Control Systems Technology

Les Fenical

Control Systems Technology

Les Fenical

THOMSON

DELMAR LEARNING

Australia • Canada • Mexico • Singapore • Spain • United Kingdom • United States

THOMSON

DELMAR LEARNING

Control Systems Technology

Les Fenical

Vice President, Technology and Trades Academic Business Unit:
David Garza

Director of Learning Solutions:
Sandy Clark

Managing Editor:
Larry Main

Executive Editor:
Stephen Helba

Senior Product Manager:
Michelle Ruelos Cannistraci

Marketing Director:
Deborah S. Yarnell

Marketing Manager:
Guy Baskaran

Director of Production:
Patty Stephan

Content Project Manager:
Christopher Chien

Senior Editorial Assistant:
Dawn Daugherty

Cover Image:
Paul Todd
(pmtdesign)

Library of Congress Cataloging-in-Publication Data:
Fenical, L. H.
  Control systems technology / Les Fenical. -- 1st ed.
    p. cm.
  ISBN 1-4018-7779-6 (hardcover : alk. paper) 1. Adaptive control systems.
  I. Title.
  TJ217.F46 2006
  629.8--dc22

2006024967

ISBN: 1-4018-7779-6

NOTICE TO THE READER

Contents

Chapter 7   Root-Locus Design   203

Preface

This book is the product of over fifteen years of teaching the subject of process and feedback control systems. Many of the notes and student handouts derived during that time are incorporated in this book with the hope that they will enhance students' understanding of what makes a system and that they will simplify the design of systems.

For the Professor/Instructor: This text is designed as a senior-level course in control system and process engineering technology in an electronics technology or electronics/electrical engineering program. More material is included than can be covered in one semester or quarter. For example, those who do not teach digital control as part of the course can ignore that part of the book. Alternatively, the content covering digital portions can be incorporated as part of continuous control design, such as PD, PI, and PID control systems.

In some chapters, the step response and frequency response are shown together. That allows students to gain a better understanding of the importance of both types of analysis in determining the response of the system and the potential for oscillation.

Discrete systems are presented as an extension of continuous systems. Students generally find discrete methods easier to understand than continuous methods, but do not realize that small changes in the coefficients of the z transfer function can cause the system response to be anything other than what they desire. Students also are not aware that an ultimately stable first-order system in the continuous time domain can become an unstable system in the discrete domain.

The concept of the root locus in the z-plane being a unit circle can sometimes cause confusion for students; but when it is presented as an extension of Euler's equation, their confusion is usually minimized. Converting the z-plane expressions to the r-plane to conduct stability calculations using the Routh table allows students to see that the Routh table is not just for continuous or discrete systems. The Routh table has implications in other areas as well, such as filter design.

Several of the chapters contain separate tutorials on the use of **MATLAB** for system simulation and analysis. Individual sections on the use of **Simulink**, the **LTI Viewer**, and the **SISO Design Tool** of **MATLAB** are also included to facilitate the design of compensators and pole placement design. In other chapters, **MATLAB** is included as part of the design and analysis process, using methods students have already learned.

For the Student: Well, you have signed up for what may be the most interesting and challenging course of your undergraduate college career.

The mathematics with which you should be familiar includes differential and integral calculus, Laplace transforms, complex variables, frequency response using Bode's methods, and discrete time functions. More than one hundred examples appear in this book, not including the **MATLAB** examples at the end of most chapters. While not designed to read like a novel, this book is designed for self-study. You should have no problem reading and understanding the concepts presented.

You will find the mathematics in this book challenging. A few of the mathematical concepts, such as transformation from the z-plane to the r-plane, are presented without a rigorous mathematical derivation. But the lack of a mathematical proof of the method does not detract from the usefulness of the concept.

You should do your best to understand the concepts and methodologies presented. The reason? After graduating, you will receive a large dose of reality when, for one of your first projects, you are asked to design some form of control system, whether it is a simple velocity control or part of a more complex system.

Acknowledgments: The author wishes to thank the staff at Thomson Learning, particularly Senior Product Managers Michelle Ruelos Cannistraci and Dawn Daugherty for their guidance in the production of this book. In addition, the author thanks everyone on the staff at Thomson Learning who contributed in many ways to the production of this book.

The author also wishes to express his gratitude to Remington College instructors Dr. Richard Washabaugh for reading and correcting the text and Dr. Met Stojanowski for checking the accuracy of the mathematics.

The author would also like to thank the following reviewers for their helpful comments and suggestions during the development stages of the text:

Sohail Anwar – Penn State University, Altoona, PA
Ray Bachnak – Texas A&M University, Corpus Christi, TX
Hal Brogerg – Indiana University Purdue, Fort Wayne, IN
A. Kent Johnson – Brigham Young University, Provo, UT
Eli Mmari – DeVry University, Calgary, Alberta
Kathleen Ossman – University of Cincinnati, Cincinnati, OH
Richard Sturtevant – Springfield Technical Community College, Springfield, MA
Ricardo Unglaub – DeVry University, Colorado Spring, CO

Finally, to my family, my son and two daughters, who have been so supportive of this and other endeavors over the last fifty years or so. The only regret is that my wife of forty-four years is not here to enjoy the fruits of this labor with me.

Part 1

Chapter 1

Introduction to Control Systems

Chapter Objectives

After completing this chapter, you should be able to:

➤ Define the concept of continuous time and discrete control systems.

➤ Define the differences between open and closed-loop control systems.

➤ Explain how feedback improves the stability of a system.

➤ Define the differences between a regulating system and a servo system.

➤ Define keywords used in the control industry, such as *setpoint, error, summing junction, plant,* and *transducer.*

1.1 Introduction

The study of control systems involves how components are integrated together to control a parameter such as velocity, torque, and pressure to force an output parameter to remain at a specific value.

A number of techniques for the analysis of linear systems are available. Some of the techniques available are **Laplace transforms, state variables,** and **frequency response methods,** which apply to what is called *classical control*. **Discrete time methods** apply to the control of a system using a digital computer.

The systems may be classified as either **single-input-single-output (SISO)** or **multiple-input-multiple-output (MIMO) systems.** If the system is a MIMO system, you will probably use state-variable analysis, along with the Laplace transform and frequency response methods.

If the system is to be controlled by a digital computer, difference equations and the methods of frequency response and state variables as applicable to difference equations will be used to completely design the system.

Whether a course is called *automatic control, feedback control, robotics,* or another name, you should understand the methods and mathematics used to analyze the systems because they are used in many engineering sciences.

1.2 Definition of Continuous and Discrete Systems

While continuous and discrete systems share many techniques, there are differences that warrant discussion of each method separately. Continuous systems do not assume any time at which measurement of the output would not be valid, but discrete systems do assume measurement at specific times between which no measurement of the output would be available.

Continuous Systems

A continuous system is one in which the condition of the system is known from instant to instant. The system is considered to be a real-time system, which means that any change in the desired output variable of the system is known instantaneously.

A change in the output variable is sensed using a transducing device that generates an analog of the variable being controlled. An example is the control of the velocity of an electric motor. A small electric generator, called a **transducer,** whose output voltage is proportional to the velocity of the motor, called the **plant,** is coupled to the motor. The transducer signal voltage is applied to another device called a **summing junction.** The summing junction compares the transducer voltage to an input reference voltage, called the **setpoint.** If there is a difference between the setpoint and the motor velocity, a voltage called the **error signal** is generated and applied to the plant control device (amplifiers and power source) to increase or decrease the velocity according to the polarity of the summing junction output voltage.

Discrete Systems

A discrete system is a system that compares the analog transducer output in the summing junction. The error signal generated is then sampled at predetermined times, using an **analog-to-digital (A/D)** converter. The sampled error signal is then sent to a digital computer where the system model, in the form of a difference equation, is stored. When the samples are different from the desired condition, the computer outputs a digital signal to a **digital-to-analog (D/A)** converter, which returns the signal to analog form and applies it to the controlling devices; and the system is corrected.

One problem with a discrete system is that if an event occurs between the samples, the event will likely be missed; and if it causes a problem in the system, the correction will be delayed. That is generally not a problem in modern systems since the sample time is very short.

There are two types of system in use: open-loop and closed-loop systems. Open-loop systems have no feedback and must be corrected manually if a change in the desired output conditions occurs. Closed-loop systems use feedback to overcome this without any human intervention.

1.3 Open-Loop Control

The concept of human control of objects is appealing since people tend to think that their ability to regulate anything is extremely accurate. Actually, people's abilities are very limited. For any device that requires speed of response, sensitivity, and accuracy, humans are not likely to be able to provide effective control since they are limited in their responses with regard to time and accuracy of adjustment.

Open-loop control generally requires human intervention to assure the proper operation of the system. An example of open-loop control is a simple pick-and-place robotic system using a stepper motor with no feedback. A computer or another device supplies a signal to a controlling device for a specific time. If there is no slippage in the motor, the system could operate indefinitely without any intervention. However, a stepper motor may slip one or two steps over time and the robotic arm will eventually run into its stops. At some point, a human will probably be required to reset the robot. Devices of this nature are used in systems where extreme accuracy is not required. Figure 1-1 shows the block diagram of an open-loop system.

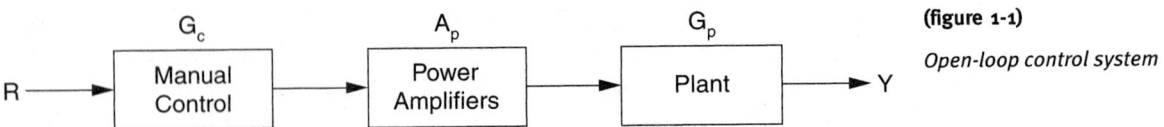

(figure 1-1)

Open-loop control system

Assume that the system is initially at rest. If the system is turned on when the load is disconnected, the system will rise to the desired level. But when a load is applied, the output will decrease from the desired level and remain at the reduced level unless corrected by some manual method, as shown in Figure 1-2.

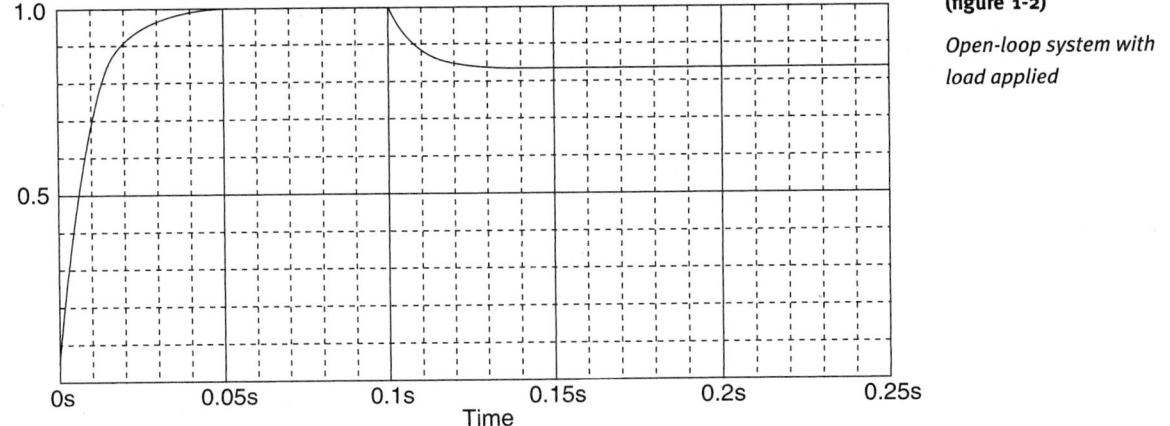

(figure 1-2)

Open-loop system with load applied

1.4 Closed-Loop Control

Closed-loop control does not require intervention to maintain the output variable at the desired condition. Rather, the system output condition is monitored continuously and is fed back to a summing junction where the desired output value is compared with the analog of the actual output value. Correction of the system occurs when the output and the desired values differ. This type of system is self-correcting and maintains its integrity over long periods of time.

Figure 1-3 shows the form of a closed-loop control system to regulate the motor velocity in the case of a motor velocity control system. G_c represents the controller (amplifiers and power source), G_p represents the plant or process (motor or other device), and H determines the percentage of the output variable fed back to the summing junction.

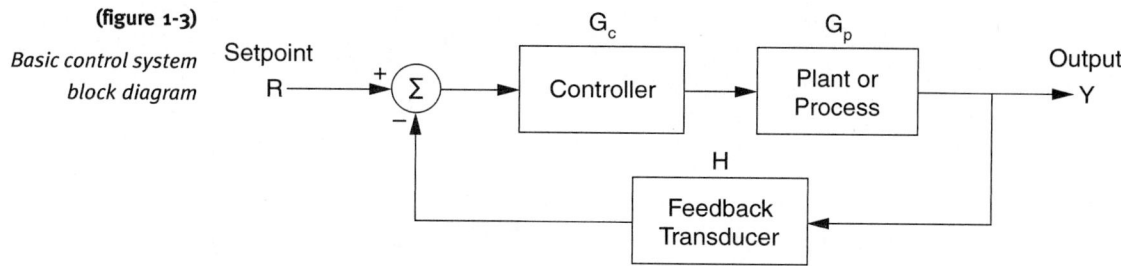

(figure 1-3)

*Basic control system
block diagram*

In all cases, specifications for the system performance are given and will relate values such as risetime, overshoot, and error allowed for the system.

Assuming the system is initially at rest, when an input reference signal is applied to $R(s)$, the motor velocity will begin to rise toward its maximum value. But at nearly the same time, a sample of the output is applied to the summing junction, reducing the signal applied to G_c. As the transducer nears equality with $R(s)$, the error signal approaches zero and the system limits itself to the value of the reference input. If a change of the output velocity to less than the setpoint value occurs, the system variable will decrease, reducing the error signal. When the output velocity is reduced, the error signal will increase and be applied to the controller, causing the system to return to the desired value. A similar operation would occur if the output velocity increased beyond the setpoint, causing the velocity to return to the desired value. The graphical output of the system is shown in Figure 1-4.

(figure 1-4)

*Closed-loop system with
load applied*

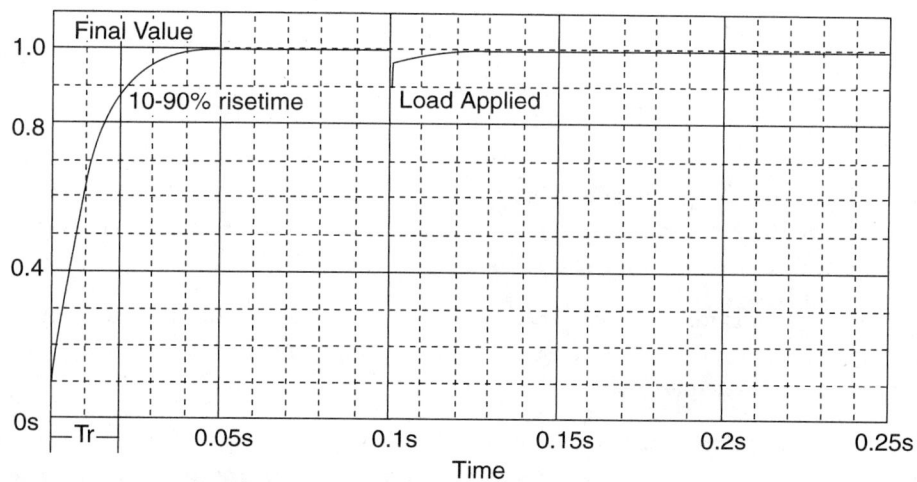

This type of system causes the motor velocity to remain at a value that is fixed by the reference value, regardless of changes in the load on the motor. This type of control system is called a **regulating system**, or simply, **regulator.**

The gain of this system is calculated from

$$\frac{Y}{R} = \frac{G_c G_p}{1 + G_c G_p H}$$

(1.1)

Remember that G_c and G_p will contain expressions that may be Laplace or frequency response functions relating the performance of the controller and plant and that H may contain an expression or may be a simple number. In that sense, there are a few different forms of response that can be produced in the system depending on the parameters calculated for the system. Three different forms of response are shown in Figure 1-5.

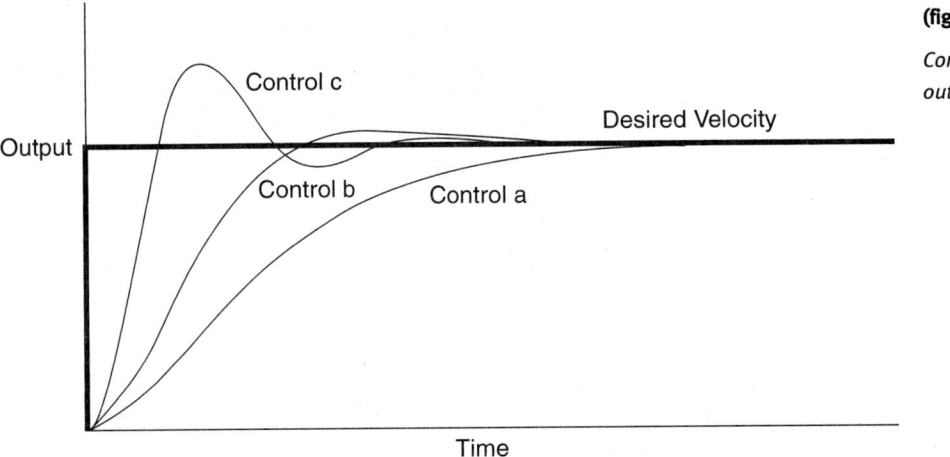

(figure 1-5)

Control system
output types

Control a shows a slow rise with no overshoot of the final value of the system. If a faster rise to the final value is required, Control b would be a better choice. Control c shows excessive overshoot, which may not be desirable in the system. From the figure, you can see that improper design or miscalculation of parameters may result in a system with excessive overshoot and may indeed become oscillatory. In that case, the system is unstable. An unstable system is not in control and is useless.

Another form of system that can be designed is one in which the reference signal is varying and the output variable must track the changing reference. That type of system is called a **servo system**.

It is worth noting here that *fast* is a relative term in control systems. Fast for mechanical systems may be from a few milliseconds to seconds, while fast for chemical mixing plants may be on the order of hours.

Discrete Systems

The discrete form of closed-loop control is similar to the continuous form with one exception: the discrete system samples the plant error signal at precise times; converts the output to a digital signal representing the condition of the plant; and sends the information to a computer, where it is determined whether a correction is needed. If a change in the output variable occurs, the output of the computer is sent to a D/A converter and is then converted to an analog signal that is supplied to the controller.

The transducer used to sample the output of the plant is an analog transducer. The output of the transducer is applied to an A/D converter, then to a zero-order-hold (ZOH) device, and is then passed to the microprocessor. If a correction is required, the computer outputs a signal that is passed to a D/A converter, which changes the signal to an analog signal. The analog signal is then applied to the controller to correct the output variable.

In general, the plant will be an analog component. Thus, the output of the computer must be converted back to an analog signal to be applied to the final control system. Figure 1-6 shows the general block form of only the discrete portion of the controller.

(figure 1-6)

Simple discrete control system

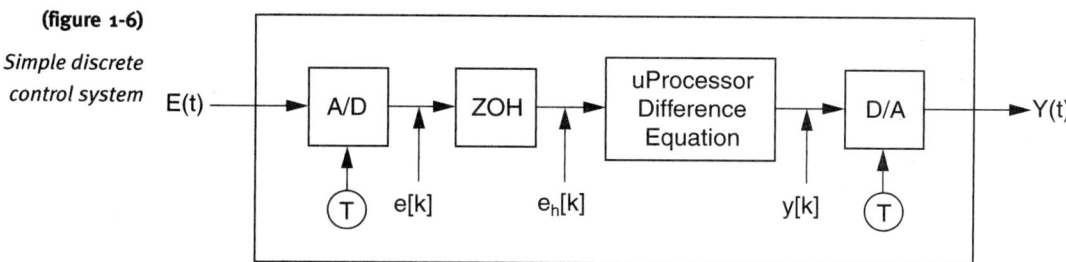

In Figure 1-6, $E(t)$ is the analog error input signal and $Y(t)$ is the continuous time output signal applied to the controller. T is the sampling time interval, $e[k]$ is the error signal sampled at intervals of kT, and $e_h[k]$ is the signal supplied to the microprocessor.

Summary

➤ Two types of control systems are open-loop and closed-loop. Open-loop control does not measure the output variable, and no correction of the actuating signal is made to cause the output to conform to the reference input. In closed-loop control, the output is measured and feedback is used to cause the output signal to conform to the reference input.

➤ Control is a process of forcing an output variable to conform to an input reference value.

➤ If the process is one in which the output is made to remain at a fixed value regardless of changes in the load on the system, it is called a *regulator*.

➤ If the system is one in which the system must track a changing input, it is called a *servo system*.

➤ Simple feedback control systems consist of the plant, which is to be controlled; an output sensor that samples the output variable; a summing junction that compares the input reference to the sampled output and generates an error signal; and a controller that causes the output variable to change according to the error signal.

➤ There are several different forms of response of a system to an input.

➤ The concept of stability is important in terms of the types of response to an input reference.

➤ The discrete form of control system and the differences between continuous and discrete control forms in terms of the block diagrams and components used for each type of system is shown.

➤ Block diagrams are useful for visualizing the system structure and the flow of signals in a system. The common forms of block diagram represent the mathematical relationships between the input and output signals in the control system.

Chapter 2

Mathematical Concepts for Control Systems

Chapter Objectives

After completing this chapter, you should be able to:

➤ Explain the underlying mathematical concepts that are essential to the design of continuous and discrete time domain control systems.

➤ Define the differential equation as a linear time invariant equation containing only derivative functions or combinations of integrals and derivative functions.

➤ Define the properties of the Laplace transform relative to the time domain for solution of complex differential equation representations of a control system.

➤ Transform the Laplace transform to the time domain, using partial fractions and the tables of Laplace transforms.

➤ Apply the basic concepts of matrix theory for representing solutions of control system parameters in state space.

➤ Apply the concepts of difference equations and conversion of continuous equations to the discrete time domain and their relation to Laplace transforms in the discrete time domain.

➤ Apply **MATLAB** to solve for the residues of a complex s-domain equations used in the retransformation of equations from the s-domain to the time domain.

➤ Use **MATLAB** to obtain the discrete time domain equation of a control system from a continuous time equation in the s-domain.

2.1 Introduction

Control system analysis requires the use of complex applied mathematics that is often difficult to solve. In order to understand the concepts presented in this book, which covers classical control theory and discrete systems, you should have taken courses that cover the subjects of complex variables, differential equations, difference equations, Laplace transforms, and z-transforms.

This chapter reviews the various mathematical methods used to calculate the required parameters of control systems.

2.2 Differential Equations

Most systems are mathematically modeled using differential equations. The limitations on differential equations for them to be linear and ordinary are that they must be of first degree and have constant coefficients. For example, a differential equation whose highest derivative is fourth order and in which any coefficients of the derivative terms are constants and not functions of the variable is a linear ordinary differential equation. A simple type of system is the series RLC circuit of electrical engineering, as shown in Figure 2-1.

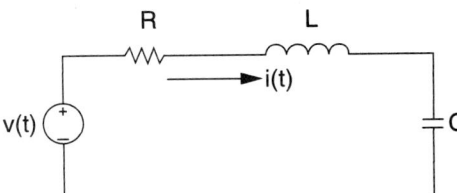

(figure 2-1)

RLC circuit for analysis

The describing equation for that circuit is

$$v(t) = Ri(t) + L\frac{di(t)}{dt} + \frac{1}{C}\int i(t)dt + v_C(0) \tag{2.1}$$

where $v(t)$ is the excitation voltage and $v_c(0)$ is the initial voltage on the capacitor at the time that $v(t)$ is applied. In reality, Equation 2.1 is an integrodifferential equation since it contains both derivative and integral terms. However, differentiating the equation once results in a second-order differential equation.

The general form for a differential equation is

$$\frac{d^n y(t)}{dt^n} + a_{n-1}\frac{d^{n-1}y(t)}{dt^{n-1}} + \quad + a_1\frac{dy(t)}{dt} + a_0 y(t) = f(t) \tag{2.2}$$

That equation is linear, is of order n, and is ordinary if the coefficients, a_n, are not functions of $y(t)$.

Generally, working with differential equations is more difficult than working with Laplace transformed functions. The differential equation is easily transformed to the s-domain by inspection. The next section develops the necessary Laplace transforms for use in the problems of this book.

2.3 Laplace Transforms for Control Systems

As in any system of mathematics, there are certain theorems and rules that are required for the proper solution of problems. Laplace transforms are no different and there are more rules than will be presented here, but the ones provided in this chapter will be sufficient for the proper solutions of all the control systems in this book.

Definition of the Laplace Transform

The Laplace transform is a method by which a complex differential equation in the time domain can be changed to an easier-to-solve algebraic equation in the complex frequency domain. The complex frequency plane is called the *s-plane*, where *s* is a complex number composed of a constant and a radian frequency. Functions of *s* in the denominator of the s-domain function are called the *poles of the system;* when they are in the numerator; they are called the *zeros of the function.* For a system to be stable, all of its poles must lie in the left half of the s-plane. Some systems will have poles in the right half of the plane but may still be stable under certain conditions. The conditions are discussed in Chapter 5, on the stability of systems. All poles in the left half of the plane are decaying exponential functions and tend asymptotically to zero with increasing time. The general form for the one-sided Laplace integral is

$$\mathcal{L}\big[f(t)\big] = \int_0^\infty f(t)e^{-st}dt = F(s) \tag{2.3}$$

When the integration is completed, the expression, which remains, is an expression in the complex frequency, or s-plane. The s-plane function $F(s)$ will not contain any values of time or time derivatives.

Inverse Laplace Transform

When transforming from the time domain to the complex frequency domain, the method of arriving at the transform is the integral shown in Equation 2.3. A method of transforming from the complex frequency domain back to the time domain is also necessary. That method is the integral shown below.

$$f(t) = \frac{1}{j2\pi}\int_{c+\infty}^{c+\infty} F(s)e^{st}\,ds = \mathcal{L}^{-1}\{F(s)\} \tag{2.4}$$

You will not use this integral, but instead will use tables of transforms, which are included in Appendix B.

Impulse (Delta) Function

The impulse, or delta, function is important in science and engineering. In many cases, the intrinsic behavior of a system can be discovered by disturbing the system in an approximately instantaneous way, that is, to apply a finite amount of energy to the system for a very short period of time. Since energy is a function of the area under a curve, it is easy to see that when the area is held constant, the energy remains the same. Figure 2-2 illustrates the approximation of an impulse.

(figure 2-2)

Approximation of the impulse function

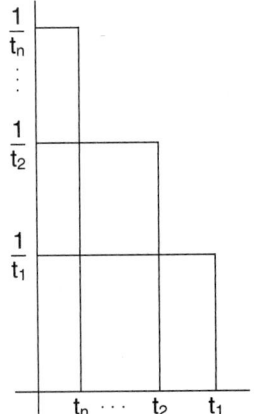

Figure 2-2 shows that the impulse can be thought of as a sequence of functions indexed as integer values whose area always equals one unit, and as $n \to \infty$, the amplitude becomes arbitrarily large, imparting one unit of energy to the system. However, the mathematical interpretation of the impulse requires more precision than this physical representation. The literature includes several methods of defining sequences of functions whose limit is the delta function.

The analysis of control systems usually assumes that the system is at rest at $t = 0$. Then some energy is applied to the system, and the response is calculated. In the case of a physical impulse, the energy is imparted to the system a short time after $t = 0$, thus the choice of this model.

The delta function is usually written as $\delta(t)$, and application of the Laplace integral yields

$$\mathcal{L}\left[\delta(t)\right] = \int_0^\infty \delta(t)e^{-st}dt$$

$$= \lim_{p \to 0}\int_0^p \frac{e^{-st}}{p}dt$$

$$= \lim_{p \to 0}\frac{-1}{ps}\left[e^{-st}\right]_0^p$$

$$= \lim_{p \to 0}\frac{1 - e^{-ps}}{ps}$$

where p is a dummy variable. Then, by application of L'Hospital's rule to both numerator and denominator

$$\lim_{p \to 0}\frac{1 - e^{-ps}}{ps} = \lim_{p \to 0}\frac{\left(\dfrac{d}{dp}\right)\left(1 - e^{-ps}\right)}{\left(\dfrac{d}{dp}\right)ps} = \lim_{p \to 0}\frac{e^{-ps}s}{s} = 1.0 \qquad (2.5)$$

Thus, the Laplace transform of the delta function is simply 1.0. Note that other methods of obtaining the impulse function are available, but this method most conveniently suits this analysis.

Unit-Step Function, *u(t)*, Transformed to the s-Plane

To show the transform method, the unit-step function $u(t)$ will be transformed first. When converting from the time domain to the frequency domain, the function converted to the frequency domain is specified using a capital letter such as $F(s)$. The unit step function in the time domain is defined by

$$u(t) = \begin{cases} 0, & \text{for } t < 0 \\ 1, & \text{for } t > 0 \end{cases} \qquad (2.6a)$$

The unit-step function has a value of 1 in the time domain, so $f(t) = 1.0$. Substitute that value in the Laplace integral and integrate between zero and infinity. The integration is

$$F(s) = \int_0^\infty (1) \cdot e^{-st}dt$$

$$F(s) = -\frac{1}{s}\left[e^{-s\infty} - e^{-s0}\right] = -\frac{1}{s}[0 - 1] = \frac{1}{s} \qquad (2.6b)$$

Transform of e^{-at}

Often you will encounter e^{-at} in your system analysis. While that is not a difficult function to evaluate, it forms part or all of the response of a system; therefore, a transform is needed for the expression.

In the case of $f(t) = e^{-at}$, you substitute that value in the Laplace integral and perform the indicated operations.

$$\mathcal{L}\{e^{-at}\} = \int_0^\infty e^{-at} e^{-st} dt$$

$$\mathcal{L}\{e^{-at}\} = \int_0^\infty e^{-(s+a)t} dt$$

$$\mathcal{L}(e^{-at}) = -\frac{1}{s+a}[e^{-(s+a)\infty} - e^{-(s+a)0}] = -\frac{1}{s+a}[0 - 1]] \qquad (2.7)$$

$$\mathcal{L}(e^{-at}) = \frac{1}{s+a}$$

Thus, if e^{-at} exists in your system analysis, the value just derived is the corresponding function in the complex frequency, or s-domain.

Transform of the Derivative Function

Given a derivative function of the form

$$\frac{df(t)}{dt} = f'(t) \qquad (2.8)$$

Using the Laplace integral

$$\mathcal{L}[f'(t)] = \int_0^\infty f'(t) e^{-st} dt \qquad (2.9)$$

The integral of this function is more rigorous than for the other forms in that you must use integration by parts to find a proper solution. The form for integration by parts is

$$\int_0^\infty u\, dv = uv\Big|_0^\infty - \int_0^\infty v\, du \qquad (2.10)$$

The only questions are which part of the equation is chosen to be u and which part is chosen to be v. One rule of thumb is to select the part of the equation that, when differentiated, will make du simpler than u. For the equation above, it is obvious that the integral of $f'(t)$ is simply $f(t)$; so the choice in this case is easy. u will be e^{-st}, and du will be $-se^{-st}$. The four variables to be used are

$$u = e^{-st} \text{ and } \frac{du}{dt} = -se^{-st}$$

$$\frac{dv}{dt} = f'(t) \text{ and } v = f(t) \tag{2.11}$$

Putting those values into the formula, you have

$$\mathcal{L}[f'(t)] = e^{-st}f(t)\Big|_0^\infty - \int_0^\infty f(t)(-se^{-st})dt = e^{-st}f(t) + s\int_0^\infty f(t)e^{-st}dt \tag{2.12}$$

This evaluation is not as difficult as it looks because s does not depend on t and can be taken outside the integral. Since the limits are zero to ∞ for both uv and the integral, each term must be evaluated between those two limits.

When uv is evaluated between the limits, $e^{-s\infty} = 0$, and the component itself is equal to zero. When the 0 limit is applied, the function becomes $f(0)$. For that reason, the value $f(0)$ can be considered any initial condition in the circuit, such as the voltage $V_c(0)$ of a precharged capacitor.

You can now operate on the integral portion of the transform. Examination of this portion shows that it is the Laplace transform of $f(t)$ multiplied by s. The transform is

$$\mathcal{L}[f'(t)] = -f(0) + sF(s)$$
$$\mathcal{L}[f'(t)] = sF(s) - f(0) \tag{2.13}$$

If only a first derivative is involved, the above equation is all that is necessary. In the case of second- and higher-order derivatives, the initial conditions at each of the derivatives must be taken into consideration. In most cases, the initial conditions will be zero and the function reduces to $s^n F(s)$. Equation 2.14 shows the second derivative form and the general derivative form.

$$\mathcal{L}[f''(t)] = s^2 F(s) - sf(0) - f'(0)$$
$$\mathcal{L}[f^n(t)] = s^n F(s) - s^{n-1}f(0) - s^{n-2}f'(0) - \cdots - f^{n-1}(0) \tag{2.14}$$

Transform of the Integral Function

The general integral form is shown in Equation 2.15.

$$\mathcal{L}\left[\int_0^{t_1}\int_0^{t_2}\cdots\int_0^{t_n} f(\tau)d\tau\, dt_1 dt_2 \cdots dt_{n-1}\right] = \frac{F(s)}{s^n} \tag{2.15}$$

Transform of sin(ωt) and cos(ωt)

The transforms for the sine and cosine functions can be developed using Euler's equation $e^{j\omega t} = \cos(\omega t) + j\sin(\omega t)$ for the sine and cosine functions

$$\sin(\omega t) = \frac{e^{j\omega t} - e^{-j\omega t}}{2j} \qquad \cos(\omega t) = \frac{e^{j\omega t} + e^{-j\omega t}}{2} \qquad (2.16)$$

The transformation for the sine function proceeds as follows:

$$\mathcal{L}\{\sin(\omega t)\} = \frac{1}{2j}\int_0^\infty \left(e^{j\omega t} - e^{-j\omega t}\right)e^{-st}dt = \frac{1}{2j}\left(\int_0^\infty e^{-(s-j\omega t)}dt + \int_0^\infty e^{-(s+j\omega)t}dt\right) \quad (2.17)$$

$$\mathcal{L}[\sin(\omega t)] = \frac{1}{2j}\left[-\frac{e^{-(s-j\omega)t}}{s - j\omega}\Big|_0^\infty + \frac{e^{-(s+j\omega)t}}{s + j\omega}\Big|_0^\infty\right] = \left[0 + \frac{1}{s - j\omega} + 0 - \frac{1}{s + j\omega}\right]$$

$$= \frac{1}{2j}\left[\frac{s + j\omega - s + j\omega}{s^2 + \omega^2}\right] = \frac{\omega}{s^2 + \omega^2} \qquad (2.18)$$

The transform for the cosine function is arrived at by a similar process:

$$\mathcal{L}[\cos(\omega t)] = \frac{s}{s^2 + \omega^2} \qquad (2.19)$$

Transform of $e^{-at}\sin(\omega t)$ and $e^{-at}\cos(\omega t)$

These functions are determined in essentially the same way as the sine and cosine functions. The development of the functions will not be shown, but you will run into them more often than not in system analysis. They are exponentially damped sine and cosine transforms. They occur only in functions that are of second order or greater.

$$\mathcal{L}[e^{-at}\sin(\omega t)] = \frac{\omega}{(s + a)^2 + \omega^2}$$

$$\mathcal{L}[e^{-at}\cos(\omega t)] = \frac{s + a}{(s + a)^2 + \omega^2} \qquad (2.20)$$

It is important that you understand the use of the Laplace integral, in almost every case, the tables will be used to put the formula in the required form and then the tables are used to retransform the function to the time domain.

Theorems Needed for the Laplace Transforms

The theorems provided here are not all of the theorems of the Laplace transform, but they are required for solution of the systems outlined in this book.

Linearity Property

The Laplace transform of constants times the sum, difference of two or more functions, is the same as the sum, or difference of the constants times the Laplace transforms of the individual functions.

$$\mathcal{L}\{[K_1 f_1(t) + K_2 f_2(t)]\} = K_1 \mathcal{L}[f_1(t)] + K_2 \mathcal{L}[f_2(t)] \tag{2.21}$$

Final-Value Theorem

This theorem is used extensively to determine the output of the system as time approaches infinity, the steady-state condition of the system. That occurs when all time-varying terms have decayed to zero. In terms of frequency, the frequency variations are approaching zero. You can then say that as time approaches infinity, if the limit of $f(t)$ exists, the frequency is approaching zero and that the limit of $f(t)$ as t approaches infinity is equal to the limit of $sF(s)$ as s approaches zero. That is

$$\lim_{t \to \infty} f(t) = \lim_{s \to 0} sF(s) \tag{2.22}$$

Initial Value Theorem

This theorem predicts the operation of the system at the time when $f(t)$ is very nearly, but not exactly equal to, zero $t(0^+)$. It is used to establish the initial value $f(0)$ for a system without going through the entire inverse transformation. It can then be shown by induction that, in the limit, when $f(t) \to f(0^+)$, the system is responding in a manner that shows any tendency toward oscillation. Thus, the limit of $f(t)$ as t approaches zero is the same as the limit of $sF(s)$ as s approaches infinity.

$$\lim_{x \to 0} f(t) = \lim_{x \to \infty} s \cdot F(s) \tag{2.23}$$

A short table of Laplace transforms and their time domain functions is shown in Table 2-1.

	f(t)	F(s)	(table 2-1)
1.	$\delta(\tau)$	1	
2.	$u(t)$	$\dfrac{1}{s}$	
3.	t	$\dfrac{1}{s^2}$	
4.	$e^{-\alpha t}$	$\dfrac{1}{s + \alpha}$	
5.	$\sin \omega t$	$\dfrac{\omega}{s^2 + \omega^2}$	
6.	$\cos \omega t$	$\dfrac{s}{s^2 + \omega^2}$	
7.	$e^{-\alpha t} \sin \omega t$	$\dfrac{\omega}{(s + \alpha)^2 + \omega^2}$	
8.	$e^{-\alpha t} \cos \omega t$	$\dfrac{s + \alpha}{(s + \alpha)^2 + \omega^2}$	
9.	$t^n u(t)$	$\dfrac{n!}{s^{n+1}}$	
10.	$t^n e^{-\alpha t}$	$\dfrac{n!}{(s + \alpha)^{n+1}}$	

Convolution in the s-Domain

In many cases, there is a product of two functions in the s-plane. An important point to keep in mind is that *the inverse transform of a product in the frequency domain corresponds to the product of convolution of the functions in the time domain.* When a product occurs in the frequency domain, the concept of convolution is applied so the correct time domain functions are determined. The convolution product is denoted as $\mathcal{L}\left[(f_1(t)f_2(t)\right] = F_1(s) * F_2(s)$ where * indicates convolution. Convolution is accomplished using a dummy variable, τ. The dummy variable allows you to determine the time domain function. The convolution integral is

$$\int_0^t f_1(\tau)f_2(t - \tau)d\tau = f_1(t) * f_2(t) \qquad (2.24)$$

Then $\mathcal{L}\{f(t)\} = \mathcal{L}\{f_1(t) * f_2(t)\} = F_1(s)F_2(s)$.

It is important to remember that convolution in the time domain corresponds to a product in the frequency domain, or $f_1(t) * f_2(t) \xleftrightarrow{\mathcal{L}} F_1(s)F_2(s)$, and that convolution in the frequency domain, or $f_1(t)f_2(t) \xleftrightarrow{\mathcal{L}} F_1(s) * F_2(s)$, corresponds to a product in the time domain. Some examples will clarify the process.

Example 2-1

Consider the product of two functions in the s-domain.

$$F(s) = \frac{1}{(s + 1)(s + 2)} = \frac{1}{s + 1} \cdot \frac{1}{s + 2} = F_1(s)F_2(s)$$

where

$$F_1(s) = \frac{1}{s + 1} \qquad\qquad F_2(s) = \frac{1}{s + 2}$$

Solution

These functions represent e^{-t} and e^{-2t}, respectively. The response is not e^{-3t}. Partial fraction expansion, which is explained in Section 2.4, would show that the actual response is

$$f(t) = e^{-t} - e^{-2t} = e^{-t}\left(1 - e^{-t}\right)$$

Let $f(t - \tau)$ replace either one of the exponents in the multiplied functions and $f(\tau)$ the other. Let e^{-t} be $e^{-\tau}$ and e^{-2t} be $e^{-2(t-\tau)}$. Then substituting those functions into the integral Equation 2.24

$$f(t) = \int_0^t e^{-\tau}e^{-2(t-\tau)}d\tau$$

Summing the two exponents, you see that

$$e^{-\tau} \cdot e^{-2(t-\tau)} = e^{-\tau} \cdot e^{(-2t+2\tau)} = e^{-2t}e^{\tau}$$

Since τ is the variable of integration and e^{-2t} is not a function of τ, e^{-2t} can be taken outside the integral, yielding a simpler integration to perform.

$$f(t) = e^{-2t}\int_0^t e^{\tau}d\tau$$

$$f(t) = e^{-2t}\left[e^{\tau}\right]_0^t = e^{-2t}\left[e^t - 1\right] = (e^{-t} - e^{-2t})$$

Example 2-2

Consider the function

$$F(s) = \frac{s}{(s + 2)(s + 4)} = sF_1(s)$$

Example 2-2 (continued)

where

$$F_1(s) = \frac{1}{(s + 2)(s + 4)}$$

then

$$f(t) = \frac{d}{dt} f_1(t)$$

Solution

$F_1(s)$ is solved in the same way as Example 2-1. Let $e^{-2t} = e^{-2\tau}$ and $e^{-4t} = e^{-4(t-\tau)}$. Then summing the exponents and removing the constant term

$$f(t) = \int_0^t e^{-2\tau} \cdot e^{-4(t-\tau)} d\tau = e^{-4t} \int_0^t e^{2\tau} d\tau$$

The integrated function is

$$f_1(t) = \frac{e^{-4t}}{2} \left(e^{-2\tau} \right) \Big|_0^t \frac{e^{-4t}}{2} \left[e^{-2t} - 1 \right] = \frac{1}{2} \left(e^{-2t} - e^{-4t} \right)$$

This function is now differentiated to form the complete convolution result.

$$f(t) = \frac{1}{2} \left[(-2e^{-2t}) - (-4e^{-4t}) \right] = -e^{-2t} + 2e^{-4t}$$

$$f(t) = 2e^{-4t} - e^{-2t}$$

2.4 Inverse Laplace Transforms Using Partial Fractions

System design usually results in a transfer function that is not ready for retransformation to the time domain. Generally, you do not use the inverse integral to determine the result. The use of tables of transforms and partial fraction expansion is used.

The first approach when working with any Laplace expression is to factor the denominator of the transfer function into its roots. In many cases, quadratic functions are a part of the expression. When this occurs, there are three possible results of the factorization of the quadratic functions in the expression:

1. If the discriminant $b^2 - 4ac > 0$, the roots of the transfer function are real and unequal.
2. If the discriminant $b^2 - 4ac = 0$, the roots of the transfer function are real and equal.
3. If the discriminant $b^2 - 4ac < 0$, the roots of the transfer function are complex conjugates.

For the methods of partial fractions, the numerator must be at least one order less than the denominator. If the numerator order is the same or is of greater order, a long division can be done to extract the proper form. The methods of partial fractions are demonstrated in Examples 2-3, 2-4, and 2-5.

Transfer Functions with Real Roots, Not Repeated

In some cases, the transfer function, when factored into its roots, has simple roots. In that case, the transfer function has the general form

$$F(s) = \frac{N(s)}{(s + s_1)(s + s_2)\cdots(s + s_n)}$$ (2.25)

The general form for determining the constants is

$$K_{s_n} = \left[(s + s_n)\frac{N(s)}{D(s)}\right]$$ (2.26)

Example 2-3

Consider the function

$$F(s) = \frac{4}{s(s^2 + 4s + 3)} = \frac{4}{s(s + 1)(s + 3)}$$

Retransform that function into the time domain, using partial fractions.

Solution

The second-order expression factors into two real and unequal roots as shown. The transfer function is then separated into its constituent functions

$$F(s) = \frac{A}{s} + \frac{B}{s + 1} + \frac{C}{s + 3}$$

The next step is to form the constants of the function by multiplying the function by the roots individually and setting s equal to the value of the constant of the multiplying root. The values generated are

$$A = \left[s\frac{4}{s(s + 1)(s + 3)}\right]_{s=0} = 1.333$$

$$B = \left[(s + 1)\frac{4}{s(s + 1)(s + 3)}\right]_{s=-1} = -2$$

$$C = \left[(s + 3)\frac{4}{s(s + 1)(s + 3)}\right]_{s=-3} = 0.6667$$

Example 2-3 (continued)

Substituting the constants into the expression

$$F(s) = 1.333\left(\frac{1}{s} - 1.5 \cdot \frac{1}{s+1} + 0.5 \cdot \frac{1}{s+3}\right)$$

The time domain function is formed using the transforms 2 and 4 from Table 2-1:

$$f(t) = 1.333\left(1 - 1.5e^{-t} + 0.5e^{-3t}\right)u(t)$$

Transfer Functions with Repeated Roots

When a transfer function is factored and shows roots of multiplicity r, the general form of the transfer function has the form

$$F(s) = \frac{N(s)}{(s+p_1)^m} \tag{2.27}$$

The general form for that type of partial fraction expansion is

$$G(s) = \frac{N(s)}{(s+p_1)^m} = \frac{A_{s1}}{s+p_1} + \cdots + \frac{B_{s2}}{(s+p_1)^{m-1}} + \frac{C_m}{(s+p_1)^m} \tag{2.28}$$

Where

$$A_{s1} = \left[(s+p_1)^m G(s)\right]\Big|_{s=-p_1}$$

$$B_{s2} = \frac{d}{ds}\left[(s+p_1)^m G(s)\right]\Big|_{s=-p_1}$$

$$\vdots$$

$$C_m = \frac{1}{(m-1)!}\frac{d^{m-1}}{ds^{m-1}}\left[(s+p_1)^m G(s)\right]\Big|_{s=-p_1}$$

Example 2-4

Consider the s-domain function

$$F(s) = \frac{s+4}{s(s^3 + 6s^2 + 12s + 8)}$$

Example 2-4 (continued)

Solution

When factored, that equation has roots of multiplicity three, as shown:

$$= \frac{s + 4}{s(s + 2)^3}$$

And is set into partial fraction form, as shown:

$$F(s) = \frac{s + 4}{s(s + 2)^3} = \frac{A}{s} + \frac{B}{s + 2} + \frac{C}{(s + 2)^2} + \frac{D}{(s + 2)^3}$$

The resulting value for A is

$$A = s \left[\frac{s + 4}{s(s + 2)^3} \right]_{s=0} = 0.5$$

The results for B, C, and D are

$$B = \left[\frac{s + 4}{s} \right]_{s=-2} = \frac{-2 + 4}{-2} = -1$$

$$C = \left[\frac{d}{ds} \left(\frac{s + 4}{s} \right) \right]_{s=-2} = \left[\frac{s \cdot 1 - (s + 4) \cdot 1}{s^2} \right]_{s=-2} = \frac{s - s - 4}{4} = -1$$

$$D = \frac{1}{2} \left[\frac{d^2}{ds^2} \left(\frac{-4}{s^2} \right) \right]_{s=-2} = \frac{1}{2} \frac{s^2 \cdot 0 - (-4) \cdot 2s}{s^4} = \frac{4s}{s^4} = \frac{4 \cdot (-2)}{16} = -0.5$$

The resulting transfer function in both the frequency and time domain is

$$F(s) = \frac{0.5}{s} - \frac{1}{s + 2} - \frac{1}{(s + 2)^2} - \frac{0.5}{(s + 2)^3}$$

$$f(t) = 0.5 - e^{-2t} - te^{-2t} - 0.25t^2 e^{-2t}$$

Transfer Functions with Complex Roots

When transfer functions contain complex conjugate roots, the function has an underdamped response. That means the system will usually contain exponentially damped sine and cosine terms. In that case, you provide a numerator for the second-order function that is one order less than the denominator.

If a function contains simple complex roots such as

$$s_{1,2} = -a \pm j\omega$$

The method of Equation 2.26 can be used; but it is often easier to form a set of simultaneous equations to determine the constants. Example 2-5 shows such decomposition.

Example 2-5

Consider the transfer function

$$F(s) = \frac{4s + 13}{s(s^2 + 6s + 13)}$$

Determine the required constants and retransform the function to the time domain.

Solution

Separating the transfer function into individual factors, the required numerator is formed as shown

$$F(s) = \frac{A}{s} + \frac{Bs + C}{s^2 + 6s + 13}$$

$$4s + 13 = As^2 + 6As + 13A + Bs^2 + Cs$$

$$= (A + B)s^2 + (6A + C)s + 13A$$

The matrix and the values for A and for $Bs + C$ are

$$A + B + 0 = 0$$

$$6A + 0 + C = 4$$

$$13A + 0 + 0 = 13$$

$$A = 1, \qquad B = -1, \qquad C = -2$$

The resultant Laplace domain function is

$$F(s) = \frac{1}{s} - \frac{s + 2}{(s + 3)^2 + 4}$$

The second-order function that results from the calculation is not correct for direct transformation to the time domain, as it does not fit the exponentially damped sine or cosine form. To determine the time domain response, the property of superposition will be used to find the required components of the second-order function. Also, the square for the second-order function will be completed to determine the exponentially damped sine and cosine terms.

$$\frac{s + 2 + 1 - 1}{(s + 3)^2 + 4} = \frac{s + 3}{(s + 3)^2 + 4} - \frac{1}{(s + 3)^2 + 4}$$

The addition and subtraction of 3 from the expression gives both the exponentially damped sine and cosine terms, but the sine term is not correct for the transform. If you multiply and divide by 1, in the form of 2/2, the result is the proper form for the sine term.

$$-\frac{1}{(s + 3)^2 + 4} \cdot \frac{2}{2} = -0.5 \frac{2}{(s + 3)^2 + 4}$$

Example 2-5 (continued)

The separate terms are then added together to form the transfer function. The transfer function is then transformed back to the time domain.

$$F(s) = \frac{1}{s} - \frac{s + 3}{(s + 3)^2 + 4} + 0.5\frac{2}{(s + 3)^2 + 4}$$

$$f(t) = \left[1 - e^{-3t}\left(\cos 2t - 0.5\sin 2t\right)\right]u(t)$$

As an aside, there is an easy way to determine if the roots of a quadratic equation are complex if the coefficient of the first term is 1.0. Take one-half the coefficient of s and square it; then subtract the value from the constant term. If the result is negative, the quadratic has real roots. If the result is positive, the square root of the remaining value is the imaginary part and one-half the coefficient of s is the real part.

If you have already factored the quadratic and have complex roots, twice the real part is the coefficient of s and the sum of the squares of the real and imaginary parts is the constant term. Those two methods can save you some time in evaluating a quadratic.

2.5 Basic Matrix Theory

Matrices are used to simplify complex systems of simultaneous equations. A matrix is composed of the variables and constants of the simultaneous equations. Matrices are used exclusively in the solution of control systems in state space, which is discussed in Chapter 10.

A simple definition of a matrix is:

Any rectangular array of numbers or functions. If a matrix has only one column, it is a column vector, or simply a vector. If the matrix has only one row, it is a row vector. Examples of the different types are as follows:

$$\begin{bmatrix} 1 & -5 & 10 \\ 6 & 0 & -1 \end{bmatrix} \quad Matrix$$

$$\begin{bmatrix} 1 \\ 4 \\ 0 \end{bmatrix} \quad Column\ Vector$$

$$\begin{bmatrix} 1 & -7 & 9 \end{bmatrix} \quad Row\ Vector$$

Matrices are in row-column format $(m \times n)$, where m is the number of rows and n is the number of columns. The fundamental form of a matrix is

$$\mathbf{A} = \begin{bmatrix} a_{11} & a_{12} & a_{13} \\ a_{21} & a_{22} & a_{23} \\ a_{31} & a_{32} & a_{33} \end{bmatrix} \tag{2.29}$$

Matrix Definitions

Some important definitions of matrices follow.

Square matrices

Square matrices are matrices that have an equal number of rows and columns; only square matrices have a determinant.

Determinant of a matrix

If the determinant of a square matrix is not zero, the matrix is said to be nonsingular. If the matrix determinant is zero, the matrix is said to be singular. If a matrix is singular, the rows of the matrix are dependent. For example, if one or more rows of a matrix are the sum or difference of the other rows, the rows are dependent. Similarly, if a column is a scalar multiple of other columns, the columns are dependent. Examples of a nonsingular and singular matrix are

$$\begin{bmatrix} 2 & 2 & 0 \\ -2 & 1 & 1 \\ 3 & 0 & 1 \end{bmatrix} \qquad \det \mathbf{A} = 12 \neq 0$$

$$\mathbf{A} = \begin{bmatrix} -1 & 3 & 0 & -6 \\ 2 & 7 & -3 & 5 \\ 1 & 10 & -3 & -1 \\ -3 & -4 & 3 & -11 \end{bmatrix} \quad \det \mathbf{A} = 0$$

In the singular matrix, Rows 3 and 4 are the addition and subtraction of Rows 1 and 2. The resulting determinant is zero.

Transpose of a matrix

The transpose of a matrix is one in which the rows and columns are interchanged. The transpose of a matrix is denoted by $\mathbf{A}^T$. (In **MATLAB**, the transpose is given by A'.)

$$\mathbf{A} = \begin{bmatrix} 3 & 6 & 2 \\ 2 & 5 & 1 \\ -1 & 2 & 4 \end{bmatrix}, \ \mathbf{A}^T = \begin{bmatrix} 3 & 2 & -1 \\ 6 & 5 & 2 \\ 2 & 1 & 4 \end{bmatrix}$$

Identity matrix

An identity matrix is a square matrix with all ones on the main diagonal and zeros in all other positions.

$$\mathbf{I} = \begin{bmatrix} 1 & 0 & 0 & 0 \\ 0 & 1 & 0 & 0 \\ 0 & 0 & 1 & 0 \\ 0 & 0 & 0 & 1 \end{bmatrix}$$

Diagonal matrix

A diagonal matrix has values only on the main diagonal.

$$\mathbf{A} = \begin{bmatrix} a_{11} & 0 & 0 \\ 0 & a_{22} & 0 \\ 0 & 0 & a_{33} \end{bmatrix} \tag{2.30}$$

Null matrix

A null matrix is one whose elements are all zero.

$$\mathbf{A} = \begin{bmatrix} 0 & 0 & 0 \\ 0 & 0 & 0 \\ 0 & 0 & 0 \end{bmatrix}$$

Multiplication of a matrix by a scalar

Multiplication of a matrix $\mathbf{A}$ by a scalar k is a multiplication of each element of the matrix by k.

$$k \begin{bmatrix} 2 & 4 \\ 3 & 7 \end{bmatrix} = \begin{bmatrix} 2k & 4k \\ 3k & 7k \end{bmatrix}$$

Rank of a matrix

The rank of a matrix is the number of linearly independent columns in the matrix or the largest nonsingular matrix within the matrix.

$$\begin{bmatrix} 1 & 0 \\ 2 & 0 \end{bmatrix} \text{ rank } = 1 \qquad \begin{bmatrix} 0 & 8 & -2 \\ 1 & 8 & 4 \end{bmatrix} \text{ rank } = 2$$

$$\begin{bmatrix} 1 & 3 & -2 \\ 2 & 0 & 5 \\ -3 & 1 & 0 \end{bmatrix} \text{ rank } = 3$$

Addition and subtraction of matrices

Matrices are added by adding like terms in each matrix. The associative and commutative laws of addition apply to matrices.

$$A = \begin{bmatrix} 3 & -4 \\ -6 & 5 \end{bmatrix} \quad B = \begin{bmatrix} -1 & 5 \\ 8 & -5 \end{bmatrix}$$

$$A + B = \begin{bmatrix} 3 - 1 & -4 + 5 \\ -6 + 8 & 5 - 5 \end{bmatrix} = \begin{bmatrix} 2 & 1 \\ 2 & 0 \end{bmatrix}$$

Matrices are subtracted by subtracting like terms in each matrix.

$$A = \begin{bmatrix} 3 & -4 \\ -6 & 5 \end{bmatrix} \quad B = \begin{bmatrix} -1 & 5 \\ 8 & -5 \end{bmatrix}$$

$$A - B = \begin{bmatrix} 3 - (-1) & -4 - 5 \\ -6 - 8 & 5 - (-5) \end{bmatrix} = \begin{bmatrix} 4 & -9 \\ -14 & 10 \end{bmatrix}$$

Multiplication of matrices

Matrix multiplications have rules that are not unlike algebra; but there are differences, especially regarding the commutative law of multiplication. For multiplication to be possible, the column value in the **A** matrix must be equal to the row value in the **B** matrix, such as $A = (m \times n)$ and $B = (n \times p)$. For example, $A \cdot B = (m \times n) \cdot (n \times p)$ is possible, but $B \cdot A = (n \times p) \cdot (m \times n)$ is not possible if $m \neq p$. The associative and distributive laws of multiplication do apply. The size of the matrix produced by multiplication is equal to the row value of the first matrix and the column value of the second matrix.

$$A = (3 \times 4) \quad B = (4 \times 2) \quad A \cdot B = (3 \times 2) \quad (B \cdot A \text{ is not possible.})$$

Further, $A \cdot B$ may or may not be the same as $B \cdot A$. However, for the matrix multiplications to be possible, the matrices must be square.

Matrices are multiplied by multiplying the top row of the matrix **A** by the columns of matrix **B** and adding the results, then multiplying the second row of matrix **A** by the columns of matrix **B** and adding the results, and so on. The following simple matrix multiplication shows the method.

Consider two matrixes

$$A = \begin{bmatrix} 4 & 7 \\ 3 & 5 \end{bmatrix}, \quad B = \begin{bmatrix} 9 & -2 \\ 6 & 8 \end{bmatrix}$$

Then

$$A \times B = \begin{bmatrix} 4 & 7 \\ 3 & 5 \end{bmatrix} \times \begin{bmatrix} 9 & -2 \\ 6 & 8 \end{bmatrix} = \begin{bmatrix} 4 \cdot 9 + 7 \cdot 6 & 4 \cdot (-2) + 7 \cdot 8 \\ 3 \cdot 9 + 5 \cdot 6 & 3 \cdot (-2) + 5 \cdot 8 \end{bmatrix} = \begin{bmatrix} 78 & 48 \\ 57 & 34 \end{bmatrix}$$

$$B \times A = \begin{bmatrix} 9 & -2 \\ 6 & 8 \end{bmatrix} \times \begin{bmatrix} 4 & 7 \\ 3 & 5 \end{bmatrix} = \begin{bmatrix} 9 \cdot 4 + (-2) \cdot 3 & 9 \cdot 7 + (-2) \cdot 5 \\ 6 \cdot 4 + 8 \cdot 3 & 6 \cdot 7 + 8 \cdot 5 \end{bmatrix} = \begin{bmatrix} 30 & 53 \\ 48 & 82 \end{bmatrix}$$

2.6 Cofactors and Adjoint of a Matrix

Cofactors of a Matrix

A cofactor matrix exists for all square matrices. The cofactor matrix elements are determinants of the factors of the matrix and have signs that are determined by addition of the row-column subscripts for each term. If the summed row and column indices are odd, the sign is negative; if they are even, the sign is positive. For example, the cofactors for a (3 × 3) matrix are

$$A = \begin{bmatrix} a_{11} & a_{12} & a_{13} \\ a_{21} & a_{22} & a_{23} \\ a_{31} & a_{32} & a_{33} \end{bmatrix}$$

$$A_{11} = \begin{bmatrix} a_{22} & a_{23} \\ a_{32} & a_{33} \end{bmatrix} \qquad A_{12} = -\begin{bmatrix} a_{21} & a_{23} \\ a_{31} & a_{33} \end{bmatrix} \qquad A_{13} = \begin{bmatrix} a_{21} & a_{22} \\ a_{31} & a_{32} \end{bmatrix}$$

$$A_{21} = -\begin{bmatrix} a_{12} & a_{13} \\ a_{32} & a_{33} \end{bmatrix} \qquad A_{22} = \begin{bmatrix} a_{11} & a_{13} \\ a_{31} & a_{33} \end{bmatrix} \qquad A_{23} = -\begin{bmatrix} a_{11} & a_{12} \\ a_{31} & a_{32} \end{bmatrix} \qquad (2.31)$$

$$A_{31} = \begin{bmatrix} a_{12} & a_{13} \\ a_{22} & a_{23} \end{bmatrix} \qquad A_{32} = -\begin{bmatrix} a_{11} & a_{13} \\ a_{21} & a_{23} \end{bmatrix} \qquad A_{33} = \begin{bmatrix} a_{11} & a_{12} \\ a_{21} & a_{22} \end{bmatrix}$$

Adjoint of a Matrix

The adjoint of a matrix A, denoted by *Adj* A, is defined as

$$Adj\ A = \begin{bmatrix} A_{ij} \text{ of det } A \end{bmatrix}_{m,n} \qquad (2.32)$$

A_{ij} is the cofactor of each element of A where i and j are the row-column indices. For a simple (2 × 2) matrix

$$A = \begin{bmatrix} a_{11} & a_{12} \\ a_{21} & a_{22} \end{bmatrix}$$

The cofactors of the matrix are $\mathbf{A}_{11} = a_{22}$, $\mathbf{A}_{12} = -a_{12}$, $\mathbf{A}_{22} = a_{11}$, from which the adjoint matrix is

$$\mathbf{adj\,A} = \begin{bmatrix} \mathbf{A}_{11} & \mathbf{A}_{12} \\ \mathbf{A}_{21} & \mathbf{A}_{22} \end{bmatrix}^T = \begin{bmatrix} a_{22} & -a_{21} \\ -a_{12} & a_{11} \end{bmatrix}^T = \begin{bmatrix} a_{22} & -a_{12} \\ -a_{21} & a_{11} \end{bmatrix}$$

More simply, the adjoint of a (2 × 2) matrix can be formed by exchanging the main diagonal elements and changing the sign of the elements off the diagonal.

Inverse of a Matrix

The multiplicative inverse of a matrix $\mathbf{A}^{-1}$ may exist if and only if the matrix A is square and nonsingular.

The inverse of a matrix is determined by

$$\mathbf{A}^{-1} = \frac{adj(\mathbf{A})}{\det(\mathbf{A})} \qquad (2.33)$$

To evaluate the inverse method, a simple (2 × 2) matrix will be used. Consider the matrix

$$\mathbf{A} = \begin{bmatrix} 1 & 4 \\ 2 & 10 \end{bmatrix} \qquad \det \mathbf{A} = 2$$

The determinant of the matrix is $[10 - 8] = 2 \neq 0$, the cofactor matrix is nonsingular, and the inverse matrix exists because the matrix is nonsingular.

The cofactors for this matrix are $\mathbf{A}_{11} = 10$, $\mathbf{A}_{12} = -2$, $\mathbf{A}_{21} = -4$, $\mathbf{A}_{22} = 1$. Setting those values in a matrix and taking the transpose of the matrix yields

$$Cofactor\ \mathbf{A} = \begin{bmatrix} 10 & -2 \\ -4 & 1 \end{bmatrix} \qquad Adj\ \mathbf{A} = \begin{bmatrix} 10 & -4 \\ -2 & 1 \end{bmatrix}$$

And the inverse matrix is

$$\mathbf{A}^{-1} = \frac{1}{2}\begin{bmatrix} 10 & -4 \\ -2 & 1 \end{bmatrix} = \begin{bmatrix} 5 & -2 \\ -1 & \frac{1}{2} \end{bmatrix}$$

Note that for a (2 × 2) matrix, the inverse is done simply by interchanging the values of the main diagonal and changing the sign of the other two values.

An example inverse of a more involved (3×3) matrix is

$$\begin{bmatrix} 2 & 2 & 0 \\ -2 & 1 & 1 \\ 3 & 0 & 1 \end{bmatrix} \qquad \det A = 12 \neq 0$$

Nonsingular

The cofactors, cofactor matrix, and inverse of the matrix are

$$A_{11} = \det\begin{vmatrix} 1 & 1 \\ 0 & 1 \end{vmatrix} = 1 \quad A_{12} = -\det\begin{vmatrix} -2 & 1 \\ 3 & 1 \end{vmatrix} = 5 \quad A_{13} = \det\begin{vmatrix} -2 & 1 \\ 3 & 0 \end{vmatrix} = -3$$

$$A_{21} = -\det\begin{vmatrix} 2 & 0 \\ 0 & 1 \end{vmatrix} = -2 \quad A_{22} = \det\begin{vmatrix} 2 & 0 \\ 3 & 1 \end{vmatrix} = 2 \quad A_{23} = -\det\begin{vmatrix} 2 & 2 \\ 3 & 0 \end{vmatrix} = 6$$

$$A_{31} = \det\begin{vmatrix} 2 & 0 \\ 1 & 1 \end{vmatrix} = 2 \quad A_{32} = -\det\begin{vmatrix} 2 & 0 \\ -2 & 1 \end{vmatrix} = -2 \quad A_{33} = \det\begin{vmatrix} 2 & 2 \\ -2 & 1 \end{vmatrix} = 6$$

$$A_{ij} = \begin{bmatrix} 1 & 5 & -3 \\ -2 & 2 & 6 \\ 2 & -2 & 6 \end{bmatrix} = Cofactor\ Matrix \qquad A_{ij}^T = \begin{bmatrix} 1 & -2 & 2 \\ 5 & 2 & -2 \\ -3 & 6 & 6 \end{bmatrix} = adj(A)$$

$$A^{-1} = \frac{1}{12}\begin{bmatrix} 1 & -2 & 2 \\ 5 & 2 & -2 \\ -3 & 6 & 6 \end{bmatrix} = \begin{bmatrix} 1/12 & -1/6 & 1/6 \\ 5/12 & 1/6 & -1/6 \\ -1/4 & 1/2 & 1/2 \end{bmatrix} = \frac{adj(A)}{|A|}$$

$$A^{-1} = \begin{bmatrix} 0.0833 & -0.1667 & 0.1667 \\ 0.4167 & 0.1667 & -0.1667 \\ -0.25 & 0.5 & 0.5 \end{bmatrix}$$

Multiplying $A \cdot A^{-1} = I$ the identity matrix, proves that this is the inverse matrix:

$$A^{-1} = \begin{bmatrix} 0.0833 & -0.1667 & 0.1667 \\ 0.4167 & 0.1667 & -0.1667 \\ -0.25 & 0.5 & 0.5 \end{bmatrix} \times \begin{bmatrix} 2 & 2 & 0 \\ -2 & 1 & 1 \\ 3 & 0 & 1 \end{bmatrix} = \begin{bmatrix} 1 & 0 & 0 \\ 0 & 1 & 0 \\ 0 & 0 & 1 \end{bmatrix}$$

Since direct division of matrices is not possible, it may still be possible to provide an equivalent, such as $A^{-1} \cdot B$:

$$A = \begin{bmatrix} 2 & 5 \\ 3 & 9 \end{bmatrix} \qquad B = \begin{bmatrix} 4 & 2 \\ 1 & 3 \end{bmatrix}$$

$$adj(A) = \begin{bmatrix} 9 & -5 \\ -3 & 2 \end{bmatrix}$$

$$A^{-1} = \frac{1}{3}\begin{bmatrix} 9 & -5 \\ -3 & 2 \end{bmatrix} = \begin{bmatrix} 3 & -1.667 \\ -1 & 0.667 \end{bmatrix}$$

$$A^{-1}B = \begin{bmatrix} 3 & -1.667 \\ -1 & 0.667 \end{bmatrix} \cdot \begin{bmatrix} 4 & 2 \\ 1 & 3 \end{bmatrix} = \begin{bmatrix} 10.333 & 1 \\ -3.333 & 0 \end{bmatrix}$$

$B^{-1} \cdot A$ is also possible, but results in a different matrix.

2.7 Exponential of a Matrix

Matrix exponentials are useful in the solution of systems of differential equations. There are several methods that can be used to form the matrix exponential, but only two will be discussed here: Taylor's series and the inverse Laplace method.

Taylor's Series Method

Taylor's series for the exponential is

$$e^x = 1 + x + \frac{x^2}{2!} + \frac{x^3}{3!} + \cdots \frac{x^n}{n!} \tag{2.34}$$

If $x = 1$, the series converges to an approximation of e^1 in about six terms. If x is a matrix A, the series becomes

$$e^A = I_n + \frac{1}{1!}A + \frac{1}{2!}A + \frac{1}{3!} + \cdots \frac{1}{n!}A^n \tag{2.35}$$

To illustrate the use of Taylor's method, a simple diagonal matrix will be used:

$$A = \begin{bmatrix} 3 & 0 \\ 0 & -2 \end{bmatrix}$$

By inspection, it is easy to show that

$$A^n = \begin{bmatrix} 3^n & 0 \\ 0 & (-2)^n \end{bmatrix}$$

where $n = 1, 2, \cdots$. Hence

$$I_n + \frac{1}{1!}A + \frac{1}{2!}A + \frac{1}{3!} + \cdots \frac{1}{n!}A^n =$$

$$\begin{bmatrix} 1 + \dfrac{3}{1!} + \dfrac{3^2}{2!} + \cdots + \dfrac{3^n}{n!} & 0 \\ 0 & 1 + \dfrac{-2}{1!} + \dfrac{(-2)^2}{2!} + \cdots + \dfrac{(-2)^n}{n!} \end{bmatrix}$$

Using the above system of equations, you see that

$$e^A = \begin{bmatrix} e^3 & 0 \\ 0 & e^{-2} \end{bmatrix}$$

Indeed, for any diagonal matrix A of this type, e^A may be determined by replacing the diagonal elements with e raised to the element value. This method becomes cumbersome when the matrix dimensions are greater than two and the matrix is not diagonal.

Inverse Laplace Method

From the Laplace transform, you understand that

$$e^{at} = \mathcal{L}^{-1}\left(\frac{1}{s-a}\right) \tag{2.36}$$

If a is replaced with a square matrix A, the natural matrix form for e^A is then

$$e^{At} = \mathcal{L}^{-1}\left(\left[sI - A\right]^{-1}\right) \tag{2.37}$$

Where I is an $n \times n$ identity matrix the same size as A.

As an example, consider a matrix where the values are not on the main diagonal:

$$A = \begin{bmatrix} 0 & 1 \\ -2 & 0 \end{bmatrix}$$

The first step is to determine $\left[sI - A\right]$:

$$\left[sI - A\right] = \begin{bmatrix} s & 0 \\ 0 & s \end{bmatrix} - \begin{bmatrix} 0 & 1 \\ -2 & 0 \end{bmatrix} = \begin{bmatrix} s & -1 \\ 2 & s \end{bmatrix}$$

The determinant of $\left[sI - A\right] = s^2$, then, is

$$e^{At} = \left[sI - A\right]^{-1} = \frac{adj\ A}{\det\left[sI - A\right]} = \begin{bmatrix} \dfrac{s}{s^2+2} & \dfrac{1}{s^2+2} \\ \dfrac{-2}{s^2+2} & \dfrac{s}{s^2+2} \end{bmatrix}$$

$$= \begin{bmatrix} \cos\sqrt{2}t & 0.7071 \cdot \sin\sqrt{2}t \\ -1.4142 \cdot \sin\sqrt{2}t & \cos\sqrt{2}t \end{bmatrix}$$

There is much more to matrices than is covered here. However, this information is sufficient for use with state variable analysis.

2.8 Difference Equations

Difference equations are used when we wish to control a system using a computer. The difference equations are arrived at through the use of Newton's method for the approximation of the derivative, or Euler's methods for the approximation of the derivative. The development of difference equations and their applications follows.

Development of the Finite Difference Equation

If you remember learning the calculus of the derivative, you started out using the step method with Δx and Δy functions. The delta functions were allowed to become smaller and smaller until the delta value was infinitesimally small and a continuous function of the variables emerged. Prior to the time the delta function became infinitesimal, you had discrete, or finite, measurements for the variables.

It is relatively easy to take a continuous system and turn it into a discrete system; all that is required is to take samples of the output of the system at specific times. When that is done, the amplitude of the individual samples will vary in height. That being true, there will always be a small amount of error associated with the pulse amplitude compared to the continuous value of the output.

The discrete system is represented by an equation that is not continuous in time. It calculates the values of the system at specific samples of time. To accomplish the analysis, we will use the backward Euler approximation. The general form for Euler's equation is

$$\frac{d}{dx}f_k(x) = \lim_{\Delta x \to 0} \frac{f_k(x) - f_{k-1}(x)}{\Delta x} \tag{2.38}$$

The sample $(k - 1)$ is the sample prior to sample k and is therefore going backward from the present sample. Note that two values of x are used. The values equate to the present value of x and the previous value of x.

You are interested only in functions of time, that is, $f(x) = f(t)$. For that, Δx is equal to the time difference between samples, or Δt. Since the time difference between samples will not change, you can say that $\Delta t = T$, the sample time. Further, since you are working with functions of time, the formula is rewritten as

$$f'(t) = \frac{f_k(t) - f_{k-1}(t)}{T} \tag{2.39}$$

That is the equation of a first-order system. To get an idea of how that type of system is produced, look at the following example of a simple first-order discrete system derived from the continuous time function in the s-domain.

Example 2-6

Consider the following first-order system, as shown in Figure 2-3.

Solution

The first step is to change the Laplace form to the time domain, which is used to determine the discrete system. The time domain output of the transfer function of the block, using a unit step input, is $v_{out} = 0.5(1 - e^{-2t})$, an exponential that rises to 0.5 volts in approximately 2.5 seconds.

Using the principles that you learned in the preceding section, you can develop an equation for $u(t)$:

$$U(s) = \frac{1}{s(s + 2)} \quad A = \left[s\frac{1}{s(s + 2)} \right]_{s=0} = 0.5 \quad B = \left[(s + 2)\frac{1}{s(s + 2)} \right]_{s=-2} = -0.5$$

$$U(s) = 0.5\left[\frac{1}{s} - \frac{1}{(s + 2)} \right]$$

$$u(t) = 0.5(1 - e^{-2t})$$

From that, you can develop the value of $e(t)$ in terms of $u(t)$, as shown in Figure 2-3.

$$u(t) = \frac{e(t)}{\dfrac{d}{dt} + 2}$$

What is required is to find the derivative of $u(t)$ alone. Using the above formula, you can see that

$$e(t) = \frac{d}{dt}u(t) + 2u(t)$$

In all cases, you will designate $u(t) = u(kT)$, the value of $u(t)$ evaluated at a constant time interval T, which is called u_k, $e(t) = e(kT)$, and is designated as e_k. You also need a past value for $u(t)$, which you will always designate u_{k-1}. The function above now becomes

$$\frac{d}{dt}u(t) = e(t) - 2u(t)$$

You can replace the derivative in the above equation with the expression for the derivative, which was derived earlier, replacing f_k with u_k. That gives the derivative function as

$$\frac{d}{dt}u_k = e_k - 2u_k$$

(figure 2-3)

Block to convert to discrete form

$$E(s) \longrightarrow \boxed{\frac{1}{s + 2}} \longrightarrow U(s)$$

$$e(t) \longrightarrow \boxed{\frac{1}{\frac{d}{dt} + 2}} \longrightarrow u(t)$$

Example 2-6 (continued)

$\frac{d}{dt} u_k$ is replaced by Equation 2.39.

$$\frac{u_k - u_{k-1}}{T} = e_k - 2u_k$$

Then, multiply through the equation by T and collect the like terms, you have

$$u_k - u_{k-1} = Te_k - 2Tu_k$$

$$u_k(1 + 2T) = Te_k + u_{k-1}$$

$$u_k = \frac{Te_k + u_{k-1}}{1 + 2T}$$

The denominator of the equation will always be a constant since T never changes its value. You are free to select any value of T, which is convenient. Keep in mind that this type of system always introduces some error in the conversion of the input signal to the discrete form, and the error is a function of T. The smaller the value of T, the less the error in the conversion will be. There are, however, tradeoffs to be made, since very small values of T take longer to convert than larger values. You must determine the amount of error that can be tolerated.

For the purposes of this analysis, you will use $T = 0.1$ second and a value of $e(t) = 1$. You should recognize that when time is zero, the value of the function is also zero since u_{k-1} does not exist yet and is also zero. Since e_k is 1, the value of Te_k is 0.1 and will never change. You can rewrite the equation in the following form:

$$u_k = \frac{0.1 + u_{k-1}}{1.2}$$

Table 2-2 shows the output of this circuit vs. the continuous time function and the error for each of the time steps.

While the first-order system is useful, it is also important to have available different discrete time functions, such as second-order or other higher-order functions.

(table 2-2)

time	u_k	e(t)	Δ	time	u_k	e(t)	Δ
0.1	0.0833	0.0906	0.0073	1.6	0.4730	0.4796	0.0066
0.2	0.1528	0.1648	0.0120	1.7	0.4775	0.4833	−0.0058
0.3	0.2106	0.2256	0.0150	1.8	0.4812	0.4863	0.0051
0.4	0.2589	0.2753	0.0164	1.9	0.4844	0.4888	0.0044
0.5	0.2991	0.3161	0.0170	2.0	0.4870	0.4908	0.0038
0.6	0.3326	0.3494	0.0168	2.1	0.4891	0.4925	0.0034
0.7	0.3605	0.3767	0.0162	2.2	0.4909	0.4939	0.0030
0.8	0.3837	0.3990	0.0154	2.3	0.4925	0.4950	0.0025
0.9	0.4031	0.4174	0.0142	2.4	0.4937	0.4959	0.0022

(table 2-2) *continued*

time	u_k	e(t)	Δ	time	u_k	e(t)	Δ
1.0	0.4192	0.4323	0.0131	2.5	0.4948	0.4966	0.0018
1.1	0.4327	0.4446	0.0119	2.6	0.4956	0.4972	0.0016
1.2	0.4439	0.4546	−0.0107	2.7	0.4964	0.4977	0.0013
1.3	0.4533	0.4629	0.0096	2.8	0.4970	0.4982	0.0012
1.4	0.4611	0.4696	0.0085	2.9	0.4975	0.4985	0.0010
1.5	0.4676	0.4751	−0.0075	3.0	0.4979	0.4988	0.0009

The general equation for determining higher-order discrete functions is

$$\frac{d^N}{dt^N} f_k \triangleq \lim_{T \to 0} \frac{\sum_{n=1}^{N+1} (-1)^{n+1} P_\Delta(N,n) f_{k-n+1}}{T^N} \tag{2.40}$$

where $P_\Delta(N,n)$ is the Nth row and n^{th} element of Pascal's triangle. For the derivation of the second-order function, see Eric W. Weisstein, "derivative," from MathWorld, A Wolfram Web Resource.

The second-order derivative is, without derivation

$$\frac{d^2}{dt^2} f_k = \frac{f_k - 2f_{k-1} + f_{k-2}}{T^2} \tag{2.41}$$

Note that the time, T, is raised to the order of the derivative. That will make the derivation of the process equations more complex. An example of the second-order function is shown in Example 2-7.

Example 2-7

Consider the following s-plane function:

$$U(s) = \frac{E(s)}{s^2 + 2s + 5}$$

Provide a discrete time function based on a sampling time of 0.1 sec.

Example 2-7 (continued)

Solution

The first step is to change the Laplace function to the time domain:

$$u(t) = \frac{e(t)}{\dfrac{d^2}{dt^2} + 2\dfrac{d}{dt} + 5}$$

$$\frac{d^2}{dt^2}u(t) + 2\frac{d}{dt}u(t) + 5u(t) = e(t)$$

In this case, you are looking for the second derivative function; so all other functions can be transposed to the right-hand side of the equation. Also, $u(t)$ becomes u_k and $e(t)$ becomes e_k.

$$\frac{d^2}{dt^2}u_k = e_k - 2\frac{d}{dt}u_k - 5u_k$$

Since that equation has both a first and second derivative, you replace the second derivative with Equation 2.40 and the first derivative with Equation 2.39.

$$\frac{u_k - 2u_{k-1} + u_{k-2}}{T^2} = e_k - 2\left(\frac{u_k - 2u_{k-1}}{T}\right) - 5u_k$$

Convert the equation to a single line by multiplying both sides by T^2:

$$u_k - 2u_{k-1} + u_{k-2} = T^2e_k - 2Tu_k + 2Tu_{k-1} - 5T^2u_k$$

To find u_k, you must collect all terms that contain u_k on one side of the equation, then isolate u_k as a common factor:

$$u_k + 2Tu_k + 5T^2u_k = T^2e_k + 2Tu_{k-1} + 2u_{k-1} - u_{k-2}$$

$$u_k(1 + 2T + 5T^2) = T^2e_k + 2u_{k-1}(T + 1) - u_{k-2}$$

and the final function is

$$u_k = \frac{T^2e_k + 2u_{k-1}(T + 1) - u_{k-2}}{(1 + 2T + 5T^2)}$$

While that is a somewhat more complex formula, it is useful for any second-order system. Only the constants will change. At this point, you would select the time function, T, and simplify the equation. Keeping in mind that e_k will be equal to 1 as before, if T is selected to be 0.1 second, then

$$u_k = \frac{0.01 + 2.2u_{k-1} - u_{k-2}}{1.25}$$

The most difficult part of implementing that function is putting u_{k-1} and u_{k-2} in arrays for use in calculating values of u_k.

Definition of the z-Transform

The z-domain is the discrete time domain's equivalent of the s-domain for continuous functions. The z-transform method allows you to obtain difference equations that are easily implemented into computer programs.

The z-transform is, comparatively, a discrete version of the Laplace transform. The Laplace transform assumes that all values of any function have been summed over all time. That is because the derivatives used are infinitesimally small. When a function is sampled in discrete time steps, the difference is no longer infinitesimally small. Thus, for this type of function, a summation in time is used. The function of time, $f(t)$, which you used for continuous time systems, now becomes $f(kT)$, where T is the sample time and k is the specific sample being taken.

For the transforms, the exponential decay e^{-at} in s becomes a geometric series in $(z - k)$; and this series will generally converge to a final value. The following equations show the similarity between the Laplace function and the discrete series.

$$\mathcal{L}[f(t)] = \int_0^\infty f(t)e^{-st}dt$$

$$\mathcal{Z}\{f(t)\} \triangleq \sum_{k=0}^\infty f(kT)z^{-k}$$

(2.42)

Derivation of the Transforms

As in the continuous time domain, the discrete time domain also has transforms that are used to convert from the z-plane to the $f(kT)$ domain. These transforms enable us to write difference equations for use with computer implemented systems.

Discrete Delta Function

This is the counterpart of the continuous delta function, but in the discrete domain. The discrete delta function is defined as

$$\delta_0(kT) = \begin{cases} 1 & k = 0 \\ 0 & \text{otherwise} \end{cases}$$

(2.43)

δ_0 is a well-defined function whose z-transform is

$$\mathcal{Z}[\delta_0(kT)] = \sum_{k-0}^\infty \delta_0(kT)z^{-k} = z^0 = 1.0$$

(2.44)

Note that, like its continuous counterpart, its value is also 1.0.

Discrete Unit Step Function, *u(kT)*

The unit step function is defined in a manner similar to that in the time domain:

$$\begin{cases} u(kT) = 0, & k < 0 \\ u(kT) = 1, & k \geq 0 \end{cases} \tag{2.45}$$

Applying the equation given above, you have

$$Z[u(kT)] = \sum_{k=0}^{\infty} 1 \cdot z^{-k} \tag{2.46}$$

The series that is generated is

$$Z[u(kT)] = 1 + z^{-1} + z^{-2} + z^{-3} + \cdots \tag{2.47}$$

Equation 2.47 is a geometric series with terms to infinity. Remembering from your math courses, the sum to infinity of a geometric series is given by

$$s_{\infty} = \frac{A}{1 - r} \tag{2.48}$$

Where A is the first term in the series (1 in this case), and r is the common ratio of the series. The common ratio is found by dividing any term in the series by the term raised to the power of the term plus 1:

$$r = \frac{z^{-1}}{z^{0}} = \frac{z^{-2}}{z^{-1}} = \frac{z^{-3}}{z^{-2}} = \cdots \tag{2.49}$$

Since $z^{0} = 1$, $r = z^{-1}$. The unit step function for discrete time then becomes

$$Z[u(kT)] = \frac{1}{1 - z^{-1}} \tag{2.50}$$

It should be clear that as z approaches infinity, z^{-1} approaches zero and the equation approaches (converges to) 1.

Discrete Exponentially Decaying Function

The exponentially decaying continuous function is $f(t) = e^{-a}$. To make the continuous function a function of discrete time, you need to add k to the exponent and change the time t to a discrete time, T. The function of time now becomes

$$f(kT) = e^{-akT} \tag{2.51}$$

To transform the function to the z-plane, you must multiply the function by $\left(z^{-k}\right)$:

$$Z[f(kT)] \;=\; \sum_{k=0}^{\infty} e^{-akT} z^{-k} \;=\; \left(\sum_{k=0}^{\infty} e^{-aT} z^{-1}\right)^{k} \;=\; \sum_{k=0}^{\infty} q^{k} \tag{2.52}$$

Where $q = e^{-aT} z^{-1}$. This series converges only if $|q| < 1.0$, which requires a > 0. Then

$$Z\left[f(kt)\right] = \sum_{k=0}^{\infty} q^{k} = \frac{1}{1-q} = \frac{1}{1-e^{-aT}z^{-1}} \tag{2.53}$$

and

$$Z(e^{-akT}) \;=\; \frac{1}{1-e^{-aT}z^{-1}} \tag{2.54}$$

While that is an exponentially decaying function, it converges to a value of 1.0 when z is very large.

Functions of the Discrete sin(ωt) and cos(ωt)

For the function $f(t) = \sin(\omega t)$, you use the Euler function

$$e^{j\omega t} \;=\; \cos \omega t + j\sin \omega t \tag{2.55}$$

To use that function, you must add the sample number k and change t to T, as follows:

$$e^{j\omega \, kT} \;=\; \cos \omega kT + j\sin \omega Kt \tag{2.56}$$

In which the sine function is the imaginary part and the cosine function is the real part of the identity.

$$Z[\sin(\omega kT)] \;=\; Im\left[Z\!\left(e^{j\omega kT}\right)\right] \tag{2.57}$$

$$Z[\cos(\omega kT)] \;=\; Re\left[Z\!\left(e^{j\omega kT}\right)\right] \tag{2.58}$$

For those transforms, you will work only with the sine function and then assume that the cosine function works similarly, which it does. The series for the sine and cosine functions is

$$Z[f(kT)]= 1 + e^{j\omega T} \cdot z^{-1} + e^{j\omega 2T} \cdot z^{-2} + e^{j\omega 3T} \cdot z^{-3} + \cdots \tag{2.59}$$

The common ratio is found similarly to the other functions that have been defined, yielding

$$Z[f(kT)] \;=\; \frac{1}{1-e^{j\omega T}z^{-1}} \tag{2.60}$$

Then using Euler's identity, you expand the denominator of the function. Also, keep in mind that you want no imaginary terms in the denominator. Therefore, you must also conjugate the denominator, as follows:

$$Z[f(kT)] = \frac{1}{1 - \cos(\omega T)z^{-1} - j\sin(\omega T)z^{-1}} \tag{2.61}$$

$$Z[f(kT)] = \frac{1 - \cos(\omega T)z^{-1} + j\sin(\omega T)z^{-1}}{1 - 2\cos(\omega T)z^{-1} + [\cos^2(\omega T) + \sin^2(\omega T)]z^{-2}} \tag{2.62}$$

Both functions are determined from the single equation by separating the real and imaginary parts. The sine term is the imaginary component

$$Z[f(jkT)] = \frac{j\sin(\omega T)z^{-1}}{1 - 2\cos(\omega T)z^{-1} + [\cos^2(\omega T) + \sin^2(\omega T)]z^{-2}} \tag{2.63}$$

Notice the $\sin^2 + \cos^2$ term. Based on trigonometry, that term always equates to a value of 1. Thus, the transform can be simplified. In addition, the j's may be canceled from both sides of the equation, yielding a real value for the sine function:

$$Z[\sin(\omega T)] = \frac{\sin(\omega T)z^{-1}}{1 - 2\cos(\omega T)z^{-1} + z^{-2}} \tag{2.64}$$

The cosine term is developed in the same manner, using the real terms in the numerator of Euler's identity:

$$Z[\cos(\omega T)] = \frac{1 - \cos(\omega T)z^{-1}}{1 - 2\cos(\omega T)z^{-1} + z^{-2}} \tag{2.65}$$

Functions of Exponentially Decaying Discrete Sine and Cosine Functions

The forms for these functions are derived similarly to the sine and cosine terms, but you must add the exponentially damped function to the equation, which gives

$$Z[e^{-akT}\sin(\omega T)] = \frac{e^{aT}\sin(\omega T)z^{-1}}{1 - 2\left[e^{aT}\cos(\omega T)\right]z^{-1} + z^{-2}} \tag{2.66}$$

$$Z[e^{-akT}\cos(\omega T)] = \frac{1 - e^{aT}\cos(\omega T)z^{-1}}{1 - 2\left[e^{aT}\cos(\omega T)\right]z^{-1} + z^{-2}} \tag{2.67}$$

Those are all of the z-transforms that you will need to solve the problems assigned in this course. Be aware that there are more z-plane functions. The development of any additional transforms is done in the same manner as those developed here.

2.9 Theorems of the z-Transforms Similar to the Laplace Transform

Like the Laplace transform, the z-transform has certain properties. They are the property of linearity; the property of advancement, which is similar to the derivative function; and the property of delay, similar to integration. Further, there exists an initial value and final-value theorem.

Property of Linearity

The property of linearity states that a constant in the time domain is a constant in the z-domain:

$$Z[K_1 f_1(kT) + K_2 f_2(kT)] = K_1 Z[f_1(kT)] + K_2 Z[f_2(kT)] \qquad (2.68)$$

An example will clarify the concept.

Example 2-8

The following time function is to be sampled at one-second intervals ($T = 1$). Find the z-transform for the function.

$$f(t) = 2 + 3e^{-t} - 3e^{-2t} \cos(6t)$$

Solution

$$f(kT) = 2 + 3e^{-kT} - 3e^{-2kT} \cos(6kT)$$

The property of linearity allows you to rewrite the function in terms of z-transforms.

$$Z[f(kT)] = Z(2) + 3Z(e^{-kT}) - 3Z[e^{-2kT} \cos(6kT)]$$

$Z(2)$ is a unit step function whose amplitude is 2.0. Each of the functions can be written in terms of its z-transform.

$$Z[f(kT)] = Z[2] = 2 \cdot \frac{1}{1 - z^{-1}} = \frac{2}{1 - z^{-1}}$$

$3Z(e^{-kT})$ is an exponentially damped function in both domains.

$$Z[f(kT)] = 3Z[f(e^{-kT})] = 3 \cdot \frac{1}{1 - e^{-1}z^{-1}} = \frac{3}{1 - 0.368z^{-1}}$$

Example 2-8 (continued)

$3Z[e^{-2kT}\cos(6kT)]$ is an exponentially damped cosine wave function in both domains.

$$Z[f(kT)] = 3Z\{f[e^{-2kT}\cos(6kT)]\}$$

$$= \frac{1 - [e^{-2}\cos(6)]z^{-1}}{1 - 2[e^{-2T}\cos(6)]z^{-1} + e^{-4T} \cdot z^{-2}}$$

$$3Z\{f[e^{-2kT}\cos(6kT)]\} = \frac{1 - 0.135 \cdot 0.96z^{-1}}{1 - (2 \cdot 0.135 \cdot 0.96)z^{-1} + 0.018z^{-2}}$$

$$= \frac{1 - 0.13z^{-1}}{1 + 0.259z^{-1} + 0.018z^{-2}}$$

There are two points to be made regarding the last function. First, values of the sine and cosine functions are always in radians/sec. Second, all constants must be included in the final equation; they are not changed by the discrete time changes. Once you have arrived at the transforms for each of the functions, you can combine them in a single equation.

$$Z[f(kT)] = \frac{2}{1 - z^{-1}} + \frac{3}{1 - e^{-T}z^{-1}} + \frac{1 - 0.13z^{-1}}{1 + 0.259z^{-1} + 0.018z^{-2}}$$

Advance Theorem

This theorem is similar to the derivative in the Laplace domain. Its primary purpose is to deal with future values of a function. The theorem implies a shift to the left of n samples. If the function is $F(Z)$, then the transform of the function shifted to the left of n samples is

$$Z[f_{k+n}] = z^{-n}F(z) \tag{2.69}$$

This transform is not used often in this course and is included only for completeness.

Delay Theorem

The delay theorem is similar to the integral property in the Laplace domain. The delay property allows the calculation of the z-transform of a discrete function, which has been shifted n samples to the right. This delay implies the past values, as did the integral in the continuous sense. If the z-transform of the original discrete function is $F(Z)$, then the z-transform of the function shifted n samples to the right is

$$f_{k-n} = Z[f(k - n)T] = z^{-n}F(z) \tag{2.70}$$

This is an extremely important property that is used extensively in the derivation of computer algorithms.

If you consider the step function of the previous example and shift the step two units to the right, the shifted function becomes

$$Z[f_{k-2}] = z^{-2} \frac{2}{1 - z^{-1}} = \frac{2z^{-2}}{1 - z^{-1}} \tag{2.71}$$

Final-Value Theorem

The final-value theorem allows the calculation of the steady-state value of a function. It is similar to the final-value theorem of the Laplace domain. This theorem refers to the value of the function as the number of samples, k, approaches infinity. Thus

$$\lim_{k \to \infty} f_k = \lim_{z \to 1} (1 - z^{-1}) \cdot F(z) \tag{2.72}$$

Once again, this formula, similar to the final value of a continuous system, is applicable only if $\lim_{k \to \infty} f(k)$ exists. An example will help clarify the process.

Example 2-9

Consider the following discrete function:

$$Y(z) = \frac{z^2 + 2}{(1 - z^{-1})(z^2 - 1.6z + 0.7)}$$

Determine the final value of the function.

Solution

Using the final-value theorem and replacing z with 1

$$\lim_{k \to \infty} y_k = \lim_{z \to 1} (1 - z^{-1}) \cdot \frac{z^2 + 2}{(1 - z^{-1})(z^2 - 1.6z + 0.7)}$$

$$\lim_{k \to \infty} y_k = \frac{z^2 + 2}{(z^2 - 1.6z + 0.7)} = 30$$

There is also an initial-value theorem; but since all initial values are considered to be zero, it will not be derived here.

The properties that have been discussed are most important for the application of computer algorithms for control systems. It is now time to discuss the concept of the inverse of the discrete z-transform.

2.10 Inverse z-Transforms

Just as the Laplace transforms have inverse transforms in the continuous time domain, the z-transforms also have inverse transforms, but in the discrete time domain. Just as you can separate the Laplace transform into its poles and zeros; you also can separate the z-transform into its pole-zero form

$$F(z) = \frac{P(z)}{(z - p_0)(z - p_1)(z - p_2)\cdots(z - p_n)} \tag{2.73}$$

where $P(z)$ is an unspecified polynomial in z.

Similar to the Laplace transform, the z-transform can also be separated into its partial fraction form; but there is a slight difference.

$$F(z) = \frac{A_0 z}{z - p_0} + \frac{A_1 z}{z - p_1} + \cdots + \frac{A_n z}{z - p_n} \tag{2.74}$$

The difference is that in all cases, a z is in the numerator of the functions. Also note that this expansion is for real roots only. When the roots are complex, you must make some adjustments, repeating the sine and cosine function for convenience.

$$f(kT) = e^{-akT}[\cos(\omega kT) + j\sin(\omega kT)]$$

$$Z\{f[e^{-akT}(\cos(\omega kT) + j\sin(\omega kt))]\}$$
$$= \frac{1 - [e^{-aT}\cos(\omega T)]z^{-1} + j[e^{-aT}\sin(\omega T)]z^{-1}}{1 - 2[e^{-aT}\cos(\omega T)]z^{-1} + e^{-2aT}z^{-2}} \tag{2.75}$$

$$Z\{f[e^{-akT}\sin(\omega t)]\} = \frac{[e^{-aT}\sin(\omega t)]z^{-1}}{1 - 2[e^{-aT}\cos(\omega T)]z^{-1} + e^{-2aT}z^{-2}}$$

$$Z\{f[e^{-akT}\cos(\omega t)]\} = \frac{1 - [e^{-aT}\cos(\omega T)]z^{-1}}{1 - 2[e^{-aT}\cos(\omega T)]z^{-1} + e^{-2aT}z^{-2}}$$

If you generalize the second-order discrete function with complex roots, you have

$$\frac{A_0 z^2 + A_1 z}{z^2 + 2Bz + C} \tag{2.76}$$

Obviously, you must define factors B and C. From the forms above, you see that B must be equal to

$$
\begin{aligned}
B &= e^{-aT} \cos(T) \\
C &= e^{-2aT} \\
e^{-aT} &= \sqrt{C} \\
\omega T &= \cos^{-1}\left(\frac{B}{\sqrt{C}}\right)
\end{aligned}
\tag{2.77}
$$

Multiplying the general equation numerator and denominator above by z^2 allows you to see how the denominator is determined. The numerator must also be simplified. It is broken into two components, one for the sine and one for the cosine. The sine function is

$$
F[\sin(\omega kT)] = \frac{\sqrt{C} \cdot \sin(\omega T)z}{z^2 - 2Bz + C}
\tag{2.78}
$$

And the cosine is

$$
F[\cos(\omega kT)] = \frac{z^2 - Bz}{z^2 - 2Bz + C}
\tag{2.79}
$$

An example will help you understand the process for complex and real roots of a z-transformed function.

Example 2-10

Determine the discrete time domain output of the following z-transform.

$$
F(z) = \frac{z^2 + z}{z^3 - 1.8z^2 + 1.42z - 0.42}
$$

Solution

The first step is to factor the denominator to determine if the roots are real or complex. Since this is an odd-ordered function, there must be at least one real root. The factored form of the equation is

$$
F(z) = \frac{z^2 + z}{(z^2 - 1.2z + 0.7)(z - 0.6)}
$$

This function is then put into partial fraction form as follows:

$$
F(z) = \frac{A_0 z^2 + A_1 z}{z^2 - 1.2z + 0.7} + \frac{A_2 z}{z - 0.6}
$$

You now form the common denominator for a new numerator and collect the like terms.

$$
(A_0 + A_1)z^3 + (-0.6A_0 + A_1 - 1.2A_2)z^2 + (-0.6A_1 + 0.7A_2)z
$$

Example 2-10 (continued)

You form a three-equation simultaneous equation from the above values, which gives

$$A_0 \qquad\qquad\qquad A_2 = 0$$
$$-0.6A_0 \qquad A_1 \quad -1.2A_2 = 1$$
$$-0.6A_1 \qquad 0.7A_2 = 1$$

$$A_0 = -4.706 \qquad A_1 = 3.823 \qquad A_2 = 4.706$$

The function now becomes

$$F(z) = \frac{-4.706z^2 + 3.824z}{z^2 - 1.2z + 0.7} + \frac{4.706z}{z - 0.6}$$

The value of ωt is $\cos^{-1}(B/\sqrt{C})$. That yields the value of 0.771 rad/sec. From before, the sine term is $C \cdot \sin(\omega T)z$; and

$$\sqrt{C}\sin(\omega T)z = \sqrt{0.7}\sin(0.771)z = 0.837 \cdot 0.697z = 0.583z$$

Continuing the analysis, remove −4.706 from the numerator and determine the values of the sine and cosine terms.

$$F(z) = -4.706\left(\frac{z^2 - 0.813z}{z^2 - 1.2z + 0.7}\right)$$

Since the cosine term must contain the factor 0.6 and you see that the numerator must be $z^2 - 0.6$ and that the sine term must contain the value 0.583z, the same technique as for separating the transform into its sine and cosine components is used here, the new transform is then

$$F(z) = -4.706\left[\frac{z^2 - 0.6z}{z^2 - 1.2z + 0.7} - \frac{0.213z}{z^2 - 1.2z + 0.7}\right]$$

$$F(z) = -4.706\left[\frac{z^2 - 0.6z}{z^2 - 1.2z + 0.7} - \frac{0.213}{0.583} \cdot \frac{0.583z}{z^2 - 1.2z + 0.7}\right]$$

$$F(z) = -4.706\left[\frac{z^2 - 0.6z}{z^2 - 1.2z + 0.7} - 0.365 \cdot \frac{0.583z}{z^2 - 1.2z + 0.7}\right]$$

You can now determine the overall transform in the discrete time domain. Recall that the value of e^{-aT} is equal to the root of C and its value is 0.837; therefore, you replace e^{-aT} with that value. From the table of transforms, you can see that the damped cosine and sine functions are

$$0.837^k \cdot \cos(0.771k) \qquad 0.837^k \cdot \sin(0.771k)$$

The value of the exponentially damped part is *4.706(0.6)^k*, and the entire function is

$$f_k = -4.706\left[(0.837)^k \cos(0.771k) + 1.719(0.837)^k \sin(0.771k) + 4.706(0.6)^k\right]$$

Example 2-10 (continued)

You can simplify that a little more by removing the factor 0.837^k:

$$F(k) = 0.837^k\left[1.719\sin(0.771k) - 4.706\cos(0.771k)\right] + 4.706(0.6)^k]$$

2.11 Solution of Differential Equations Using z-Transforms

Just as in the continuous time domain, where the differential equations of a system must be transformed to the s-plane, the differential equations of a system must be converted to the z-plane for application to computer driven systems.

Difference Equations

Difference equations are often used to approximate the differential equation. The differential equation is always in the time domain, so the difference equation should approximate the same response as the differential equation except with a small error. As a simple example of the approximation, you will examine the series RC circuit. Example 2-11 shows the voltage equation from classical theory.

Example 2-11

Consider the following time domain equation:

$$v_s(t) = i(t)R + \frac{1}{C}\int i(t)dt$$

Transform the equation to the discrete time domain.

Solution

Differentiating both sides of the equation yields the differential equation

$$\frac{dv_s(t)}{dt} = R\frac{di(t)}{dt} + \frac{i(t)}{C}$$

From the previous difference equations, the derivative functions are

$$\frac{dv(t)}{dt} = \frac{v_k - v_{k-1}}{T} \qquad \frac{di(t)}{dt} = \frac{i_k - i_{k-1}}{T}$$

Rewriting the differential equation in terms of the difference gives

$$\frac{v_k - v_{k-1}}{T} = R \cdot \frac{i_k - i_{k-1}}{T} + \frac{i_k}{C} = \frac{i_k R}{T} - \frac{i_{k-1}R}{T} + \frac{i_k}{C}$$

Example 2-11 (continued)

Collecting all terms in i_k and taking T as a common factor yields

$$\frac{v_k - v_{k-1}}{T} = \frac{i_k}{T}\left[R + \frac{T}{C}\right] - \frac{i_{k-1}R}{T}$$

Then canceling all of the possible Ts yields

$$v_k - v_{k-1} = i_k\left[R + \frac{T}{C}\right] - i_{k-1}R$$

Solving for i_k results in

$$i_k = \frac{1}{R + \dfrac{T}{C}} \cdot \left[v_k - v_{k-1} + i_{k-1}R\right]$$

The term $1/[R + T/C]$ is a constant, which does not change when you transform the equation to the z-domain. To transform a difference equation to the z-domain, there are some procedures to follow:

$$u_k = U(z) \qquad e_k = E(z)$$
$$u_{k-1} = U(z)z^{-1} \quad e_{k-1} = E(z)z^{-1}$$
$$u_{k-2} = U(z)z^{-2} \quad e_{k-2} = E(z)z^{-2}$$

The quantities u and e can be any values, such as voltage and current. You can then say

$$I(z) = i_k \qquad i_{k-1} = I(z)z^{-1} \qquad V(z) = v_k \qquad v_{k-1} = V(z)z^{-1}$$

You can now rewrite the equation in terms of the new functions.

$$I(z) = \left[\frac{1}{R + \dfrac{T}{C}}\right]\left[V(z) - V(z)z^{-1} + RI(z)z^{-1}\right]$$

As you can see, $I(z)$ and $V(z)$ are both common to at least two of the terms in the equation. Rewriting the equation and solving for $I(z)$ results in

$$I(z) - I(z)R\left[\frac{1}{R + \dfrac{T}{C}}\right]z^{-1} = V(z)\left[\frac{1}{R + \dfrac{T}{C}}\right](1 - z^{-1})$$

$$I(z) = V(z)\left[\frac{(1 - z^{-1})}{R + \dfrac{T}{C}}\right] \cdot \frac{1}{1 - \left[\dfrac{R}{R + \dfrac{T}{C}}\right]z^{-1}}$$

Example 2-11 (continued)

If the previous equation is retransformed into the discrete time domain, you can use it to calculate the current in the circuit for any individual time sample.

If $v_s = 5$-volt step, $R = 1$ MegOhm and $C = 1$ μF.

$$I(z) = \frac{5}{1 - z^{-1}} \cdot \left[\frac{(1 - z^{-1})}{10^6 + \frac{0.1}{10^{-6}}} \right] \cdot \frac{1}{1 - \left[\frac{10^6}{10^6 + \frac{0.1}{10^{-6}}} \right] z^{-1}}$$

$$I(z) = 4.55 \cdot 10^{-6} \cdot \frac{1}{1 - 0.909 z^{-1}}$$

The inverse transform of that is

$$I(z_k) = 4.55 \cdot 10^{-6} \cdot (0.909)^k$$

2.12 MATLAB Operations on Functions, Matrices, and Discrete Functions

Operations are available as built-in functions in **MATLAB** for many of the forms of analysis you have learned in this chapter. Of particular usefulness is the M-file, which can be written for often-used mathematical calculations, then saved and used whenever needed.

MATLAB is matrix-oriented; that is, most functions are entered in the form of a matrix. While the usual mathematical calculations of addition, subtraction, multiplication, and division can be carried out in the command window, those same functions can be carried out for matrices.

MATLAB also has many built-in functions that can produce the familiar forms of an expression or equation in both the s- and z-domains. In most cases, you will enter the necessary coefficients in matrix form, often as a row vector. To enter the coefficients, use square brackets for the vector—$x = [a \ b \ c \ ... \ ]$.

Simple MATLAB Operations

MATLAB is a powerful tool to help you in the design of control or process systems. Almost all functions of a system can be evaluated using **MATLAB**, including the output of the system over either continuous or discrete time. Some of the concepts will be introduced here, and others in later chapters.

MATLAB can be used to check the answers you arrive at in the problems at the end of this chapter and all other chapters that follow, and, if you arrive at the correct constants to enter, the answers will be the same.

Use of MATLAB to Evaluate Partial Fractions

The partial fraction expansion capabilities of **MATLAB** will be used to show that the methods used for Examples 2-3 and 2-4 produce the same results. For some of the conventions used when entering values, vectors, and matrices in **MATLAB**, see Appendix A.

For Example 2-3

For Example 2-3, enter the vectors as shown below (the % indicates a comment).

$n = 4;$ %definition of the numerator
$d = \begin{bmatrix} 1 & 4 & 3 & 0 \end{bmatrix};$ %definition of the denominator
$\text{residue}(n,d)$ %call for partial-fraction values

Note that the numerator is entered as a number without square brackets. That is acceptable when the numerator is a pure number.

The output of the calculation is 1.3333, –2.0, and 0.6667, the same as for the manual method. The one thing to remember is that the coefficients A, B, and C are in reverse order, with A being the bottom coefficient.

For Example 2-4

The form for this expansion is the same as for the previous example:

$n = \begin{bmatrix} 1 & 4 \end{bmatrix};$
$d = \begin{bmatrix} 1 & 6 & 12 & 8 & 0 \end{bmatrix};$
$\text{residue}(n,d)$

The output of this calculation is, –0.5, –1, –1, 0.5, the same as for the example. Note that the coefficients are in reverse order. Be careful when applying the coefficients to the final function.

Use of MATLAB for Convolution

MATLAB is useful for convolution also. In this case, convolution is used to multiply polynomials. For instance, given two polynomials

$$p_1 = s^2 + 4s + 7 \qquad p_2 = s^3 + 6s^2 + 11s + 15$$

The *conv(a,b)* command can be used to multiply the two polynomials and return the coefficients of s for the product.

p1 = $\begin{bmatrix} 1 & 4 & 7 \end{bmatrix}$; p2 = $\begin{bmatrix} 1 & 6 & 11 & 15 \end{bmatrix}$;
conv$(p1,p2)$

ans =
    1   10   42   101   137   105

The polynomial is

$$P(s) = s^5 + 10s^4 + 42s^3 + 101s^2 + 137s + 105$$

Deconvolution is the process of polynomial division. However, you must have at least one of the polynomials that make up the polynomial multiplication, as well as the product. For instance, if you have *p1* as above and the polynomial, you can recover the other polynomial, using $\begin{bmatrix} q,r \end{bmatrix}$ = deconv(p1, p2), where *q* is a quotient and *r* is the remainder after the division takes place, as follows:

p1 = $\begin{bmatrix} 1 & 4 & 7 \end{bmatrix}$; p = $\begin{bmatrix} 1 & 10 & 42 & 101 & 137 & 105 \end{bmatrix}$;
$\begin{bmatrix} q,r \end{bmatrix}$ = deconv$(p,p1)$

That returns the quotient of the division of *p* by *p1*.

q =
    1   6   11   15
r =
    0   0   0   0   0   0

The remainder, *r*, is zero, as expected; and the quotient is

$$p(s) = s^3 + 6s^2 + 11s + 15$$

Addition and Subtraction of Matrices

To add matrices, define each matrix as shown; then define the sum of the matrixes with another matrix and the sum of the matrixes entered.

A = $\begin{bmatrix} 1 & 3 & -5; & 2 & 6 & 0; & 4 & 4 & 1 \end{bmatrix}$ *% definition of a (3×3) matrix; notice the*
B = $\begin{bmatrix} 2 & 5 & 7; & 3 & 6 & 9; & -5 & 0 & 1 \end{bmatrix}$ *% semicolons between the row values.*
C = A + B   *% add matrix A to matrix B.*

The value of C is

C = $\begin{bmatrix} 3 & 8 & 2; & 5 & 12 & 9; & -1 & 4 & 2 \end{bmatrix}$

Note that the matrices are entered by typing in the row values separated by a semi-colon, and the output is presented in the same way.

Subtract the matrix:

$D = A - B$

The value of D is

$D = \begin{bmatrix} -1 & -2 & -12; & -1 & 0 & -9; & 9 & 4 & 0 \end{bmatrix}$

Multiplication and Division of Matrices

Since the previous matrices meet the criteria for multiplication, you can multiply them both ($A*B$ and $B*A$), which will result in different matrices.

$E = A * B$

Where * is the multiplication symbol in **MATLAB**. The value of E is

$E = \begin{bmatrix} 36 & 23 & 39; & 22 & 46 & 68; & 15 & 44 & 65 \end{bmatrix}$

$F = B*A$

The value of F is

$F = \begin{bmatrix} 40 & 64 & -3; & 51 & 81 & -6; & -1 & -11 & 26 \end{bmatrix}$

Note that the value of the matrixes is not the same.

Divide the matrices:

$G = A/B$

The value of G is

$G = \begin{bmatrix} 16 & 12.8333 & -1.5; & 15.333 & 11.778 & -1.333; & 7.667 & -5.722 & -1.667 \end{bmatrix}$

Matrix Exponentials

MATLAB can be used to produce the matrix exponential in two ways: using the commands *exp(matrix)*, which performs an element-by-element matrix exponentiation; and *expm(matrix)*, which uses the Padé approximation to the exponential. The matrices returned are numeric, based on the power of the matrix. If a square matrix is

$$A = \begin{bmatrix} 1 & 0 & 0 \\ 0 & 3 & 0 \\ 1 & 0 & -2 \end{bmatrix}$$

In element-by-element exponentiation, *e* is raised to the power indicated by each matrix element as follows:

$\exp(A)$

ans =

2.7183	1.0000	1.0000
1.0000	20.0855	1.0000
2.7183	1.0000	0.1353

The matrix exponential expm(a) returns a different matrix:

$\text{expm}(a)$

ans =

2.7183	0	0
0	20.0855	0
0.8610	0	0.1353

Notice that the main diagonal elements are the same, which is true for any triangular matrix, but the values above and below the diagonal are different.

Creation of MATLAB M-Files

An M-file is a reusable file that can contain variables, expressions, and analysis and plot types. If there are calculations that you make often, the M-file is a convenient way to store the equations for calculation. There are two types of M-files:

➤ Scripts, which do not accept input arguments or return output data. They operate in the workspace environment.

➤ Functions, which do accept input arguments and can return output data. Any variables in the function are local to the function.

An M-file may be made using any text editor, but it must not contain any special characters or formatting functions of the editor. Only straight text is allowed.

Scripts

A script file contains all of the parameters required for analysis. This type of file executes when the filename is typed in the command window.

To invoke the internal editor, type either

edit

or

edit filename

If the filename does not exist, you are prompted to create a new M-file. If you do not use the internal editor, make sure you use *filename.m* when you save the file.

An example of a simple script that produces the residues of a third-order s-domain transfer function (which will be saved by the name *resid.m*) produced using the **MATLAB** editor is

%This file produces the residues of a third-order equation with real roots

%not repeated. The file is fixed; and to obtain other residues, this file

%must be opened and the vectors changed.

$$A = \begin{bmatrix} 1 & 6 \end{bmatrix};$$
$$B = \begin{bmatrix} 1 & 6 & 11 & 6 \end{bmatrix};$$
$$C = \text{residue}(A,B)$$

If you type *help filename,* the first three lines of the file are displayed, such as

help resid

To run the file, make sure you have the proper path name for the file. To change the path, use the following method:

cd('c:/directory1/directory2)

To save the file, you should change the path to a folder in which you can keep M-files. If you do not, the next time **MATLAB** is updated, the files are likely to be erased.

When you run the file, it will always produce the same values unless you edit the file to have the transfer function you need.

Script M-files are useful, but they are limited in scope since changing the output requires changing the vectors in the M-file. A more utilitarian file is the function file.

Functions

The use of a function in an M-file allows you to enter parameters in the command window and then execute the file. The keyword *function* is usually the opening line for this type of M-file. The help lines follow the function keyword but can come before the function argument, as shown:

function   output argument = function name(input arg1, input arg2, …)

The name of the M-file should be the same as the function name.

An example of a function M-file which allows you to enter the numerator, denominator, and sampling time of a transfer function in the continuous time form and calculates the continuous and discrete forms of transfer function. Transfer functions are discussed in Chapter 4. For now, you can consider it as a simple gain: $A = V_{out} / V_{in}$.

function sysc = xfer(n, d, t)

% M-file to generate a transfer function in both

% continuous and discrete time. Enter the numerator and denominator in

% continuous form; then enter a sampling time, $t > 0$ to produce the discrete %functions.

sys = $tf(n,d)$
sysd = $c2d(sys,t)$

Note that there is no separation between the first and second lines, but there is between the last line of the help lines and the first line of the function. The space ends the help for the function. To call this M-file, enter n, d, and t in this manner:

$n = \begin{bmatrix} 1 & 15 \end{bmatrix}$;
$d = \begin{bmatrix} 1 & 8 & 11 & 15 \end{bmatrix}$;
$t = 0.5$;

Then enter

xfer(n, d, t)

Both the continuous and discrete transfer functions are displayed in the command window.

If you type *help xfer*, the help lines will be displayed on the command window screen.

Prompting for input

The M-file can also prompt for keyboard input. To prompt for input from the user, use the following method:

name = input('prompt string')

For the above file, enter the prompted information at each prompt. (Enter only the vector for the requested information, not the name of the vector.) Note that the function is now empty at the calling line. The file is modified as follows:

function sysc = xfer()

% M-file to generate a transfer function in both

% continuous and discrete time. Enter the numerator and denominator in

% continuous form; then enter the sampling time, $t > 0$.

n = input('Enter numerator ')

d = input('Enter denominator ')

t = input('Enter sampling time ')

sys = $tf(n,d)$

sysd = c2d(sys,t)

To activate the file, call the file using

xfer( )

Then follow the prompts as follows:

Enter numerator [1 15] in the form

$n = 1$ 15

Enter denominator [1 6 11 15] in the form

$d = 1$ 6   11   15

Enter a sampling time of 0.1 second

$t =$ 0.1

The output is, of course, the same as for the original M-file.

As you can see, **MATLAB** is very useful for calculations and matrix manipulations. **MATLAB** can be used for a host of other applications, as well as for matrix operations and partial fractions. More information on using **MATLAB** is given in Appendix A.

Summary

➤ In this chapter we have shown the methods of solving linear time invariant differential equations using both continuous time and discrete time methods.

➤ We have developed the concepts of the Laplace transform, convolution, matrices, and discrete mathematical models as applied to linear systems.

➤ Several examples have shown the use of partial fractions to produce the residues of s-domain equations for three conditions of the quadratic function.

➤ In this chapter we examined methods of retransformation from the Laplace domain to the time domain.

➤ You learned the concepts of matrices and matrix operations, including addition, subtraction, multiplication, and inversion (which replaces division for matrices).

➤ This chapter examined the methods of obtaining difference equations from continuous equations for computer simulation.

➤ Properties of the z-transform with related functions in the s-domain were examined.

➤ You examined the use of **MATLAB** in the solution of various types of functions, such as partial fractions and matrices, and transfer functions.

➤ **MATLAB** was used to obtain discrete time transfer functions from continuous time Laplace transforms.

➤ Fundamental concepts of M-files and functions in **MATLAB** for determining transfer functions and output graphs were shown and examples given.

Problems

P2.1   Determine which of the following differential equations are linear time invariant.

a. $5\dfrac{d^3x(t)}{dt^3} + 15\dfrac{d^2x(t)}{dt^2} + 10\dfrac{dx(t)}{dt} + 15x(t) = 0$

b. $\dfrac{d^3x(t)}{dt^3} + 4x^2\dfrac{d^2x(t)}{dt} + 3x(t) = r(t)$

c. $\dfrac{d^2x(t)}{dt^2} + 5\dfrac{dx(t)}{dt} + 6x(t) = r(t) + \dfrac{dr(t)}{dt}$

d. $\left(\dfrac{d^2x(t)}{dt}\right)^{1.5} + 3\dfrac{dx(t)}{dt} + 5x(t) = 0$

e. $\dfrac{d^3x(t)}{dt^3} + 5\cos(3t)\dfrac{d^2x(t)}{dt^2} + 5\dfrac{x(t)}{dt} + 5x(t) = 3r(t)$

f. $\dfrac{d^4x(t)}{dt^4} + 6\dfrac{d^3x(t)}{dt^3} + 11\dfrac{d^2x(t)}{dt^2} + 6\dfrac{dx(t)}{dt} + 6x(t) = 4r(t) + 2\dfrac{dr(t)}{dt}$

P2.2   For the following differential equations, find the Laplace transforms.

a. $\dfrac{d^2x(t)}{dt^2} + 3\dfrac{dx(t)}{dt} + 2x(t) = 2r(t)$

b. $\dfrac{d^3x(t)}{dt^3} + 6\dfrac{d^2x(t)}{dt^2} + 11\dfrac{dx(t)}{dt} + 6x(t) = 6r(t) + 6\dfrac{dr(t)}{dt}$

c. $4\dfrac{d^4x(t)}{dt^4} + 6\dfrac{d^3x(t)}{dt^3} + 12\dfrac{d^2x(t)}{dt^2} + 18\dfrac{dx(t)}{dt} + 24x(t) = 12r(t) + 6\dfrac{dr(t)}{dt} + 3\dfrac{d^2r(t)}{dt^2}$

P2.3 For the following Laplace transfer functions, determine the time domain functions. Use whatever computer program you have available to do any partial fraction expansions.

a. $\dfrac{2s + 6}{s^2 + 8s + 12}$

b. $\dfrac{9}{s^2 + 11s + 18}$

c. $\dfrac{6s + 6}{s^3 + 6s^2 + 11s + 6}$

d. $\dfrac{41}{s^3 + 13s^2 + 81s + 205}$

e. $\dfrac{2s + 9}{s(s^2 + 3s + 9)}$

f. $\dfrac{s^2 + 6s + 15}{s^3 + 8s^2 + 24s + 15}$

P2.4 For the following two transfer functions, determine the time domain output.

a. $F(s) = \dfrac{s^2 + 4s + 3}{s(s^2 + 5s + 6)}$

b. $F(s) = \dfrac{s^3 + 12s^2 + 47s + 60}{s(s^3 + 8s^2 + 19s + 12)}$

P2.5 For the following functions, use convolution to find the time domain function.

a. $\dfrac{s^2}{(s + 3)(s + 5)}$

b. $\dfrac{s + 2}{s^2 + 6s + 5}$

c. $\dfrac{12}{(s + 1)(s + 3)(s + 4)}$

P2.6 Plot the poles and zeros of the following functions on the s-plane. Mark the poles with an **x** and the zeros with an **o**. (The s-plane is a Cartesian coordinate graph with axes of $\pm \sigma$, and $\pm j\omega$.)

a. $F(s) = \dfrac{6(s + 2)}{s(s^2 + 7s + 12)}$

b. $F(s) = \dfrac{s - 5}{s^2(s^2 + 8s + 25)}$

c. $F(s) = \dfrac{s^2 + 16}{s^4 - 24s^2 + 400}$

d. $F(s) = \dfrac{s^2 + 6s + 13}{s(s^3 + 7s^2 + 25s + 39)}$

P2.7   For the following matrices, carry out the indicated operations.

a. $\begin{bmatrix} 1 & -5 \\ 2 & 6 \end{bmatrix} + \begin{bmatrix} -3 & 5 \\ 6 & -6 \end{bmatrix}$

b. $\begin{bmatrix} 3 & 4 & -5 \\ 6 & 0 & 9 \\ 4 & -5 & 10 \end{bmatrix} - \begin{bmatrix} -4 & 0 & -6 \\ 8 & 9 & 15 \\ 7 & -5 & 0 \end{bmatrix}$

c. $\begin{bmatrix} \dfrac{3}{s} & \dfrac{1}{s+5} & 4 \\ 5 & 9 & \dfrac{1}{s} \end{bmatrix} + \begin{bmatrix} 1 & 0 & -5 \\ \dfrac{2}{s+8} & 2s & 1 \end{bmatrix}$

P2.8   For the matrices shown, calculate the determinants. Use any calculator or computer program available for the determinants. If the determinant cannot be formed, indicate why. If the matrix is singular, so indicate.

a. $\begin{bmatrix} 6 & 5 & 0 \\ 3 & 9 & 8 \\ 4 & -6 & 9 \end{bmatrix}$

b. $\begin{vmatrix} 1 & 2 & 3 \\ 4 & 5 & 6 \\ 7 & 8 & 9 \end{vmatrix}$

c. $\begin{bmatrix} 1 & -5 & 8 \\ 2 & 0 & 9 \end{bmatrix}$

d. $\begin{bmatrix} 2 & -2 & 2 & 4 \\ 6 & 10 & 18 & 12 \\ 14 & -5 & 0 & 9 \\ 13 & 12 & 6 & -5 \end{bmatrix}$

P2.9   Determine the transpose of the matrices in Problem 2.8.

P2.10   Form the adjoint of the following matrices.

a. $\begin{bmatrix} 2 & 8 \\ 10 & 50 \end{bmatrix}$

b. $\begin{vmatrix} 3 & 8 & 4 \\ 6 & 9 & 12 \\ 0 & 15 & 5 \end{vmatrix}$

P2.11   Determine the inverse of the matrices in Problem 2.9 if they exist.

P2.12   For the following matrices, determine the exponential matrix, using the inverse Laplace method.

a. $\begin{bmatrix} 1 & 0 \\ 0 & 1 \end{bmatrix}$

b. $\begin{bmatrix} 1 & 1 \\ 0 & -4 \end{bmatrix}$

c. $\begin{bmatrix} 1 & 0 & 0 \\ 0 & 1 & 0 \\ 1 & 0 & 1 \end{bmatrix}$

P2.13   Perform the partial fraction expansion of the following functions. Then convert the continuous time functions to z-transforms, using the z-transform table.

a. $F(s) = \dfrac{1}{s + 5}$

b. $F(s) = \dfrac{3}{s^2 + 4s + 3}$

c. $F(s) = \dfrac{6}{s^2 + 6s + 9}$

d. $F(s) = \dfrac{9}{s^2 + 6s + 13}$

e. $F(s) = \dfrac{s + 6}{s^3 + 6s^2 + 11s + 15}$

f. $F(s) = \dfrac{s^2 + 2s + 1}{s^3 + 3s^2 + 4s + 2}$

P2.14   For the following Laplace functions, determine a difference equation, using 0.1 second as the sampling time.

a. $F(s) = \dfrac{1}{s^2 + s + 1}$

b. $\dfrac{s + 4}{s^2 + 5s + 4}$

c. $F(s) = \dfrac{s + 12}{s^3 + 8s^2 + 18s + 12}$

P2.15   Perform the partial fraction expansions for the following z-transform functions.

a. $F(z) = \dfrac{2}{z(z^2 - 5z + 6)}$

b. $F(z) = \dfrac{z - 2}{z(z - 1)(z^2 - 6z + 9)}$

c. $F(z) = \dfrac{z^2 - 7z + 10}{(z - 1)^2 (z^2 - 6z + 13)}$

P2.16   Find the inverse z-transforms $f(k)$ of the following functions. Use partial fraction expansions as necessary. Then use the z-transform table.

a. $F(z) = \dfrac{z}{z - 0.7z + 0.1}$

b. $F(z) = \dfrac{5}{(z - 1)(z^2 - 3z + 2)}$

P2.17   Find the discrete time function for the following z-transform.

$$F(z) = \dfrac{z^2 - z}{(z - 0.6)(z^2 - 1.2z + 0.7)}$$

P2.18   For the following transfer function, transform the equation to the discrete time domain.

$$i(t) = \dfrac{v(t)}{R} + \dfrac{1}{L}\int v(t)dt$$

MATLAB Problems

M2.1   Write a script M-file that converts the following $F(s)$ to the discrete domain. Use $T_s = 0.15$ sec.

$$F(s) = \dfrac{12}{s^3 + 6s^2 + 11s + 24}$$

Show the output of the system for a unit step input.

M2.2   Write a function type of M-file that allows input of $F(s) = N(s)/D(s)$ and the sampling time and then converts the function to the discrete time domain and plots the output for a unit step input.

Chapter 3

Control System Devices and Components

Chapter Objectives

After completing this chapter, you should be able to:

➤ Describe the concept of linear, lumped component models such as resistance, capacitance, inductance, torque, velocity, inertia, and damping.

➤ Associate the parameters of electrical and mechanical components to make a composite model of an electric motor for simulation.

➤ Calculate the requirements for gears and gear trains.

➤ Calculate the inertia, damping of gears, and gear ratios and torque transmission.

➤ Determine gear ratios for optimum transfer of torque to a load.

➤ State the three methods of heat transfer, and describe the concept of thermal resistance and time constant.

➤ Describe the difference between compressible and incompressible fluids, and describe the concept of hydraulic resistance and time constant.

➤ Design simple opamp circuits for use in control systems.

➤ State the reason for using V-I and I-V conversion circuits in current loops for transmission of electrical signals over long distances.

➤ Describe types of transducers used in control systems to convert mechanical motion, temperature, and fluid flow to electrical signals.

3.1 Introduction

Control systems use many different types of components: electrical, mechanical, fluid valves, and others. In developing a strategy for control of a plant (the device being controlled), the dynamic and static mathematical model for the system must be developed.

In general, the model will be a continuous time model; but it may also be a discrete model to facilitate computer control. Most of the devices used are continuous in nature; therefore, when they are adapted for computer control, sampling interfaces are used to convert the continuous function to a discrete function. An example is the A/D and D/A converters for input to and output from a computer. The model developed is often a composite of the plant and controlling devices that form a complete system.

The level of input to a system does not generally affect the response of the system in the sense that the normalized response curve will be the same for any input level. For that to be true, the elements must exhibit a linear characteristic. For example, a resistor has a voltage-to-current ratio (resistance) that is independent of the voltage applied. Further, some components may exhibit a derivative or integral component that defines another linear function. For that reason, linear differential equations are used to describe the system. From that point of view, linearity requires that if $r(t)$ produces $y(t)$, then $Kr(t)$ produces $Ky(t)$.

3.2 Lumped, Linear Component Models

Lumped linear components are components that are individual units of a system, as opposed to distributed components where the parameters are distributed along a path. The resistor is an example of a lumped parameter and the coaxial transmission line is an example of a distributed parameter device.

Electrical and Mechanical Elements

Passive electrical components are the resistor, capacitor, and inductor shown in Table 3-1, with their linear differential equation relationships and their Laplace equivalents. The Laplace components are shown without initial conditions, which will be the usual case for your investigations. The components are called **passive** because they can only dissipate or store energy; they cannot generate energy of their own. The resistor dissipates energy in the form of heat and is similar to friction (damping) in a mechanical system. The inductor is similar to mass in a mechanical system in that it stores **kinetic energy** in a magnetic field. The capacitor is similar to elasticity in a mechanical system because it stores **potential energy** in an electric field. The units of the components are ohms, henrys, and farads, respectively, and are constants.

(table 3-1)

Component	f(t)	F(s)
	$e(t) = R \cdot i(t)$ $i(t) = G \cdot v(t),\ G = \dfrac{1}{R}$	$E(s) = I(s) \cdot R$ $I(s) = E(s) \cdot G$
	$e(t) = L\dfrac{di(t)}{dt}$ $i(t) = \dfrac{1}{L}\displaystyle\int_0^t v(t)dt + i(0)$	$E(s) = sLI(s)$ $I(s) = \dfrac{E(s)}{sL}$
	$e(t) = \dfrac{1}{C}\displaystyle\int_0^t i(t)dt + e(0)$ $i(t) = C\dfrac{dv(t)}{dt}$	$E(s) = \dfrac{I(s)}{sC}$ $I(s) = sCE(s)$

Passive mechanical components are divided into two types based on the motion they produce: translational or rotational. For each type, the first element is a viscous damper. For translational motion, the damper is usually an oil-filled cylinder and piston, with holes in the piston to allow a restricted flow path for the fluid to go from one side of the piston to the other. For the rotational damper, the resistance to motion is a load torque.

In many cases, the friction will vary in direct proportion to the translational or angular velocity. The symbol B is used to describe the friction for both cases, but the units of measure are different. For translational motion, the friction is defined as force per unit velocity; for rotational motion, the friction is torque per unit angular velocity.

The other two components of mechanical systems store energy as kinetic or potential energy, respectively. The mass and spring are described in terms of displacement $(x$ or $\theta)$ where the zero position is the point where force or torque is zero. The components and their relationships are shown in Tables 3-2 and 3-3.

(table 3-2)

	$f(t) = Bv(t) = B\dfrac{dx(t)}{dt}$	$F(s) = BV(s) = sBX(s)$
	$f(t) = M\dfrac{dv(t)}{dt} = M\dfrac{d^2x(t)}{dt^2}$	$F(s) = sMV(s) = s^2MX(s)$
	$f(t) = K\displaystyle\int_0^t v(t)dt + f(0) = Kx(t)$	$F(s) = K\dfrac{V(s)}{s} = Kx(s)$

(table 3-3)

B (symbol) T(t), ω(t)	$T(t) = B\omega(t) = B\dfrac{d\theta(t)}{dt}$	$T(s) = B\Omega(s) = s\Theta(s)$
J (symbol) T(t), ω(t)	$T(t) = J\dfrac{d\omega(t)}{dt} = J\dfrac{d\,\theta(t)}{dt^2}$	$T(s) = sJ\Omega(s) = s^2\Theta(s)$
K (symbol) T(t) ω(t)	$T(t) = K\int_0^t \omega(t)dt + T(0) = K\theta(t)$	$T(s) = \dfrac{K\Omega(s)}{s} = K\Theta(s)$

Electrical Models

As the tables clearly show, there are similarities between the equations. In most cases, the equations are the same; only the units and variable names are different. For instance, the voltage developed across a resistor is a function of the resistance and current and represents the electrical equivalent of friction. The friction damping is a function of the damping constant and the velocity in a mechanical system.

Definitions of the system constants are:

1. **Capacitance.** The element of a system that stores energy in an electric field; it is analogous to storing potential energy.
2. **Inductance.** The element of a system that stores energy in a magnetic field; it is analogous to storing kinetic energy.
3. **Resistance.** The element of a system that opposes the flow of current in the system.

A simple, series RLC electrical circuit and its describing equation are shown in Figure 3-1 and Equation 3.1.

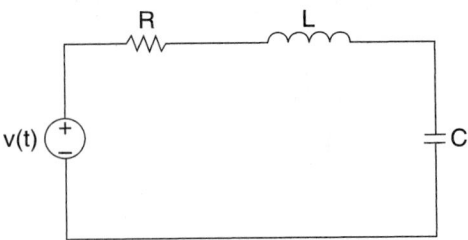

(figure 3-1)

RLC circuit for analysis

$$v(t) = Ri(t) + L\frac{di}{dt} + \frac{1}{C}\int_0^t i(t)dt + v_C(0) \qquad (3.1)$$

An example of an electrical system will show the use of the parameters discussed.

Example 3-1

For the circuit shown in Figure 3-2, write the differential equations for the input voltage and output voltage, using the capacitor as the output device.

$$v_C(t) = 20\int i(t)\,dt + v(0)$$

$$v_i(t) = 5i(t) + 2\frac{di(t)}{dt} + 20\int i(t)\,dt + v(0)$$

Solution

The differential equations are cumbersome to work with, so to simplify the solution, the equations are converted to the s-plane with zero initial conditions.

$$V_o(s) = \frac{20I(s)}{s}$$

$$V_i(s) = 5I(s) + 2sI(s) + \frac{20}{s}I(s)$$

$$V_i(s) = I(s)\left(\frac{5s + 2s^2 + 20}{s}\right) = I(s)\left(\frac{2s^2 + 5s + 20}{s}\right)$$

The equation for the voltage across the capacitor is given by

$$V_o(s) = V_i(s)\left(\frac{20}{s^2 + 5s + 20}\right)$$

Since $I(s)$ is common to all terms, it cancels and you have the equivalent of the voltage divider equation for the circuit. Clearly, the output will be equal to the input after some time, and the circuit will contain exponentially damped sine and cosine terms. The time domain response of the circuit for a unit step input is shown in Figure 3-3.

(figure 3-2)

V_{out} *for RLC circuit*

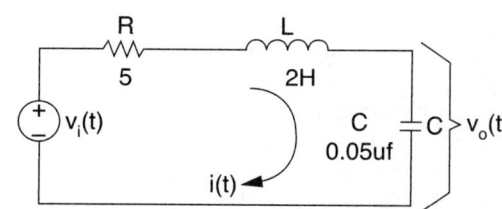

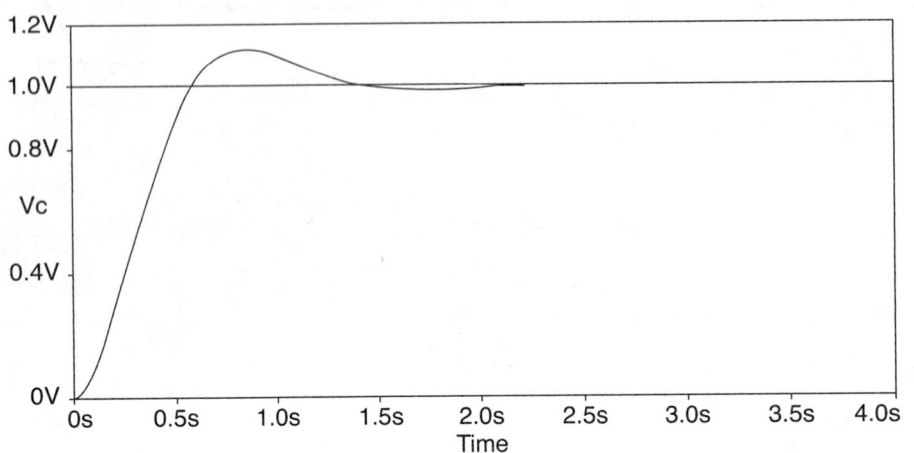

(figure 3-3)

Step response V_{out} for RLC circuit

Mechanical Models

Lumped mechanical component models have describing equations as shown in the Tables 3-2 and 3-3. Note the similarity of the model equations to the equations of the electrical models, the only thing that changes is the parameter names, but the forms are identical.

Translational Motion

Translational motion is defined as "motion in a straight line." To describe the motion, several variables are defined, such as the mass in the system, the damping in the system, and the distance traveled. The equation describing force applied to a system is given in Equation 3.2, and the system is shown in Figure 3-4.

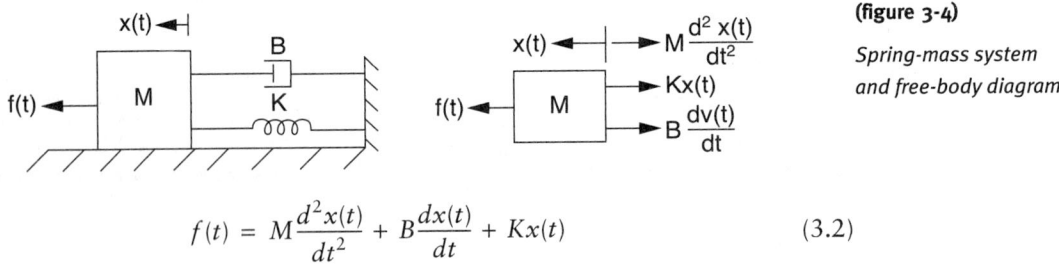

(figure 3-4)

Spring-mass system and free-body diagram

$$f(t) = M\frac{d^2x(t)}{dt^2} + B\frac{dx(t)}{dt} + Kx(t) \tag{3.2}$$

Definitions of the system constants are as follows:

Friction. Friction is the equivalent of resistance in an electrical system. Three types of friction are encountered in mechanical systems:

   Coulomb friction is constant with velocity; only the sign of the frictional force changes with a reversal of direction.

Static friction opposes the start of motion from a static position; and must be overcome before motion can start. Once the body is in motion, this friction vanishes and the friction of motion takes over.

Viscous friction varies linearly with the force applied and the velocity of the body.

Mass. The element in a mechanical system that stores the kinetic energy of the system.

Spring. An element whose linear deflection is proportional to the applied force; the spring stores the potential energy of the system and is analogous to the capacitor in an electrical system.

The sum of all forces applied to a mechanical system is:

$$\sum f(t) = M\alpha \tag{3.3}$$

Mass. The force applied to a mass is given by

$$f(t) = M\alpha = M\frac{d^2x(t)}{dt^2} = M\frac{dv(t)}{dt} \tag{3.4}$$

Where force is given in pounds $\left(\text{slug} \cdot \text{ft/s}^2\right)$ or Newton's $\left(\text{kg} \cdot \text{m/s}^2\right)$, M is in slugs or kilograms and x is distance in inches, feet, or meters. When weight is given in pounds, you must convert to slugs or convert the mass to the International System (SI) in kilograms. The slug is the gravitational constant in the United States Customary System (USCS) and is 32.167 ft/s^2 (32.2 ft/s^2 for convenience). The gravitational constant in the SI is 9.807 m/s^2. To convert from weight in pounds to mass in slugs, use Equation 3.5.

$$M = \frac{W}{g} = \frac{W(lbs)}{32.2} \tag{3.5}$$

Damping. The force applied in a mechanical system that is proportional to the rate of change of motion in a system; it is given by

$$f(t) = B\frac{dx(t)}{dt} \tag{3.6}$$

This force increases with increasing velocity and may be either positive or negative depending on the direction of motion relative to the reference, or zero, position. In the SI system, the damping is expressed in N · m/s; in the USCS system, lb · ft/s.

Spring. Capable of storing potential energy within the limits of its elasticity; the defining equation for the force on a spring is

$$f(t) = Kx(t) \tag{3.7}$$

Where the force is given in the SI system as N/m and in the USCS system as lb/ft (lb/in).

The response of a spring-mass system is the sum of all forces acting on the system. For an impulse input, the system equation is developed as follows:

$$\delta(t) = M\frac{d^2x}{dt^2} + B\frac{dx}{dt} + kx$$

$$1 = Ms^2X(s) + BsX(s) + kX(s)$$

$$X(s) = \frac{1}{\sqrt{kM}} \cdot \frac{\sqrt{\frac{k}{M}}}{s^2 + \frac{B}{M}s + \frac{k}{M}}$$

(3.8)

An example will help clarify the use of those elements in a system.

Example 3-2

Given a weight of 20 lb, a spring constant of 25 lb/ft, and damping of 0.5 lb/ft/sec, determine the equation of motion for the spring-mass system shown in Figure 3-4.

Solution

First, convert the weight to mass.

$$M = \frac{W}{g} = \frac{20}{32.2} = 0.6211 \text{ slug}$$

Then substitute all values into Equation 3.8.

$$X(s) = \frac{1}{\sqrt{25 \cdot 0.6211}} \cdot \frac{\sqrt{\frac{25}{0.6211}}}{s^2 + \frac{0.5}{0.6211}s + \frac{25}{0.6211}} = 0.254 \cdot \frac{6.344}{s^2 + 0.81s + 40.25}$$

Factoring the denominator shows that there are two complex roots of the expression. The lack of an s term in the numerator indicates that there will be only a sine term in the equation.
Completing the square and solving for the terms

$$X(s) = 0.254 \cdot \frac{6.344}{(s + 0.405)^2 + 40.09} \cdot \frac{6.332}{6.332} = 0.254 \cdot \left[\frac{6.332}{(s + 0.405)^2 + 40.09}\right]$$

$$x(t) = 0.254e^{-0.405t}\sin(6.332t)$$

Rotational Motion

Rotational motion is defined as "motion about a fixed axis." The motion is given by the sum of all torques applied to the system, as shown in Equation 3.9.

$$\sum T_A(t) = J\alpha$$

(3.9)

The torques applied to a rotational system are as follows:

Damping torque. The element of a rotational system that opposes the rotation; similar to viscous friction in a translational system. The damping torque is

$$T_B(t) = B\frac{d\theta}{dt} \tag{3.10}$$

Inertia. The element of a rotational system that stores the kinetic energy of a system; for a simple cylindrical mass, the inertia, which is called the **moment of inertia,** is defined as "the tendency to rotate about an axis" and is given by

$$J = \frac{1}{2}Mr^2 \tag{3.11}$$

The torque applied to the mass is

$$T_M(t) = J\frac{d^2\theta}{dt^2} = J\frac{d\omega}{dt} \tag{3.12}$$

Torsional spring. Similar to a linear spring in that it has a spring constant, K, but the torque is the product of the spring constant and the angular displacement; the describing equation for the torsion spring is

$$T_S(t) = K\theta(t) \tag{3.13}$$

A general form for a torsion spring, mass, damper system is

$$\sum T(t) = T_M + T_B + T_S = J\frac{d^2\theta(t)}{dt^2} + B\frac{d\theta(t)}{dt} + K\theta(t) \tag{3.14}$$

$$= J\ddot{\theta} + B\dot{\theta} + K\theta(t)$$

A good example of a rotational system is the electric motor, since motors are used in many process control systems. The example will develop an expression for the motor's rotational velocity vs. applied voltage. The permanent magnet (PM) DC motor is simple to characterize from its electrical and mechanical characteristics taken from a data sheet.

3.3 PM DC Motors

The PM type of motor has a permanent magnet to provide the magnetic field for the wound rotor. Control of this type of motor is achieved by changing the applied armature voltage. The construction of a simple PM motor is shown in Figure 3-5.

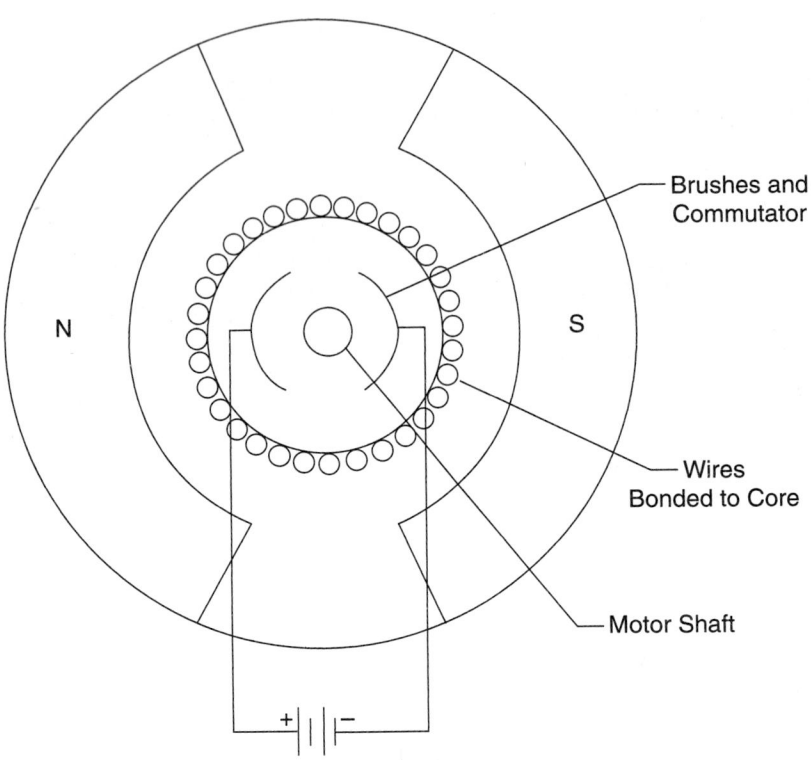

(figure 3-5)

Basic permanent magnet DC motor

Assuming the battery produces a north magnetic pole in the armature, it is clear that there is a repulsing force between the magnet's north pole and the armature north pole and an attractive force between the armature north pole and the magnet's south pole. As the armature rotates, the brushes switch the polarity of the voltage in the armature, maintaining the repulsive and attractive forces. The PM motor can be reversed by changing the polarity of the battery.

Motor Equation

The electrical and mechanical equivalents of a PM DC motor are shown in Figures 3-6 and 3-7, respectively.

The electrical parameters are as follows:

R_a = Armature resistance in ohms

L_a = Armature inductance in henrys

K_b = Motor back EMF constants in volts/rad/sec

K_T = Torque constant in oz-in/A or N-m/A

(figure 3-6)

Electrical component of motor

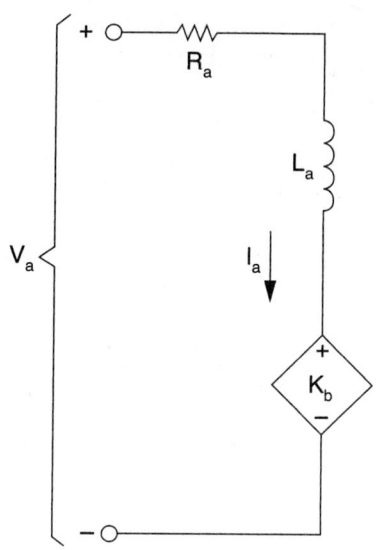

(figure 3-7)

Mechanical component of motor

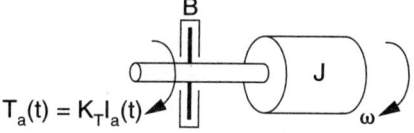

The mechanical parameters are as follows:

J = Rotor inertia in oz · in · s² or N · m · s²
B = Viscous damping of rotor in oz · in or N · m
ω = Rotational velocity in rad/sec

As shown in the previous figures, the differential equations for the two sets of parameters are

$$v_a(t) = i_a(t)R_a + L_a \frac{di_a(t)}{dt} + K_b\omega$$

$$T_a = K_T i_a(t) = J\frac{d\omega}{dt} + B(\omega)$$

The equations transformed to Laplace form, without initial conditions, are

$$V_a(s) = I_a(s)\left[sL_a + R_a\right] + K_b$$

$$T_a = K_T I_a(s) = \Omega(s)\left[sJ + B\right]$$

After determining an expression for $I_a(s)$ in the mechanical system, and replacing $I_a(s)$ in the electrical expression, you develop the following expression:

$$I_a(s) = \frac{\Omega(s)[sJ + B]}{K_T}$$

$$V_a(s) = \frac{\Omega(s)[sJ + B]}{K_T}\left[sL_a + R_a\right] + K_b(s)$$

$$\frac{\Omega(s)}{V_a(s)} = \frac{K_T}{(sJ + B)(sL_a + R_a) + K_bK_T}$$

$$\frac{\Omega(s)}{V_a(s)} = \frac{K_T}{s^2L_aJ + \left(L_aB + R_aJ\right)s + (K_bK_T + R_aB)}$$

Use of a block diagram to characterize the motor components is the clearest way to visualize how the components of the motor are related. The block diagram of a simple motor is shown in Figure 3-8.

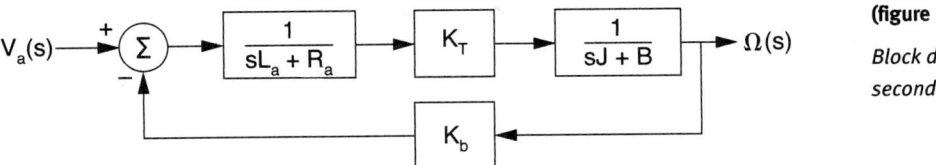

(figure 3-8)

Block diagram of second-order motor

Notice the block, K_b, which spans from output to a summing block. This block represents the back emf generated when the armature rotates in the field of the magnet. The back emf is subtracted from the applied voltage. When the resultant voltage reaches a value that causes enough current to flow to balance the forces required by the motor and its load, the back emf becomes static. Changes in load will cause the back emf to increase or decrease; and depending on whether the load increases (decreases) and the back emf decreases (increases), the motor will be returned to stable operation. Block diagrams are covered in depth in Chapter 4. The expression for the motor velocity vs. applied armature voltage developed from the block diagram for the motor is

$$\frac{\Omega(s)}{V_a(s)} = \frac{K_T / L_aJ}{s^2 + \left(\dfrac{R_a}{L_a} + \dfrac{B}{J}\right)s + \dfrac{K_bK_T + R_aB}{L_aJ}}$$

An example of a servomotor used in a control system is developed in Example 3-3.

Example 3-3

A 12-volt DC servomotor has the following characteristics:

$K_T = 1.94$ oz-in/A $\qquad\qquad B = 1.464 \cdot 10^{-4}$ oz · in/(rad/sec)

$K_b = 1.3 \cdot 10^{-2}$ V/(rad/sec) $\qquad J = 1.4 \cdot 10^{-4}$ oz-in-s^2

$R_a = 3.1$ ohms $\qquad\qquad RPM_{max} = 7847$ rpm

$L_a = 1.57 \cdot 10^{-3}$ H

Determine the expression for the rotational velocity vs. armature voltage for the motor. Then determine the rotational velocity for an input of $5u(t)$.

$$\frac{\Omega(s)}{V_a(s)} = \frac{\dfrac{1.94}{1.57 \cdot 10^{-3} \cdot 1.4 \cdot 10^{-4}}}{s^2 + \left(\dfrac{3.1}{1.57 \cdot 10^{-3}} + \dfrac{1.464 \cdot 10^{-4}}{1.4 \cdot 10^{-4}}\right)s + \dfrac{1.94 \cdot 1.37 \cdot 10^{-2} + 3.1 \cdot 1.464 \cdot 10^{-4}}{1.57 \cdot 10^{-3} \cdot 1.4 \cdot 10^{-4}}}$$

$$= \frac{8.826 \cdot 10^6}{s^2 + 1976s + 1.23 \cdot 10^5}$$

Solution

The final-value theorem is used to determine the output of the system for a step input. The final value occurs when all time-varying components have become zero, and the output is

$$\omega(t) = 5 \cdot \frac{8.826 \cdot 10^6}{1.23 \cdot 10^5} = 358.8 \text{ rad/sec}$$

The answer generated is in rad/sec and must be converted to rpm. The final value corresponds to 3426 rpm.

DC Motor—Torque-Speed Curves

For any type of motor to rotate, the inertia of the rotor assembly must be overcome. A small amount of torque must be expended for that to occur. This torque may or may not be specified in the motor data sheets as the motor constant, T_M. This torque is not usually significant relative to the torque required for the load.

PM DC motors have linear torque speed curves; that is, the torque generated is a function of the applied voltage and the rotational velocity of the motor. The torque generated is derived from the fact that the back emf generated in opposition to the applied voltage rises to a value such that the voltage across the armature winding produces the necessary current for the torque required, according to Equation 3.15.

$$T_m = \frac{K_T(V_a - K_e\omega)}{R_a} = K_T I_a \qquad (3.15)$$

Note that the armature voltage is not the applied voltage; it is reduced by the back emf.

The slope, m, of the curves is the derivative of Equation 3.15.

$$m = \frac{dT_m}{d\omega} = \frac{R_a(-K_T K_e) - 0}{R_a^2} = -\frac{K_T K_e}{R_a} \tag{3.16}$$

A general form of torque-speed curve for the equations is shown in Figure 3-9.

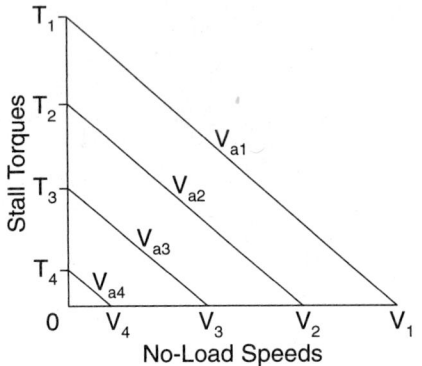

(figure 3-9)

No-load speed curves for DC motor

For the torque-speed curves shown, the applied armature voltages are, $V_{a1} > V_{a2} > V_{a3} > -V_{a4}$.

Often, the motor speed is too fast for an application and must be reduced for the system to function correctly. To change the motor speed, a set of gears is often used.

3.4 Gears

In all cases, gears either have a set of teeth around the gear periphery or are machined into a shaft in such a way that the shaft meshes with a mating gear. Among the uses of gears are changing motion from rotational motion to translational motion, as in the rack and pinion used in automobile steering systems, increasing or decreasing rotational velocity, increasing torque, and changing the direction of rotation of a shaft. There are several general types of gears:

➤ A **spur gear** is the most common form of gear; its teeth are around the perimeter of the gear.

➤ In a **pinion gear**, the teeth are often machined on the end of a shaft.

➤ A **rack gear** has a flat piece of metal with meshing teeth machined on the surface; it is used to convert rotary motion to linear motion.

➤ A **worm gear** is a spiral form of gear that rotates about a fixed axis, causing rotary motion in a mating gear.

All gears share certain common properties:

$$r_1\theta_1 = r_2\theta_2 \qquad r_1 N_2 = r_2 N_1 \qquad T_1\theta_1 = T_2\theta_2 \qquad (3.17)$$

In Equation 3.17, r is the radius of the gear, N is the number of teeth in the gear, T is the torque of the gear, ω is the rotational velocity of the gear, and θ is the angle the gear moves through. Figure 3-10 shows the relationships applied to a two-gear system.

(figure 3-10)

Rotation of gears

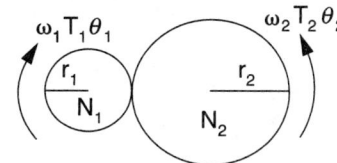

From the relationships, two spur gears coupled, with Gear 1 having teeth N_1 and Gear 2 having teeth N_2, will rotate through the same distance; that is, if Gear 1 moves 2 inches, Gear 2 will move 2 inches. When Gear 1 moves θ_1 degrees, Gear 2 will move some portion of θ_1 degrees. The relationship of motion is

$$\frac{\theta_1}{\theta_2} = \frac{N_2}{N_1} \qquad (3.18)$$

N_2 / N_1 is the gear ratio, N, where N_1 is the input gear. The velocity and acceleration ratios are also defined from the gear ratio by taking the first and second time derivatives of the gear ratio.

$$N = \frac{\theta_1}{\theta_2} = \frac{\dot{\theta}_1}{\dot{\theta}_2} = \frac{\ddot{\theta}_1}{\ddot{\theta}_2} \qquad (3.19)$$

Gear efficiency is very high, so little power is lost in translating power from a motor to an output shaft. Assuming little power loss in the gear system, the work done at the input and output are approximately the same. Work in a rotational system is the product of torque and angle, and the power transmitted is the product of torque and angular velocity.

$$P = T\dot{\theta}_1 = T\dot{\theta}_2 \qquad (3.20)$$

Considering the previous equation and considering work to be the same for each gear, then substituting $N\dot{\theta}_2 = \dot{\theta}_1$ for $\dot{\theta}_1$

$$T_1 N\dot{\theta}_2 = T_2\dot{\theta}_2$$
$$N = \frac{T_2}{T_1} \qquad (3.21)$$

Showing the torque ratio also to be a function of the gear ratio, N.

The highest practical gear ratio for two spur gears is approximately 10. If larger ratios are needed, combinations of gears are used in what is called a *gear train*. The overall ratio is the product of the gear ratios. Gears are identified by a parameter called the *pitch of the teeth*. Pitch is defined as "the number of teeth on a 1 in. diameter gear"; some common values are 12, 32, 48, and 64.

Backlash is a parameter that all gear systems have; it is the ability of a gear to move with no corresponding motion of its mating gear. It is defined as "the play between two gears at the point of mesh." The play is usually 0.001 to 0.005 inches.

Example 3-4

A motor that develops 2.0 oz-in of torque at a speed of 750 rpm is coupled to a load by means of a gear train. The output shaft is to rotate in the same direction as the input shaft. N_1 = 12 teeth (pinion shaft gear), N_2 = 36 teeth, N_3 = 90 teeth. Determine the torque and speed of the output shaft. See Figure 3-11.

Solution

The overall gear ratio N is

$$N = \frac{N_2}{N_1} \cdot \frac{N_3}{N_2} = \frac{N_3}{N_1} = \frac{90}{12} = 7.5$$

The load torque is the product of the gear ratio and the input torque.

$$T_L = T_M \cdot N = 2 \cdot 7.5 = 15 \, \text{oz-in}$$

The output rotational velocity is given from the following equations:

$$\omega_M = \frac{d\theta_M}{dt}, \qquad \omega_L = \frac{d\theta_L}{dt}$$

For this case, ω_M = 750 rpm, N = 7.5, and

$$\dot{\theta}_L = \frac{\dot{\theta}_M}{N} = \frac{750}{7.5} = 100 \, \text{rpm}$$

Note that a middle gear was required to make the output shaft rotate in the same direction as the input shaft. That occurs in all designs where the output shaft must rotate in the same direction as the input shaft.

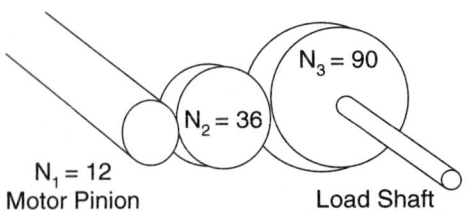

$N_3 = 90$

$N_2 = 36$

$N_1 = 12$
Motor Pinion

Load Shaft

(figure 3-11)

Gear train for Example 3-4

Inertia of Gears

Gears have inertia due to their mass. The torque produced is the same as in the mechanical equivalents shown before. The torque is the product of the inertia, J, and the acceleration of the mass.

$$T = J\frac{d^2\theta}{dt^2} = J\ddot{\theta} \qquad (3.22)$$

The effect of the inertia at the load on the input shaft is reflected inertia.

$$\ddot{\theta}_1 = \frac{T_1}{J_1} \quad \text{or} \quad J_1 = \frac{T_1}{\ddot{\theta}_1}$$

J_1 and T_1 are the inertia and torque at the input shaft. Consider the total inertia to be that reflected from the output load.

$$T_1 = \frac{T_2}{N} \quad \text{and} \quad \ddot{\theta}_1 = \ddot{\theta}_2 \cdot N$$

Substitute that into the equation for J_1.

$$J_{21} = \frac{\dfrac{T_2}{N}}{N\ddot{\theta}_2} = \frac{T_2}{N^2\ddot{\theta}_2}$$

Where J_{21} is the inertia reflected from the output to the input shaft. Since $T_2 / \ddot{\theta}_2 = J_2$, the reflected inertia can be rewritten as

$$J_{21} = \frac{J_2}{N^2}$$

For Figure 3-10, J_{21} is the inertia reflected from the output gear to the input gear. If the inertia associated with the drive system is included, the total inertia is

$$J_T = J_1 + J_{21} = J_1 + \frac{J_2}{N^2}$$

Damping of Gears

Friction, or damping, is present in all gear systems, as in all other systems. The damping torque using Equation 3.12 is

$$B = \frac{T}{\dot{\theta}} \qquad (3.23)$$

In addition, the reflected damping is

$$T_{LR} = \frac{T_{LF}}{N} \qquad (3.24)$$

T_{LF} is the friction torque of the load.

Example 3-5

The inertia of the load is $7.5 * 10^{-3}$ oz-in-s^2, and the armature inertia of the motor is $2.5 * 10^{-4}$ oz-in-s^2. Determine the total inertia in the gear system.

Solution

Setting the inertia of the gears to zero, what is the inertia being reflected from output to input?

$$J_{21} = \frac{7.5 \cdot 10^{-3}}{(7.5)^2} = 1.33 \cdot 10^{-4} \ \text{oz-in-s}^2$$

The total inertia of the system is simply the sum of the two inertias, or

$$J_T = 2.5 \cdot 10^{-4} + 1.33 \cdot 10^{-4} = 3.83 \cdot 10^{-4} \ \text{oz-in-s}^2$$

Gear Ratio Selection

Servomotors are usually high-speed, low-torque motors. To turn a specified load, it is important to select gears that provide the maximum torque and acceleration for the load. Torque values follow the formulas so far developed. The load friction torque reflected is

$$B_{21} = \frac{B_2}{N^2} \qquad (3.25)$$

T_{LR} is the reflected load friction torque. The motor speed for a given load speed is

$$\dot{\theta}_M = N\dot{\theta}_L \qquad (3.26)$$

Various possible gear ratios exist that will accelerate the load. In the selection of gear ratio, ratios are assigned arbitrarily for the motor and load. The torque and motor speed are calculated using Equations 3.21 and 3.22, respectively. Then the data points are plotted on the speed-torque curve of the motor. All points lying below the speed-torque curve and above the load curve are usable gear ratios. If none of the points lies below the curve, the motor is inadequate and a new motor must be selected.

The torque available for acceleration of the load can be determined from the speed-torque and load curve. The gear ratio that provides the maximum accelerating torque occurs when the distance between the load torque curve and the motor torque curve is a maximum.

Example 3-6

A motor is to drive a torque load of 100 oz-in at a speed of 200 rpm. Determine the range of gear ratios that satisfy that requirement and the gear ratio that provides maximum accelerating torque.

Solution

First, assign specific gear ratios. That is done by calculating the end values of the curve from the formulas and then choosing ratios between the two points.

$$N = 2 \quad rpm = 200 \qquad T_M = 50.00 \text{ oz-in} \left(\text{endpoint of graph}\right)$$
$$N = 5 \quad rpm = 1000 \qquad T_M = 20.00 \text{ oz-in}$$
$$N = 10 \quad rpm = 2000 \qquad T_M = 10.00 \text{ oz-in}$$
$$N = 15 \quad rpm = 3000 \qquad T_M = 6.67 \text{ oz-in}$$
$$N = 20 \quad rpm = 4000 \qquad T_M = 5.00 \text{ oz-in}$$
$$N = 25 \quad rpm = 5000 \qquad T_M = 4.00 \text{ oz-in} \left(\text{endpoint of graph}\right)$$

The speed-torque curve for the motor and load are shown in Figure 3-12.

From the figure, it appears that the optimum gear ratio is 10:1 since only 10 oz-in is required to accelerate the motor and there is 20 oz-in for accelerating the load.

(figure 3-12)

Speed-torque curve for Example 3-6

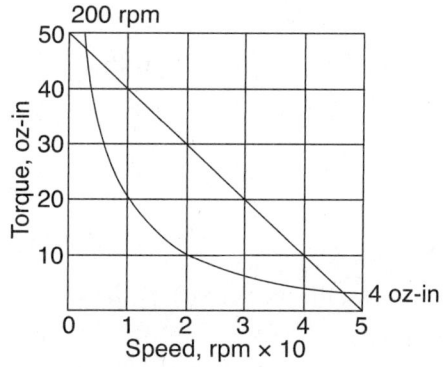

3.5 Temperature

Temperature is one of the most often measured parameters in control systems. The maintenance of constant temperature and the detection of over or under temperature conditions is extremely important.

In general, there are three ways that heat can be transferred from one body to another:

➤ **Convection** is brought about by movement of liquid or air.

➤ **Conduction** is brought about by atomic and molecular absorption and transfer.

➤ **Radiation** involves bodies emitting their characteristic spectral line.

Two laws of thermodynamics are applicable to the study of thermal systems. The first is the law of **conservation of energy**. It says that there will not be more energy output from a system than input into the system, and the best one can hope for is to break even on energy. The second law, called **entropy**, states that some energy will be lost in work done on the system or in the flow of heat in the system. It states that the system will not even have an energy output equal to the energy input. Some energy input to the system will be lost doing work in the system. There is always less energy to use than the energy that is put into the system (efficiency less than 100 percent).

Thermal systems generally require that only two parameters be measured: the thermal resistance, R, in $°F/(Btu/s)$, and the thermal capacitance, C, in Btu/ft. R and C are analogous to the resistance and capacitance of an electrical circuit.

In addition, two heat-specific parameters need to be defined. They are the rate of heat flow, q, in Btu/s. (One **Btu** is the quantity of heat required to raise 1 pound of water 1°F). The **specific heat** of a material, S, in $(Btu/lb)/°F$, is the heat capacity of a material per unit mass. In general

$$q = C\frac{dT}{dt} \qquad C = WS \qquad (3.27)$$

W is the weight of the body, and S is the specific heat of the material.

Keep in mind that heat flows from the hottest part of the system to the coolest part of the system, not the other way around. The rate of heat flow is determined by the **thermal resistance** and is usually considered linear. Thus

$$R = \frac{T_1 - T_2}{q} \qquad °F \,/\, (Btu/s) \qquad (3.28)$$

Where $(T_1 - T_2)$ is the change in temperature of the medium.

In general, q has three forms: q_i, the input heat; q_s, the stored heat; and q_o, the output heat. For thermal equilibrium, then

$$q_i = q_s + q_o \qquad (3.29)$$

The total amount of input heat must equal the sum of the stored and output heat.

Example 3-7

Consider an insulated tank of water heated by an electric element. Develop the differential equation for the water temperature T_w if the heat flow from a heating element is q_h. Assume that the water is at uniform temperature and that there is no heat storage in the tank insulation. Transform the differential equation to the s-domain.

Solution

You develop the differential equation in the following manner, where T_a is the ambient temperature outside the tank.

The heat stored by the water is

$$q_s = C\dot{T}_w \qquad \dot{T} = \frac{dT_w}{dt}$$

Where C is the thermal capacity of the water. The heat loss through the insulation (no storage in the insulation) is

$$q_o = \frac{T_w - T_a}{R}$$

The total heat to be supplied by the heater q_h, which conforms to q_i in Equation 3.27, is the sum of the heat stored by the water and the heat lost through the insulation.

$$q_H = C\dot{T}_w + \frac{T_w - T_a}{R}$$

Where

$$q_w = \frac{T_w - T_a}{R}$$

Rearranging the equation to include the quantity q_w

$$q_h - \frac{T_w - T_a}{R} = C\dot{T}_w$$
$$Rq_h - T_w + T_a = RC\dot{T}_w$$
$$Rq_h + T_a = RC\dot{T}_w + T_w$$

That is a first-order linear differential equation where T_w is the variable of interest.

There is another difficulty with the equation. To put the equation into the s-domain, the function T_a on the right side of the equation does not fit. You need to define an equivalent resistance that includes the ambient temperature. The equivalent resistance is

$$R_{eq} = \frac{Rq_h + T_a}{q_h}$$

Substitution of that equivalent into the differential equation yields

$$R_{eq}\, q_h = RC\dot{T}_w + T_w$$

Example 3-7 (continued)

Then transforming the equation to the s-domain with T_w as the output variable

$$R_{eq} Q_h(s) = RCsT_w(s) + T_w(s) = T_w(s)(RCs + 1)$$

$$\frac{T_w(s)}{Q_h(s)} + \frac{R_{eq}}{RCs + 1} = \frac{R_{eq}}{\tau s + 1}$$

Since R and C are equivalent to the electrical parameters, $\tau = RC$ is the thermal time constant.

3.6 Fluids

Fluid systems occur often in control systems. They are used for things such as maintaining the fluid level in a tank so the tank remains full under changing conditions of outflow from the tank, and maintaining a constant pressure in a pipe system under conditions of varying numbers of outlets on the pipe system opening and closing.

The operation of fluid systems operate based on a set of laws set forth by Blaise Pascal in 1648. Pascal's law is as follows:

> *Pressure applied to an incompressible fluid is transmitted everywhere through the fluid undiminished.*

Some definitions used in fluid systems are:

Compressible fluid. Any fluid whose container size can be changed and can retain the same amount of the fluid, i.e., air and any other gas.

Incompressible fluid. Any fluid whose container size cannot be changed and retain the same amount of fluid.

Mechanical advantage. The gain (or loss) in force over the applied force.

Pressure. The force per unit area of container.

Fluid Fundamentals

In a container of fluid, the pressure is least at the top of the fluid and increases linearly to the bottom of the container.

Consider that a container 10 inches tall has an area of 1 square inch and is filled to the top with a fluid whose force is 1.5 lb/in². At the top of the container (excluding atmospheric pressure), the pressure is zero; but at the bottom, it is 15 lb/in². At a distance of 5 inches, the pressure is 7.5 lb/in².

If atmospheric pressure $\left(\sim 14.7\ \text{lb/in}^2\right)$ is included, the pressure is 14.7 lb at the top and 29.7 lb at the bottom. It is clear that the system is dependent on the area and the pressure applied to the fluid.

Hydraulic Systems

Hydraulic systems use incompressible fluids to multiply a small force applied at one point to move a heavier weight at a different point over some distance.

In a complex system, a pipe connects two cylinders with pistons. The area of the first cylinder is 1.0 in², and the area of the second cylinder is 100 in². A 100 lb weight is placed on the second cylinder.

If a force of 1 pound is applied to the piston of the first cylinder, the total force on the piston area of the second cylinder is 100 lb, or 1 lb/in². If the 1 lb force moves the piston in the first cylinder a distance of 10 inches, the 100 lb weight will move 0.1 inches.

To accomplish the movement, the volume change must be the same in both cylinders.

$$V_1 = V_2 \tag{3.30}$$

And since pressure is force per unit area

$$P = \frac{F_1}{A_1} = \frac{F_2}{A_2} \tag{3.31}$$

You see can that 1 lb/1 in² = 100 lb/100 in². You are interested in equal volumes

$$V = A_1 \cdot H_1 = A_2 \cdot H_2 \tag{3.32}$$

Since volume is the area of the cylinder times the height of the cylinder, $A = \pi r^2 H$, where H = height. You can rearrange the formula slightly and make some simple determinations.

$$\frac{A_1}{A_2} = \frac{H_2}{H_1} \tag{3.33}$$

The system is considered to be a simple lever system in which force is multiplied. This gives rise to mechanical advantage, which is given by

$$MA = \frac{A_2}{A_1} = \frac{H_1}{H_2} \tag{3.34}$$

For the values above, the distance H_2 is

$$H_2 = \frac{A_1 H_1}{A_2} = \frac{1 * 10}{100} = 0.1\ \text{inches}$$

And the mechanical advantage is

$$MA = \frac{100}{1} = \frac{10}{0.1} = 100$$

The resulting advantage is being able to move a 100 lb weight by the application of 1 lb force. The price for that advantage is in the distance the 100 lb weight moves—in this case, 0.01 inches for each inch the first cylinder's piston travels.

Practical Fluid Systems

The previous calculations neglect the characteristics of the fluid being used. When water is the fluid, the formulas above work as they are. On the other hand, if oil or some other fluid is used, you must take into account the density of the fluid, the volumetric flow rate, and the frictional component of the fluid.

For fluid systems, volumetric flow rate is q, pressure is p, hydraulic resistance is R, hydraulic capacitance is C (the potential energy of the liquid), h is the head (the height of a fluid in a container), and hydraulic inertance is I (the equivalent of electrical inductance).

Hydraulic systems are composed of storage tanks, pumps, pistons, valves, and flow restrictors. All of those devices have parameters that are taken into account in the design of a hydraulic system. A typical form of flow restriction is the valve shown in Figure 3-13.

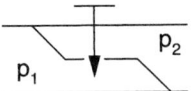

(figure 3-13)

Flow restricting valve

The pressure on the input side is greater than the pressure on the output side, giving rise to a pressure differential $(p_1 - p_2)$. The difference between the two pressures is a function of R and q.

$$p_1 - p_2 = Rq \qquad (3.35)$$

Where R is the hydraulic resistance and q is the volumetric flow rate. The larger R is, the greater the pressure differential is for a given rate of flow, q. The value R can be determined by

$$R = \frac{dh}{dq} \qquad (3.36)$$

Where h is the head of the fluid expressed in units of length or the height of fluid in a tank.

Fluids flow in infinitesimally thin layers in a pipe. As long as the inside of the pipe is smooth and the flow rate is slow, all of the layers flow at the same rate. The flow is called **laminar** flow. As the flow rate increases, the boundary layers (the layers closest to the wall of the pipe) tend to flow more slowly than the central layers due to the friction of the fluid and imperfections in the pipe wall. When that occurs, the flow is said to be **turbulent** flow. If the flow is laminar, R can be expressed as

$$R = \frac{h}{q} \tag{3.37}$$

In 1883, scientist Osborne Reynolds measured flow rates in various types and sizes of pipes and vessels. He was attempting to prove or disprove Darcy's law for laminar flow. Through his experiments, Reynolds was able to relate the velocity of flow to the diameter, density, and viscosity of a fluid, which gives a unitless number that increases with flow rate and defines the transition between laminar flow and turbulent flow. The number is called the *Reynolds number*, R_E, not to be confused with the real part of a complex number. The Reynolds number is calculated from

$$R_E = \frac{\rho v D}{\mu} \tag{3.38}$$

Where ρ is the density of the fluid in kgm^3, v is the velocity of the fluid in m/s, D is the diameter of the pipe in meters, and μ is the viscosity of the fluid in $(N \cdot s) / m^2$.

Reynolds numbers less than 2000 indicate laminar flow, numbers between 2000 and 3000 indicate partially turbulent flow (turbulence at the boundary layers), and numbers greater than 3000 indicate that the flow is turbulent.

For instance, consider a tank filled with fluid and a valve as the output device. The rate of flow into the valve will be greater than the rate of flow out of the valve. Thus, the pressure differential is also given by

$$\Delta p = (q_1 - q_2)^{\frac{1}{\alpha}} = \frac{dv}{dt} \tag{3.39}$$

The range of α is between 1.0 and 2.0. When flow is laminar, α is 1.0. For purposes here, consider that flow is laminar and that the Reynolds number is less than 600.

Volume is equal to the cross-sectional area times the height of the container. The differential flow rate is

$$q_1 - q_2 = \frac{d(A \cdot h)}{dt} = A \cdot \frac{dh}{dt} \tag{3.40}$$

If the pressure differential is defined as p, then

$$p = h\rho g \tag{3.41}$$

Where ρ is the density of the liquid and g is the acceleration due to gravity. From that equation

$$q_1 - q_2 = A\frac{d\left(\dfrac{p}{\rho g}\right)}{dt} = \frac{A}{\rho g}\frac{dp}{dt} \tag{3.42}$$

If the liquid is assumed incompressible, the hydraulic capacitance is

$$C = \frac{A}{\rho g} \tag{3.43}$$

$$q_1 - q_2 = C\frac{dp}{dt}$$

By integration of that formula, the pressure differential p is found.

$$p = \frac{1}{C}\int (q_1 - q_2)dt \tag{3.44}$$

The hydraulic inertance of the system is given as the difference between two forces, which are the result of applying pressure to move a mass of liquid.

$$F_1 - F_2 = p_1 a - p_2 a = \left(p_1 - p_2\right)a \tag{3.45}$$

a is the acceleration of the fluid, which is the rate of change of velocity, $dv\ /\ dt$, and

$$(p_1 - p_2)a = ma = m\frac{dv}{dt} \tag{3.46}$$

The mass of a fluid is given as the length of fluid in a specific area multiplied by the density of the liquid. The length of fluid is measured between two points in the pipe or tank in which the fluid is moving. The mass is then

$$m = AL\rho \tag{3.47}$$

If L is the length of the mass, then

$$(p_1 - p_2)A = AL\rho\frac{dv}{dt} \tag{3.48}$$

The inertance I is $L\rho\ /\ A$, q is the area of the vessel times the velocity in the vessel $q = A \cdot v$, and

$$p_1 - p_2 = I\frac{dq}{dt} \tag{3.49}$$

Example 3-8

For a tank of cross-sectional area A with input of fluid q_i, and output of fluid q_o, determine a transfer function in the s-domain for the system to maintain a constant flow out of the tank.

Solution

To maintain a constant flow out, the input to the tank must equal the outflow from the tank plus any losses due to resistance. Thus

$$q_i = RA\frac{dq_o}{dt} + q_o$$

$$Q_i(s) = RAsQ_o(s) + Q_o(s) = Q_0(s)(ARs + 1)$$

From which you form the transfer function of the system

$$\frac{Q_o(s)}{Q_i(s)} = \frac{1}{ARs + 1} = \frac{1}{\tau s + 1} = \frac{1/AR}{s + \dfrac{1}{AR}}$$

τ, the system time delay, is equal to $1/\alpha$, where α is $1/AR$.

Example 3-9

The single tank shown has a nominal operating head of 4 ft and a normal 0.125 ft³/sec outflow. The cross-sectional area of the tank is 10 ft², as shown in Figure 3-14.

Determine the differential equation for the outflow as a function of time. Then determine the transfer function of the system in the s-domain.

Solution

For turbulent flow, the resistance is given as

$$R = \frac{dh}{dq} = \frac{2h}{q}$$

Since you are looking at the smaller orifice, you are interested in the resistance at the output of the tank, q_o.

$$R = \frac{2h}{q_o} = \frac{2 \cdot 4}{0.125} = 64 \text{ sec/ft}^2$$

Since $dh = R\,dq$. To convert that function to a function of time

$$\frac{dh}{dt} = R\frac{dq}{dt}$$

Then, since

$$q_i - q_o = A\frac{dh}{dt} = AR\frac{dq_o}{dt}$$

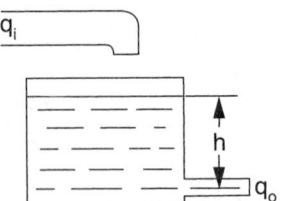

(figure 3-14)

Tank for Example 3-9

Example 3-9 (continued)

The required input flow is found to be

$$q_i = q_o + AR\frac{dq_o}{dt}$$

Since the area of the tank is 10 ft², the value of $AR = 10 * 64 = 640$ and

$$q_i = q_o + 640\frac{dq_o}{dt}$$

That is the differential equation of the tank and the required input. The transfer function of the system in the s-plane for q_o/q_i is

$$Q_i(s) = Q_o(s) + 640sQ_o(s) = Q_o(s)(1 + 640s)$$

$$\frac{Q_o(s)}{Q_i(s)} = \frac{1}{640s + 1} = \frac{\frac{1}{640}}{s + \frac{1}{640}} = \frac{1.5625 * 10^{-3}}{s + 1.5625 * 10^{-3}}$$

3.7 Review of Opamps

The **opamp** is an electronic device that can perform mathematical operations such as integration, differentiation, summation, averaging, differencing, and other operations, as well as amplify a signal input. Most control systems use combinations of opamps, sensors, and power amplifiers to provide the necessary functions of control and power supply to the system.

Inverting and Noninverting Amplifiers

In its simplest form, the opamp is an amplifier. As in other amplifying devices, it has an inverting mode and a noninverting mode of operation. Each of the forms has its own amplification formula. Both types of amplifier are shown in Figure 3-15.

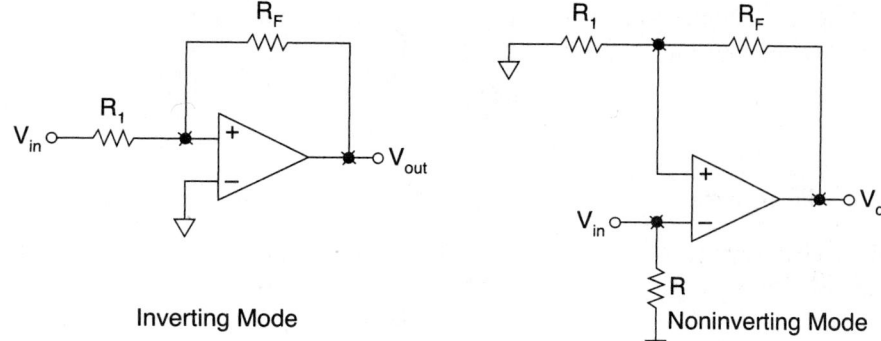

(figure 3-15)

Basic opamp configurations

Inverting Mode

Noninverting Mode

The formulas for the two types of amplifier are

$$A_{V(inv)} = -\frac{R_F}{R_1} \qquad\qquad A_{V(ni)} = \frac{R_F}{R_1} + 1 \qquad\qquad (3.50)$$

They are simple gain block forms that can replace any gain block with a pure number as its value.

Opamp Differentiator

The opamp differentiator will take the time derivative of its input. The form of the differentiator is shown in Figure 3-16.

(figure 3-16)

Ideal opamp differentiator

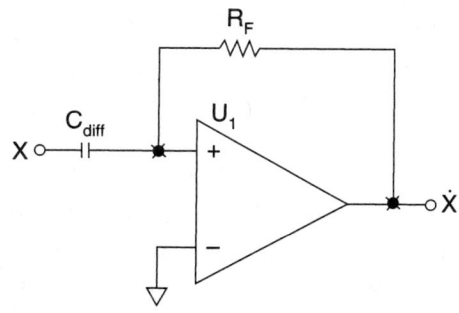

The gain equation for Figure 3-16 is

$$A_V = -\frac{R_{diff}}{\dfrac{1}{C_{diff}}} = -R_{diff}C_{diff} \qquad\qquad (3.51)$$

And the output voltage is

$$V_{out} = -R_{diff}C_{diff}\frac{dV_{in}(t)}{dt} \qquad\qquad (3.52)$$

Since $R \cdot C$ is time, the value is denoted as T. In performing the differentiation, the output voltage is given by

$$V_{out} = V_{in} \cdot \frac{T}{t} \tag{3.53}$$

Note that the use of that type of circuit places a zero in the system.

While the circuit will work, it is not practical in the sense that it is extremely sensitive to short time duration pulses that are considered noise. The reactance of the capacitor is very small for those inputs, and the gain of the amplifier approaches the open-loop gain of the opamp. To prevent that from happening, a resistor must be inserted in series with the capacitor. That limits the gain at high frequencies to some finite value. The circuit is shown in Figure 3-17.

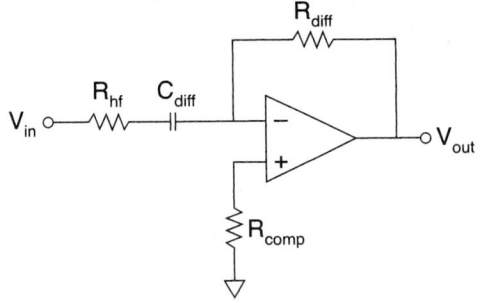

(figure 3-17)

Practical opamp differentiator

If you include the resistor R_{hf}, the gain equation becomes

$$A_V = -R_{diff}C_{diff} \frac{j\omega_c}{1 + j\dfrac{\omega}{\omega_c}} \tag{3.54}$$

Once again, the question arises: What is the value of the high frequency for gain limiting? As a rule of thumb, you can start by letting the high frequency equal about 10 times the differentiating frequency. You must also take into account the slew rate of the opamp for differentiators since you may need fast risetimes for the input signal to differentiate properly.

Opamp Integrator

The opamp integrator is used to sum the area under a voltage vs. time curve. In control systems, the integrator is used to promote zero error in the system. In other systems, it is used to generate linear ramp voltages or to generate sine-cosine outputs and other outputs that are the time integral of an input. The general form of the integrator is shown in Figure 3-18.

(figure 3-18)

Ideal opamp integrator

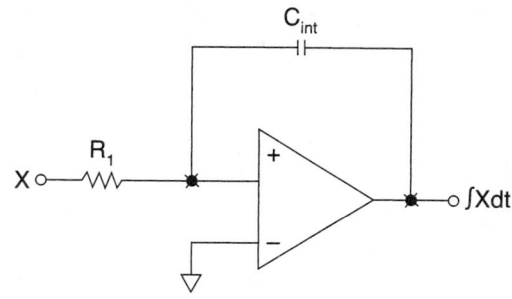

That is the simplest form of integrator, but it is not the one usually used. The reason for not using the simple form is that it is a true integrator and would have extremely high gain for low frequency inputs. An offset voltage between the inputs would also produce a railing out of the integrator. One form that is often used is shown in Figure 3-19.

(figure 3-19)

Practical opamp integrator

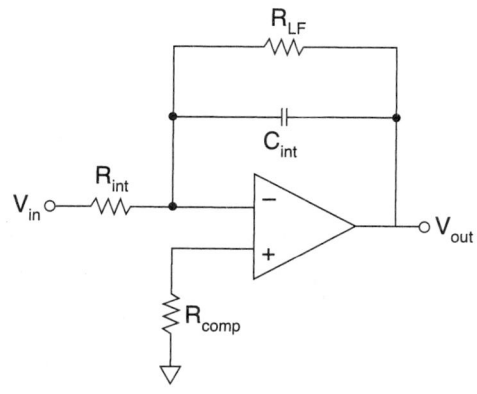

That form of amplifier limits the low frequency gain generally to about 10.

For the theory of integration, the simple integrator of Figure 3.19 is used. The input is V_{in}, and the output is V_{out}. The output voltage formula can be determined from the inverting amplifier gain equation.

$$V_{out} = -\frac{1}{R_{int}C_{int}} \int v_{in}\, dt = -\frac{1}{R_{int}C_{int}} \cdot V_{in} \cdot t + K \qquad (3.55)$$

K represents an initial condition for the capacitor. In most cases, that condition is (or should be) zero. Note that the gain of the amplifier is fixed at $-1\,/\,R_{int}C_{int}$ and the output voltage is a function of the time that an input signal is present. Since RC also is time and is fixed, you can designate it to be T. The amplifier's output voltage equation then becomes

$$V_{out} = -V_{in} \cdot \frac{t}{T} \qquad (3.56)$$

That is the equation of a true integrator. The problem is that at low frequencies, the output voltage wants to be very large and the amplifier rails out. The solution is the addition of a gain-limiting resistor. When you add the resistor, the gain equation becomes

$$A_V = \frac{\dfrac{R_{lf}}{j\omega C_{int}}}{\dfrac{R_{lf} + \dfrac{1}{j\omega C_{int}}}{R_{int}}} = \frac{R_{lf}}{R_{int}} \cdot \frac{1}{j\omega C_{int} R_{lf} + 1} = \frac{R_{lf}}{R_{int}} \cdot \frac{1}{1 + j\left(\dfrac{\omega}{\omega_c}\right)} \tag{3.57}$$

Where ω_c is the cutoff frequency of the RC network in the feedback loop. From the previous equation, when ω is zero, the gain equation reduces to that of the common inverting amplifier, limiting the low frequency gain to a reasonable value. One question remains: What should ω_{lf} be? In all cases, ω_{lf} should be less than the integrator frequency, ω_{int}. Both frequencies are determined from similar equalities.

$$\omega_{int} = \frac{1}{R_{int} C_{int}}, \qquad \omega_c = \frac{1}{R_{lf} C_{int}} \tag{3.58}$$

The circuit will act as a true integrator between ω_c and ω_{int} but will have limited gain outside that range. As a rule of thumb, you might select ω_{lf} to be one-tenth the integrating frequency. That makes calculation of the resistors simple and generally gives a maximum gain of ten outside the integrating range.

You should also pay attention to the slew rate of the opamp. The slew rate of an opamp is the rate of change of the amplifier between two levels of voltage. It is given in V / μs and, in that sense, determines the maximum frequency the opamp can amplify without excessive distortion of the signal occurring. If the integrating frequency is more than about 1000 Hz, you may need a slew rate higher than that of the μa741, which is 0.5 V / μs.

You can specify what some input signals will produce for outputs in differentiators and integrators. For the sake of simplicity, you will use some standard signals; they are the sine wave, the unit step, the unit ramp, and parabolic signals. The signals and their outputs are shown in Table 3-4. The table is for the differentiator, but it is also used for integrators by reversing input for output.

(table 3-4)

Signal Type	Output
Sine wave	−Cosine wave
Square wave	−Triangle wave
$u(t)$	$-tu(t)$

(table 3-4) *continued*

Signal Type	Output
$tu(t)$	$-\dfrac{t^2}{2}u(t)$
$\dfrac{t^2}{2}u(t)$	$-\dfrac{t^3}{6}u(t)$

Opamp Summing Amplifiers

The summing amplifier is used when there is more than one voltage to be applied to a single amplifier. The summing amplifier may have as many inputs as necessary. The concept is to add together all of the voltages of the inputs. Assuming that there are n input voltages—V_1, V_2, V_3, V_4, and V_n—that are summed to arrive at an output voltage V_O according to the output voltage formula

$$V_O = -\left(A_1 V_1 + A_2 V_2 + A_3 V_3 + \ldots + A_n V_n\right) \tag{3.59}$$

The given values in Equation 3.59 are constant and all values may be either positive or negative. A diagram of the general form of the summing amplifier for n voltage inputs is shown in Figure 3-20.

(figure 3-20)

Opamp summing amplifier

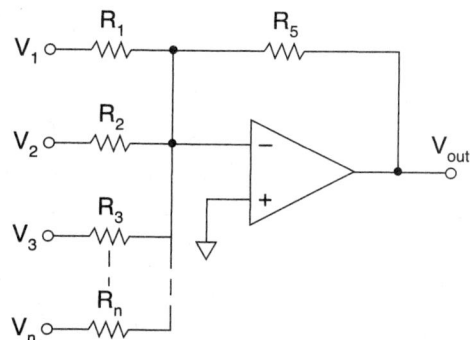

In terms of that amplifier, the analysis uses the same equation as all other opamp amplifiers, but individually for each input. In the case of unequal resistor values, the compensating resistor is the parallel combination of all of the resistors in the circuit and the feedback resistor.

Opamp Difference Amplifier

The differencing amplifier is used in all control system designs as the summing junction for the system to output the error signal to the controllers. The circuit for the differencing amplifier is shown in Figure 3-21.

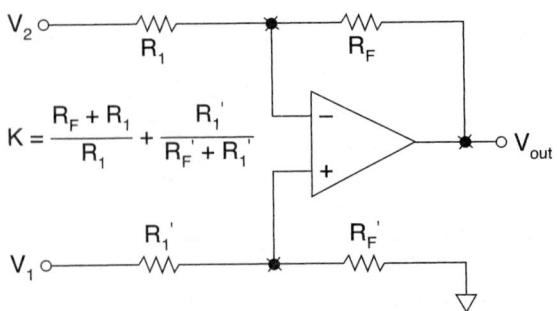

(figure 3-21)

Opamp differencing amplifier

If all resistors are equal in value, the gain of the system is one. If the resistors are not all equal, the gain of the circuit is as shown in Figure 3-21.

Instrumentation Amplifier

There are several forms of instrumentation amplifier, and each has a specific use. For this section, only the high input impedance, high common-mode-rejection-ratio (CMRR) form will be considered. The circuit of that type of amplifier is shown in Figure 3-22.

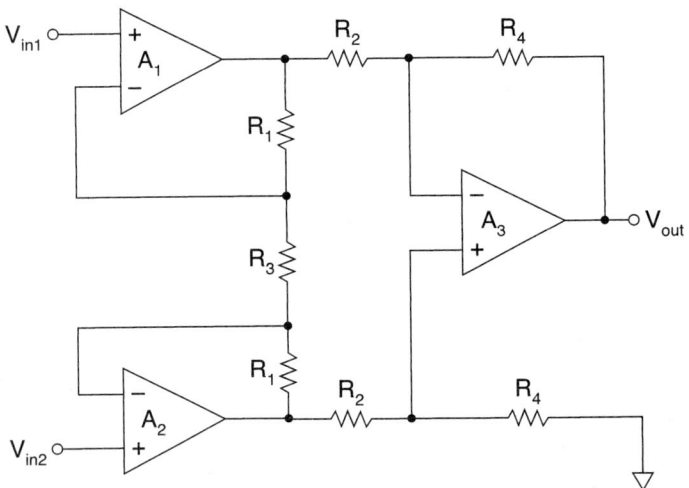

(figure 3-22)

High input Z, high CMRR instrumentation amplifier

For that amplifier, common-mode signals cancel in the input amplifiers A_1 and A_2. With careful consideration of the components and their tolerances in the output amplifier, the CMRR can be made to approach that of a single opamp. The overall gain of this amplifier is given by

$$A_V = \left(1 + \frac{2R_1}{R_3}\right)\left(\frac{R_4}{R_2}\right) \tag{3.60}$$

Note that the differential inputs are at the noninverting input, which indicates high input impedance, while the output is a differencing amplifier with low output impedance that approximates an ideal voltage source.

Voltage-to-Current and Current-to-Voltage Conversion

In many cases, the analog of the output variable is a voltage that is measured at a remote location. If that is the case, it is necessary to transmit the sensor information over some distance to the controlling devices. If the distance is long (say, more than 1,000 feet), the resistance of the transmission wires would attenuate the available voltage. In a current-driven system, even if the voltage is attenuated, the current remains the same. In addition, since voltage requires a high impedance to develop across, the chances of noise developing on the transmission line in an electrically noisy environment are high. It is better to convert the measurement to a current, in which low impedance is used, so noise will be less likely to cause problems. The current is then transmitted over a transmission line and converted back to a voltage at the receiving end.

Two standard hard-wired current loops are commonly used:

➤ 4–20 mA current loop

➤ 10–50 mA current loop

The most common is the 4–20 mA current loop.

Voltage-to-Current Converter

The general form of the voltage-to-current converter is shown in Figure 3-23.

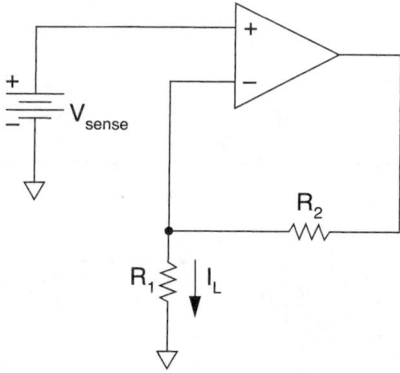

The voltage-to-current converter resembles a noninverting opamp amplifier with its attendant high input impedance, which will not load the sensor being used. The current signal taken is not at the output of the amplifier; it is the current in the resistor R_1. The current in R_1 is then passed via a transmission line to a current-to-voltage converter at the receiving end.

This circuit's principle of operation is simple. In the noninverting mode, the voltage at both of the inputs must be the same; therefore, the current through the resistor R_1 is

$$I_L = \frac{V_{sense}}{R_1} \tag{3.61}$$

V_{sense} is the sensor output voltage. If the sensor is capable of a 4–20 V output, the circuit can be a simple voltage follower, with a resistance of 1 $K\Omega$ for R_1. If not, the voltage must be conditioned so the voltage meets the required value.

Example 3-10

A certain locomotive axle speed sensor has a voltage output of 3–15 V depending on its rotational velocity. Since locomotives are an extremely noisy environment electrically speaking, it is necessary to convert the voltages to current and transmit the current to a current-to-voltage converter at the locomotive speedometer. Design a circuit to perform that function.

Solution

First, compare the ratio of the voltage sensor output to the ratio of the currents. If they are the same, the job is simple; if not, a more complicated design is required. In this case, they are the same. Each has a ratio of 5:1. The voltage range of the axle generator is within the acceptable range of most opamps, which is usually approximately equal to the supply voltages.

This design can use a unity gain voltage follower configuration with a resistor to provide the 4–20 mA range. The current range is calculated using either the 4 mA or 20 mA value and the voltage of the generator. Thus

$$R_1 = \frac{3}{0.004} = \frac{15}{0.020} = 750\Omega$$

The circuit is shown in Figure 3-24, and the output current vs. input voltage is shown in Figure 3-25.

Keep in mind that this is a very simple example meant to show the method of converting voltages to currents for transmission to a receiving current-to-voltage amplifier.

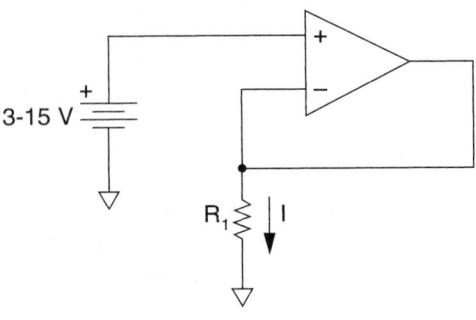

(figure 3-24)

V-I converter circuit for Example 3-10

(figure 3-25)

4–20 mA output of V-I converter for Example 3-10

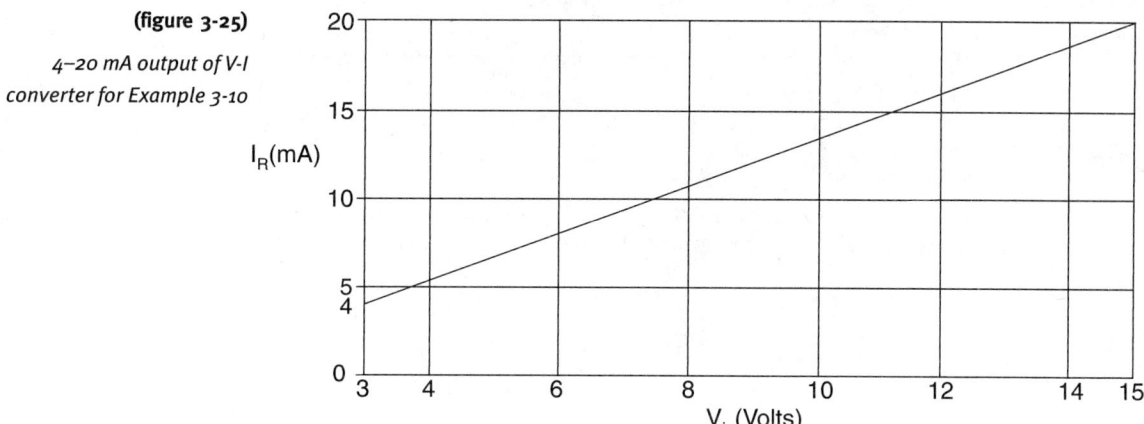

(figure 3-25)

4–20 mA output of V-I converter for Example 3-10

Current-to-Voltage Converter

The current-to-voltage converter is also a simple circuit in that the opamp has only a feedback resistor. The current entering the inverting input must be the same as the current in the feedback resistor since the inverting terminal is forced to zero. The circuit is shown in Figure 3-26.

(figure 3-26)

Current-to-voltage converter

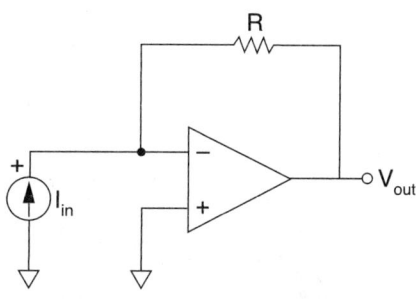

That circuit's principal of operation is also simple since the current applied to the input is the same as the current in the feedback resistor. Therefore, the output voltage is given by

$$V_{out} = I_{in} \cdot R \qquad (3.62)$$

It is to be noted that the circuit shown is a simple example and would be part of a larger, more complex system with all of the proper precautions taken (for example, input offsets negated).

Example 3-11

For the V-I converter in Example 3-10, provide an I-V converter to provide a 3–15 V output. Then show the two circuits mated and their output. (Neglect wire losses.)

Solution

The circuit of the I-V converter is shown in Figure 3-27, and the output of the circuit is shown in Figure 3-28. Calculation of the required resistance yields a resistor the same value as that of the V-I converter, as follows:

$$R = \frac{3}{0.004} = \frac{15}{0.02} = 750\Omega$$

The total circuit for the V-I, I-V conversion is shown in Figure 3-29, and the output is shown in Figure 3-30.

The break in the line between the two 750 W resistors indicates that there is some distance between the V-I converter and the I-V converter. The first resistor is not a part of the I-V converter circuit.

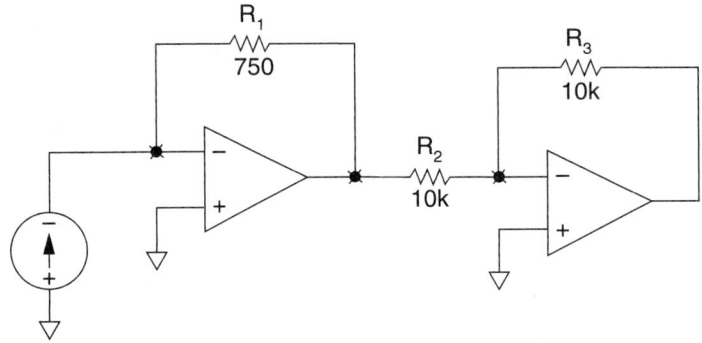

(figure 3-27)

I-V converter and inverter for Example 3-11

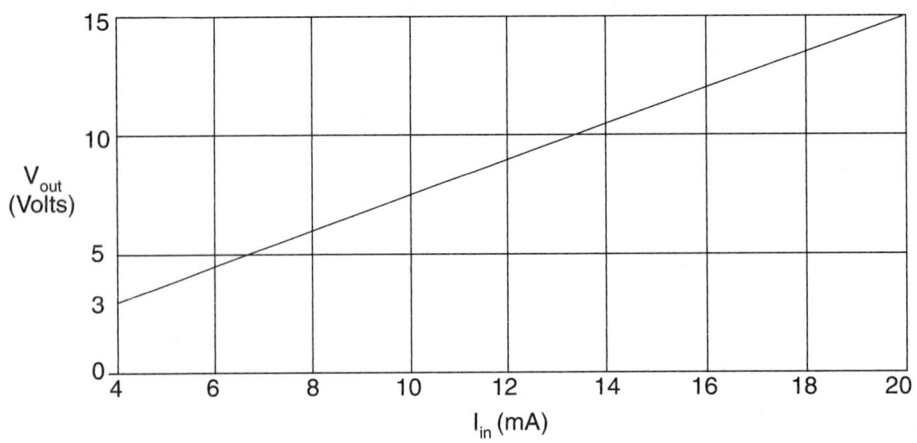

(figure 3-28)

Voltage output of I-V converter for Example 3-11

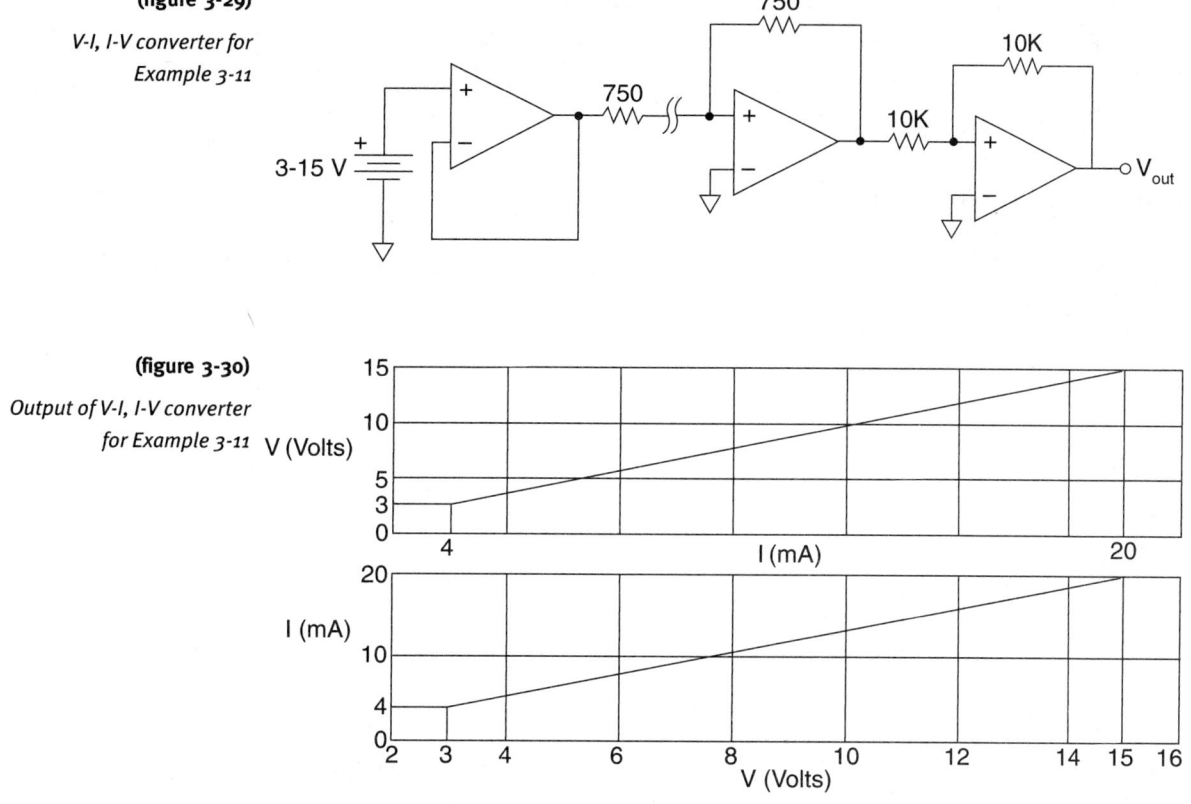

(figure 3-29)

V-I, I-V converter for Example 3-11

(figure 3-30)

Output of V-I, I-V converter for Example 3-11

3.8 Transducers

Transducers are devices used to measure various parameters of a control system, such as velocity, position, pressure, temperature, and flow. Transducers often convert the measured parameter to a voltage or a change in voltage, but not always. Take the case of fluid flow in a pipe. A change in pressure causing more or less flow can be directed directly to the diaphragm of a regulating valve by tapping the pipe and transmitting the change in pressure to the valve, causing it to close or open as needed.

DC Tachometer Generator

Tachometers are, functionally, motors in reverse. Moving a coil of wire in a magnetic field induces a voltage in the coil; alternately, passing a magnet over a coil will produce the same result. That is the principle of a tachometer.

Figure 3-31 shows a basic form of DC generator.

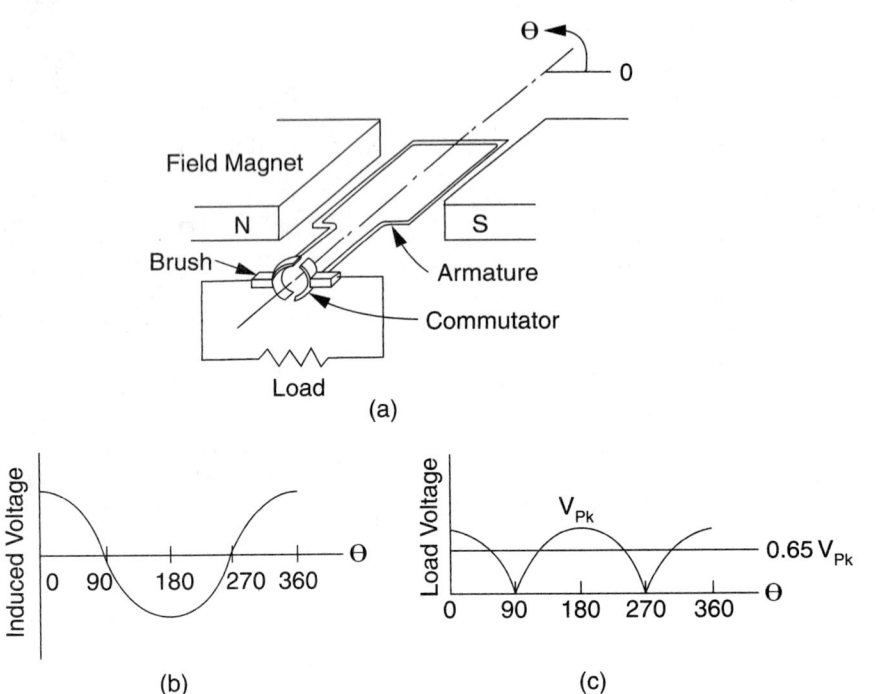

(figure 3-31)

DC tachometer generator

As the armature is rotated through 360°, Figure 3-31b shows that the magnetic field induces a sinusoidal voltage in the windings of the coil. The commutator is a segmented ring of metal that maintains the proper polarity of the induced voltage. Figure 3-31c shows the polarity of the voltage measured across the load due to the commutator. Electrical contact of the load with the commutator is maintained by using carbon brushes. The DC output voltage of this system is $V_{dc} = 0.65\ V_{Pk}$. If more segments and coils are added to the commutator, the dc output voltage rises, giving a voltage closer to the peak of the induced voltage. If the number of segments and coils is doubled, the output voltage rises to $V_{dc} = 0.9\ V_{Pk}$.

The output voltage of the DC tachometer can be predicted using the formula

$$V_g = K\phi\omega \tag{3.63}$$

ϕ is the magnetic flux of the magnets. ω is the rotational velocity of the generator armature. K is the generator constant, which is related to the number of turns in the coil windings, the number of parallel paths for current, and the number of coils.

In a PM generator, the electric field is constant, as are all of the other parameters other than velocity; K can be transformed into K_g, the generator constant; and the output voltage equation becomes

$$V_g = K_g\omega \tag{3.64}$$

Temperature Transducers

Three fundamental types of temperature-measuring devices are in common use:

➤ Thermocouples

➤ Thermistors

➤ Resistance Temperature Detectors (RTDs)

Each has a specific range of temperature measurement. The thermocouple has the largest range, the RTD the next largest, and the thermistor the smallest range.

Keep in mind that there are other methods of measuring temperature, such as infrared methods, standard mercury thermometers, and optical color measurements. Standard thermometers are least accurate in that they rely on human interpolation to read and adjust temperatures. Standard thermometers are often used as a visual indication of the temperature of a batch of material; but the temperature is controlled by a system of amplifiers with thermocouples, thermistors, or RTDs.

A very important parameter in temperature measurement is the linearity of the measuring device. The most linear of the three is the RTD, the thermocouple is the next most linear, and the thermistor is the least linear.

Thermocouples

When any two dissimilar metals are joined together, a potential difference exists between the metals. The reason for the potential difference is that different types of metals have different numbers of free electrons available. For instance, if iron wire has 100 free electrons and copper wire has 150 free electrons, when the two wires are welded together, there is an excess of 50 electrons at the joint. Therefore, a potential difference exists across the two wires. The voltage difference is very small since there are only a few electrons to produce the voltage. The charge on an electron is $1.6 \cdot 10^{-19}$ C; but while the voltage is small, it is measurable.

As temperature is increased, the number of free electrons increases in both wires, creating a larger voltage across the wires. The measured voltage of the thermocouple is called the *Seebeck voltage* in honor of the first scientist to measure the potential difference of two dissimilar metals.

Usually, the voltage generated by a thermocouple must be "conditioned"; that is, it must be amplified. Usually, some form of filtering is required to prevent noise voltages from getting to the controlling devices.

The metals used and their operating temperature range classify thermocouples. A few popular types of thermocouples are as follows:

➤ **Type J** is made of iron and copper-nickel (constantin) wire. The range of temperature measurement is approximately −50°C to about 900°C. The maximum output voltage is approximately 70 mV.

➤ **Type K** is made of nickel-chromium and nickel-aluminum wire. The range of temperature measurement is approximately −200°C to 1250°C. The maximum output voltage is approximately 60 mV.

➤ **Type N** is made of nickel-chromium-silicon and nickel-silicon-magnesium wire. The range of temperature is approximately −270°C to 1300°C. The maximum output voltage is approximately 50 mV.

Other specialized thermocouple forms are available for higher or lower temperatures.

One of the most vexing problems is that the thermocouple output voltage cannot be measured directly since connecting a voltmeter or another device automatically creates another thermocouple junction. A specialized method, such as inserting a compensating thermocouple in the measurement circuit to subtract the connection voltage, is commonly used.

Thermocouples are nonlinear devices. Over limited ranges, they are almost linear; but at the extremes of their range, they are curved and have equations that are exponential.

Thermistors

Thermistors are semiconductor resistors whose resistance depends on the temperature of the environment of the thermistor. They usually show a large change in resistance for a small change in temperature. They are usually a negative temperature coefficient (NTC) device, which means the resistance decreases as temperature increases. Positive temperature coefficient (PTC) thermistors are available. In all cases, the semiconductor form of thermistor is referenced to absolute zero (0 K).

The thermistor is often used in a Wheatstone bridge circuit for measurement, as shown in Figure 3-32.

The thermistor is a nonlinear device, which, if used over a limited temperature range, can be used as shown. Otherwise, linearizing methods must be used so the thermistor will output a linear change in voltage for a linear change in temperature.

The range of thermistor resistors is from a few thousand ohms to a few hundred thousand ohms. The general voltage change in a circuit is 10 mV/K. Designing the bridge circuit is not as easy as calculating the balance of the bridge since the thermistor exhibits a specific resistance at 0°C (273 K). Further, the thermistor self-heats current flows in the device. The bridge circuit is usually calibrated using an ice bath to provide the 0°C reference.

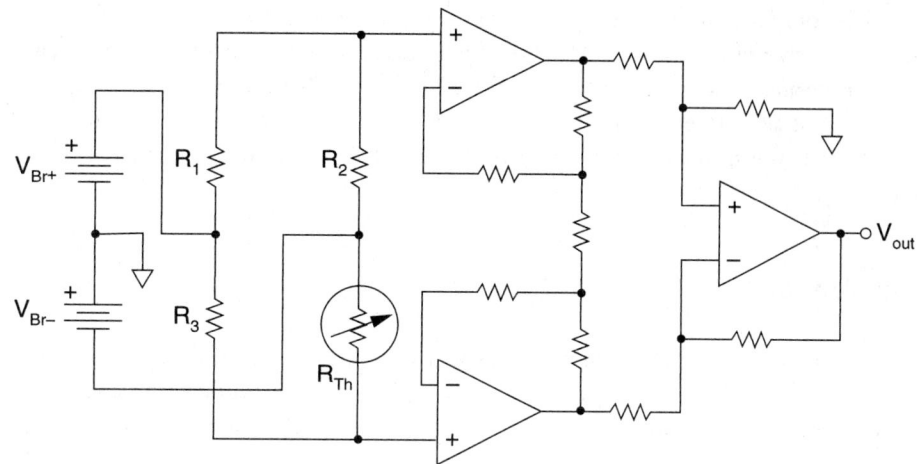

The form of instrumentation amplifier shown has very high input impedance and unity or higher gain depending on the resistors in the output amplifier. The other resistors in the bridge are usually 0.1 percent or 1 percent tolerance values.

RTDs

The RTD is a very accurate temperature-measuring device. Most RTDs are made of platinum wire, with a range of −200°C to 850°C. Platinum is used because of its linear resistance change with temperature. However, being a metal, the wire is prone to self-heating; and many times the system must be compensated to remove the effects of the self-heating. This type of system is also often used in a Wheatstone bridge configuration with an instrumentation amplifier, like the one illustrated in Figure 3-32. Since the change in voltage is very small, the amplifier usually has a very large gain (1000 or more) to provide large output voltages for direct application to control system amplifiers.

Two standards are used for calculating the resistance change of an RTD: the U.S. standard and the European standard. The standards are close but produce slightly different changes in resistance vs. temperature. The U.S. standard is used exclusively in the United States. The value of the resistance parameter is

$$\alpha = 0.00385 \ \Omega \ / \left(\Omega \ / \ °C \right) \tag{3.65}$$

The temperature correction factor, α, is determined by measuring the RTD at two temperatures: 0 and 100°C. At 100°C, the resistance of the RTD is 138.5 Ω; and at 0°C, the resistance is 100 Ω. The difference is 38.5°C. That number is divided by the value at 0°C, then by the temperature difference, such that

$$\alpha = \frac{(138.5 - 100) \ \Omega}{\dfrac{100 \ \Omega}{100°C}} = \frac{38.5}{10^4 (°C)} = 0.00385 \ / \ °C \tag{3.66}$$

The RTD nominal resistance is 100 Ω. That resistance is low and requires low voltages to operate; otherwise, self-heating of the wire occurs and the temperature measurement suffers. Filtering is not as essential for the RTD because of the low resistance. Noise voltages must contain enough power to induce a voltage in the measuring circuit, but that is usually not the case. To induce 1 millivolt in the RTD, the noise voltage requires 10 μA in the wire.

Fluid Control

Fluid controls are required when it is necessary to maintain the level of fluid in a tank, the pressure in a cylinder, the flow of fluid in a pipe, or other applications. It is therefore necessary to understand the concept of fluid resistance, quantity of flow and laminar vs. turbulent flow. This section will address these and other quantities of fluid dynamics.

Control Valves

Control valves are used to regulate the quantity, q, of fluid flowing in a pipe. Standard valves have a handle, a valve seat and plug, and a screw mechanism to determine the amount that the valve is opened with a single turn. The handle of the control valve can be manual or can be driven by an electric motor, in which case the valve opening is regulated by an electrical control system that uses a flow meter to determine the rate of flow in the pipe and to output an electrical signal proportional to flow.

Another form of control valve is the feedback-regulating valve. This type of valve can sample upstream or downstream flow to maintain constant flow in the outlet pipe. This valve is shown in Figure 3-33.

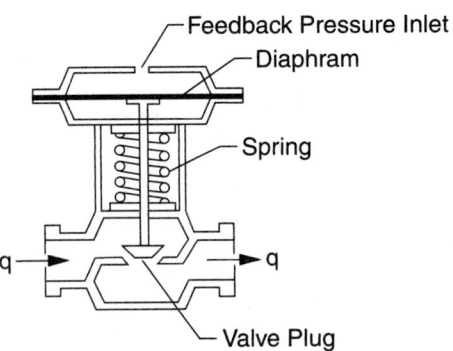

(figure 3-33)

Pressure regulating valve

In either case, the general valve constants are the same and the flow through the valve is defined by

$$q = KA\sqrt{\frac{2g\Delta p}{\rho}} \qquad (3.67)$$

Where K is the valve flow coefficient, A is the area of the pipe size, ρ is the density of the fluid in the pipe, Δp is the pressure differential across the valve, and g is the acceleration due to gravity.

The flow coefficient K and the area A are different for various types and sizes of valves. Consequently, a size coefficient C_v has been formulated by combining several of the terms of Equation 3.66. C_v is defined by the number of gallons of water per minute that flow through a completely open valve with a pressure drop of 1 psi. Thus

$$q_L = C_v\sqrt{\frac{\Delta p}{G}} \qquad (3.68)$$

Where q_L is the liquid flow-rate and G is the specific gravity of the fluid relative to water.

The size coefficient must be determined by actual test. However, for some types of valves, the coefficient may be approximated by the square of the valve size multiplied by 10.

Linear Position Transducers

Linear position transducers delineate translational motion to an electrical signal that is used to indicate position of various devices such as the movement of a piston in a cylinder, the movement of a lever system or other device. This section explains the type of transducer used for this type of measurement.

Linear Variable Differential Transformer (LVDT)

It is often necessary to convert translational motion to an equivalent electric voltage. There are a number of ways to accomplish the conversion. One popular method is by the use of a linear variable differential transformer, or LVDT. The LVDT consists of three coils of wire wound on a nonmagnetic form with a low reluctance core. Two of the coils are secondaries; the other is the primary coil. The primary coil is centered between the two secondaries. When the core is centered in the coils, the output voltages of the secondaries is equal, but opposite in phase, and the net output voltage is zero.

As the core moves in the form, a certain number of coil windings are affected by the proximity of the core, and they generate different output voltages.

The transformer wiring diagram is shown in Figure 3-34.

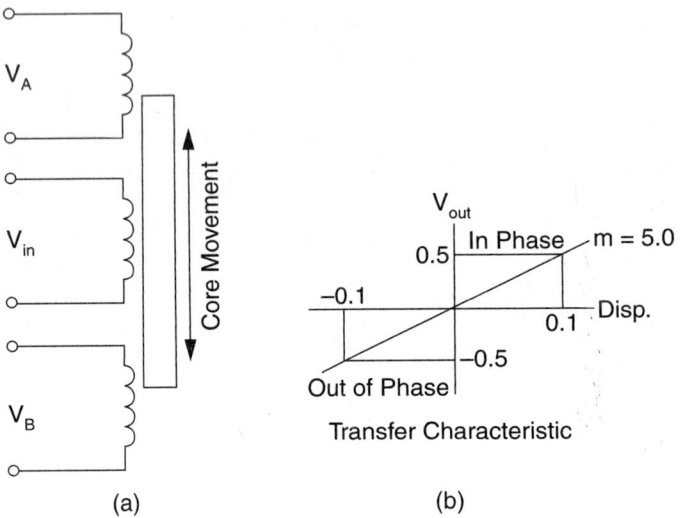

(figure 3-34)

LVDT and transfer characteristic

(a)   (b)

When the LVDT is used in the configuration of Figure 3-34, it is called *ratiometric wired*. The output voltage in this case is given by

$$D = m\frac{(V_A - V_B)}{(V_A + V_B)}$$
(3.69)

Where D is the displacement of the core and m is the slope of the displacement-voltage curve.

If the LVDT is wired as shown in Figure 3-35, it is said to be in the *open-wired* configuration, and the output voltage is proportional to the displacement of the cores

$$V_{out} = mD$$
(3.70)

(figure 3-35)

Open-wired configuration for an LVDT

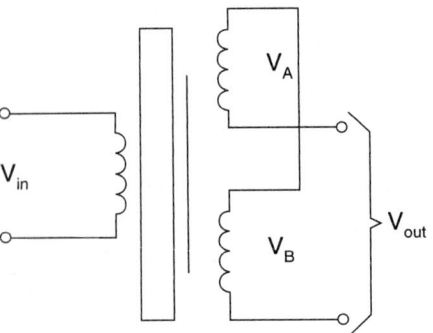

The input frequency range suitable for LVDTs is specified by the manufacturers, but usually ranges between 2 and 20 KHz.

Summary

➤ Electrical, mechanical, temperature, and fluid components all have analogous equations.

➤ Mechanical quantities such as inertia, torque, and rotational velocity are used with the electrical quantities of resistance, inductance, and damping to form an expression as to how a motor responds to an input voltage.

➤ Motors rotate at relatively high velocities. The concept of gears and driving a load with a gear train and the selection of the proper gear ratio impact optimum torque and acceleration.

➤ Operational amplifiers are components used in systems to provide various forms of gain. When used with motors in a system, opamps provide the mathematical functions needed, such as integration and differentiation, as well as proportional gain to alter (within a limited range) the response of a motor to an input voltage.

Different types of sensors are used to convert linear, rotational motion, and temperature to electrical voltages.

Problems

P3.1   Write the loop equations for currents I_1 and I_2 in the network shown below.

(figure P3-1)

Network for Problem P3.1 analysis

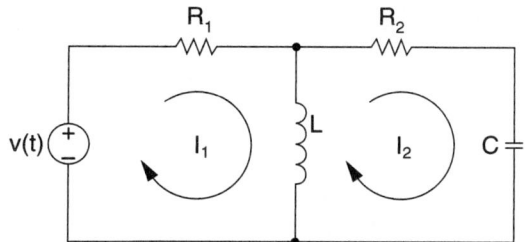

P3.2   Write the loop equations for the currents in the circuit shown, after the switch is thrown. Convert the equations to the s-plane without initial conditions. Show the voltage across R_2 as a ratio of polynomials.

(figure P3-2)

Network for Problem P3.2 analysis

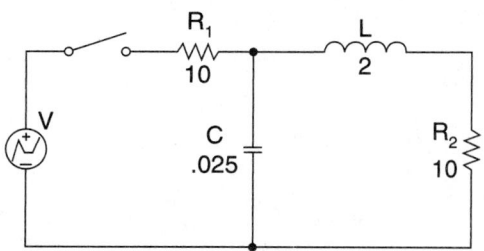

P3.3   Write the differential equations of the systems shown below.

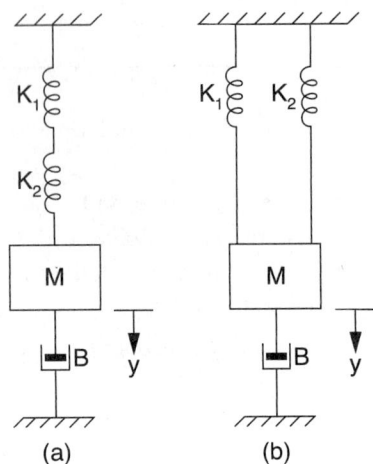

(figure P3-3)

Spring-mass systems for
Problem P3.3

P3.4   Write the torque differential equation for the rotational system shown
below. Convert the differential equation to the s-plane without initial conditions. Then provide a transfer function relating $\Theta(s) \,/\, T(s)$.

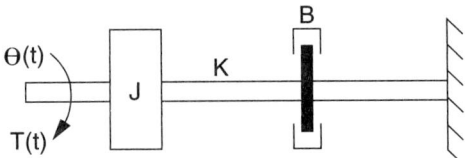

(figure P3-4)

Torsion system for
Problem P3.4

P3.5   The system shown has the following parameters: $B_1 = B_2 =$
5 oz-in/(rad/sec), $J_1 = J_2 = 8$ oz-in-s^2, $K = 8$ oz-in.

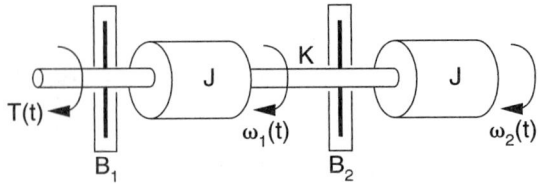

(figure P3-5)

Torsion system for
Problem P3.5

Write the differential equations of motion without initial conditions.

P3.6 For the DC motor specifications given below, determine the motor transfer function and the rpm for a 48-volt step input.

$$J = 0.005 \text{ oz-in-s}^2, \; K_b = 0.15 \text{ oz-in/krpm},$$

$$K_T = 22 \text{ oz-in/A}, \; R_a = 5.2 \text{ ohms},$$

$$L_a = 3 \text{ mH}, \; K_e = 15.9 \text{ V/krpm}$$

P3.7 The motor in Problem P3.6 has a maximum stall torque of 80 oz-in at an armature voltage of 48 volts. The motor is to be used in a constant speed conveyor system. The load is 400 oz-in, and the speed is to be 50 rpm. Make a load-torque curve for the motor and plot various gear ratios vs. motor torque, determine the optimum gear ratio for the load.

P3.8 A thermal system has a resistance $R = 2°F/(Btu/s)$, input temperature $T_1 = 100°F$, and heat flow $q = 10$ Btu/s. Determine T_2.

P3.9 An iron-constantin thermocouple has a sensitivity of 0.063 mV/°C. To measure the internal temperature of a motor, the thermocouple is placed at an internal point. The thermocouple measures 2.15 mV. Determine the temperature of the motor.

P3.10 An insulated tank contains 500 lb of water at 40°F. Water at 80°F flows into the tank at the rate of 100 lb/min. Assume that perfect mixing occurs and that heat loss through the insulation is negligible. The outflow is also 100 lb/min. Write the differential equation for the temperature of the water in the tank. Then convert the equation to the s-domain.

P3.11 A flow restricting device operates as $q = h^{1/n}$. Calculate the hydraulic resistance of the device if $n = 3$.

P3.12 A control valve has the following parameters: $K = 7.6$, $A = 2 \text{ in}^2$, $\rho = 78 \text{ lb/ft}^3$, $\Delta p = 8$ psi, and $g = 32.\text{ft/s}^2$. Determine q for the valve.

P3.13 A tachometer has an output of 5V/1000 rpm. Calculate the output voltage in $V/(\text{rad/sec})$.

P3.14 An LVDT is wired in the ratiometric mode. If $V_B = 0.5V_A$, determine m if $D = 0.25$ in.

P3.15 For the summing form of amplifiers shown, determine the output voltage.

(figure P3-15)

Summing amplifier system for Problem P3.15

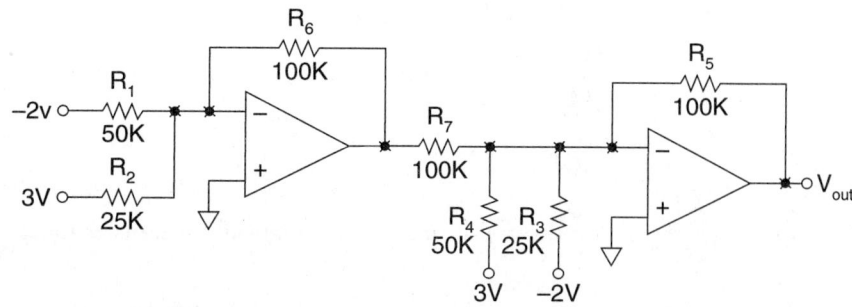

P3.16   Provide the input-output transfer function for the following opamp circuit.
         Keep in mind that the voltages at the inputs are equal.

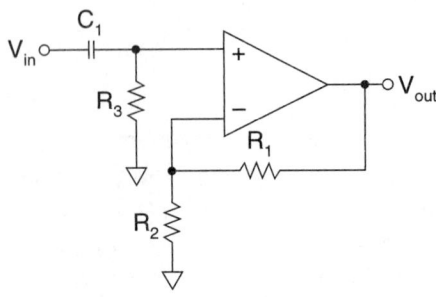

(figure P3-16)

Noninverting opamp amplifier for Problem P3.16

P3.17   Provide the input-output transfer function for the following opamp circuit.

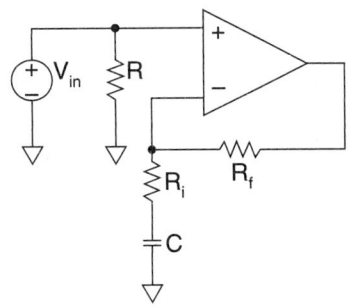

(figure P3-17)

Noninverting opamp amplifier for Problem P3.17

P3.18   For the circuit shown, show that if $R_L = R$, the current in R_L is given
         by $I_L = \dfrac{V_I}{R}$.

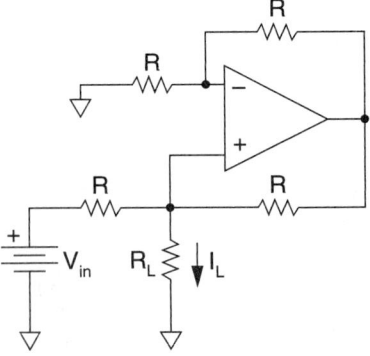

(figure p3-18)

Opamp current source for Problem P3.18

P3.19   For the opamp circuit shown, determine the input-output transfer function.

(figure p3-19)

Opamp circuit for P3.19

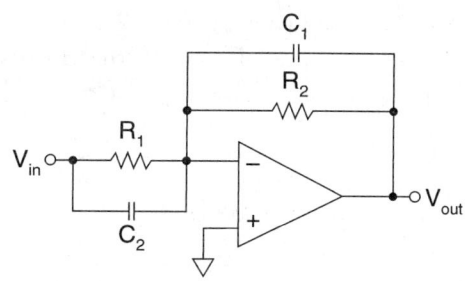

P3.20   A transducer has an output voltage range of 5–25 V. The voltage is to be converted to a 4–20 mA current for transmission over a long line. Provide a circuit for the conversion.

P3.21   Design an I-V converter to convert the current to a 2–10 V output voltage.

P3.22   The differential equation of a one-degree-of-freedom gyroscope is

$$H\omega = J\frac{d^2\theta}{dt^2} + B\frac{d\theta}{dt} + K\theta$$

Where H is the momentum of the gyroscope mass, ω is the radian velocity of the mass, J is the moment of inertia, B is the damping, and K is the spring constant of a spring attached to the spin axis.

a. Determine a transfer function for the Laplace transforms of ω and θ and show that the steady-state output is proportional to the magnitude of a constant rate input. That type of gyro is called a *rate gyro*.

b. Remove the restraining spring $(K = 0)$ and show that the output is proportional to the integral of the input rate. That type of gyro is called an *integrating gyro*.

P3.23   A tachometer generator is often used to measure the speed of the output shaft of a motor. The generator provides an output voltage that is proportional to the velocity of the shaft speed, and R_L is provided to facilitate the measurement of the voltage. For the armature circuit shown, find the transfer function relating $V_{out}(s)$ to shaft speed if $K_e = 0.5$ volts/rad/sec and the shaft is rotating at 150 rad/sec. The measured output voltage is 70 V. Calculate the ratio of load resistance to armature resistance.

(figure P3-23)

Network for Problem 3.23

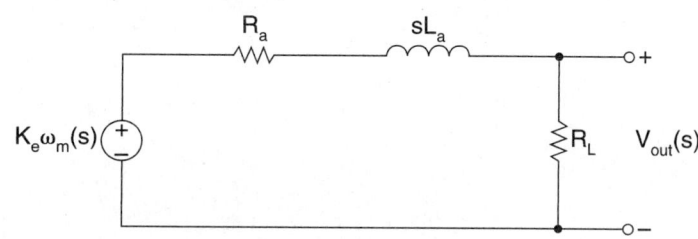

P3.24   A simple accelerometer is shown in the figure below. The position y of the mass M is proportional to the acceleration of the case. Determine the transfer function of the input acceleration $A(s)$ $\left(a = d^2x \; / \; dt^2 \right)$ to the output $Y(s)$.

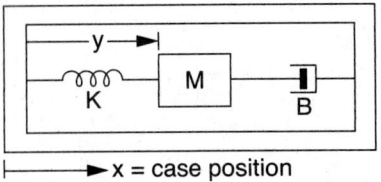

(figure P3-24)

Spring-mass system for Problem P3.24

Chapter 4

Transfer Functions from Block Diagrams and Signal Flow Graphs

Chapter Objectives

After completing this chapter, you should be able to:

➤ Represent various forms of systems by block diagrams or signal flow graphs.

➤ State the differences between block diagrams and the several forms of signal flow graphs.

➤ State what a transfer function represents.

➤ Be able to use the methods of this chapter to reduce block diagrams or signal flow graphs to transfer function form.

➤ Use **MATLAB** and **Simulink** to produce transfer functions and simulations of block diagram and SFG representations of control systems.

4.1 Introduction

Block diagrams and signal flow graphs are used universally to convey information about a system and its parameters. Whether the system is continuous or discrete, a transfer function is obtained from the type of diagram used. There are some differences in the block diagram used for a continuous system compared to a discrete system, which will become apparent in Chapter 11, which discusses discrete control. Signal flow graphs are used with state-variable solutions for systems because it is easy to envision the required matrices from state equations.

4.2 What Is a Transfer Function?

Simply put, a transfer function is the ratio of output to input of a system. It is how the system transfers its input to its output. While that sounds like simple gain, the transfer function generally gives the output in rad/volt, gal/min, or other terms.

There are two forms of transfer function: open-loop and closed-loop. The difference between them will become apparent shortly.

The transfer function is often given in the form of a numerator and denominator composed of factored polynomials. The numerator of the transfer function contains the zeros of the system, and the denominator contains the poles of the system. The transfer function also provides data for analysis of stability, frequency response, rise/settling time, and other parameters.

In Chapter 1, a simple transfer function for a continuous system was shown. It is repeated here for convenience.

$$\frac{Y}{R} = \frac{G_c G_p}{1 + G_c G_p H} \tag{4.1}$$

It should be clear that the output of the system will be less than 1 depending on $G_c H$. What comprise G_c, G_p, and H are their mathematical components, which are the subject of this chapter.

4.3 Fundamentals of Block Diagrams

Block diagrams, in general, use only two types of blocks to describe a system. The blocks are:

➤ **Gain block,** which multiplies its input by what is contained inside the block. The block may have only one input and one output and may have a gain less than, greater than, or equal to unity. The gain block can also contain expressions that may be transfer functions.

➤ **Summing junction,** which algebraically sums its inputs and may have as many inputs as necessary, but only one output. The inputs to the summing junction are labeled with their respective polarities. Plus signs indicate no inversion of the applied signal; minus signs indicate inversion of the applied signal.

Figure 4-1 shows the blocks.

(figure 4-1)

Blocks used in block diagrams

```
        B
        │
      + ▼
A ──+──▶ Σ ──▶ D = A + B − C      A ──▶ [ Gc ] ──▶ B = Gc ±A
        ▲
      − │
        C
   Summing Junction                    Gain Block
```

Canonic Forms

There are two fundamental forms of control system block diagrams. All control systems can be reduced to one of the two forms.

Form 1

Canonic Form 1 is shown in Figure 4-2. It is clear from the figure that 100 percent of the output is fed back to the inverting input of the summing junction. G_c is the gain in the forward path, the open-loop gain, with no feedback applied.

(figure 4-2)

Canonic Form 1, feedback system

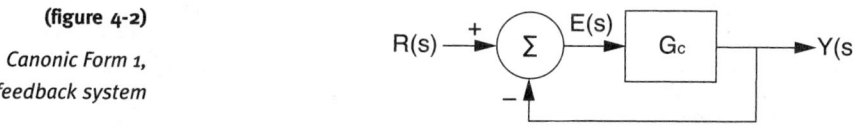

The input is called the **setpoint,** or **reference input** $R(s)$. It is the desired condition for the output. The signal $E(s)$, the difference signal between the input and output, is called the **error signal.** The transfer function of this form is constructed with some simple mathematical manipulation.

Figure 4-2 shows that the output is applied to the inverting summing junction input and the reference (setpoint) signal is applied to the noninverting input. The error signal is the difference of the setpoint and output, $R(s) - Y(s)$. The transfer function $Y(s)/R(s)$ is developed in the following way:

$$Y(s) = G_c(s)R(s) - G_c(s)Y(s)$$

$$Y(s)\left[1 + G_c(s)\right] = G_c(s)R(s)$$

$$\frac{Y(s)}{R(s)} = \frac{G_c(s)}{1 + G_c(s)}$$

(4.2)

The transfer function shows that the output of this system is always less than 1. However, if G_c is much greater than 1, the output is considered to be unity. This form is often called the *unity gain form.*

Form 2

In this form, the output is processed through a feedback block $H(s)$, which conditions the output to the proper level before it is fed back to the summing junction. The block diagram is shown in Figure 4-3.

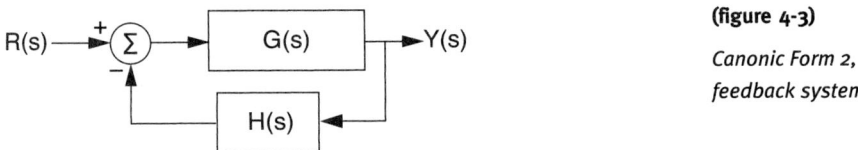

(figure 4-3)

Canonic Form 2, feedback system

The development of the transfer function for this form is identical to that of Form 1 except that the error signal is $E(s) = R(s) - H(s)Y(s)$.

The transfer function, using the same procedure as Equation 4.2, is

$$\frac{Y(s)}{R(s)} = \frac{G_c(s)}{1 + G_c(s)H(s)}$$

(4.3)

Block Diagram Algebra

To determine the transfer function, the principles of the two blocks are applied. An example of a transfer function developed from the first form will simplify the process. In all cases, the total forward path gain is the gain without feedback, and is the numerator of the transfer function. The feedback loop gains are the denominator of the transfer function.

Example 4-1

For the system shown in Figure 4-4, determine the transfer function.

Solution

Applying the derivation of Equation 4.2

$$Y(s) = G(s)R(s) - G(s)Y(s)$$

$$Y(s) = \frac{3}{s^2 + 4s + 3}R(s) - \frac{3}{s^2 + 4s + 3}Y(s)$$

$$Y(s)\left[1 + \frac{3}{s^2 + 4s + 3}\right] = \frac{3R(s)}{s^2 + 4s + 3}$$

$$\frac{Y(s)}{R(s)} = \frac{\dfrac{3}{s^2 + 4s + 3}}{1 + \dfrac{3}{s^2 + 4s + 3}} = \frac{\dfrac{3}{s^2 + 4s + 3}}{\dfrac{s^2 + 4s + 3 + 3}{s^2 + 4s + 3}}$$

$$\frac{Y(s)}{R(s)} = \frac{3}{s^2 + 4s + 6}$$

From the final value theorem, allowing all terms containing s to approach zero, the output of this system is one-half the input.

(figure 4-4)

*Feedback system for
Example 4-1*

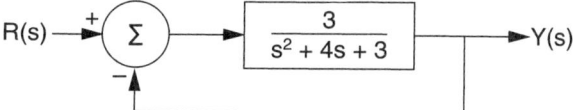

(figure 4-4)

*Feedback system for
Example 4-1*

Example 4-2

An example of the second form shows the reduction in gain controlled by feedback. The feedback supplies a signal that is less than the full output. In this case, the $H(s)$ block is a pure number. It may also be Laplace expressions or another form of expression. The block diagram is shown in Figure 4-5.

(figure 4-5)

*Control system for
Example 4-2*

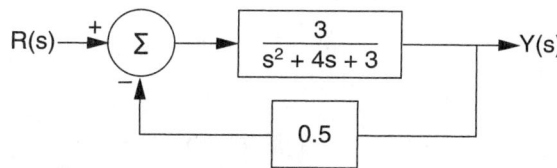

Example 4-2 (continued)

Solution

Starting again with the error signal

$$E(s) = R(s) - 0.5Y(s)$$

$$Y(s) = E(s) = R(s) - 0.5Y(s)$$

$$Y(s) = \frac{3}{s^2 + 4s + 3}R(s) - \frac{1.5}{s^2 + 4s + 3}Y(s)$$

$$Y(s)\left[1 + \frac{1.5}{s^2 + 4s + 3}\right] = \frac{3}{s^2 + 4s + 3}R(s)$$

$$\frac{Y(s)}{R(s)} = \frac{\dfrac{3}{s^2 + 4s + 3}}{\left[1 + \dfrac{1.5}{s^2 + 4s + 3}\right]} = \frac{\dfrac{3}{s^2 + 4s + 3}}{\dfrac{s^2 + 4s + 3 + 1.5}{s^2 + 4s + 3}}$$

$$\frac{Y(s)}{R(s)} = \frac{3}{s^2 + 4s + 4.5}$$

Since only 50 percent of the output is used for feedback, it is apparent that the output of this system is greater than the output of the first form. From the final-value theorem, the output of this system will be two-thirds the input.

Examples of More Complex Systems

Example 4-3

In this example, the generic terms $G_x(s)$ will be used for the system components. The block diagram is shown in Figure 4-6.

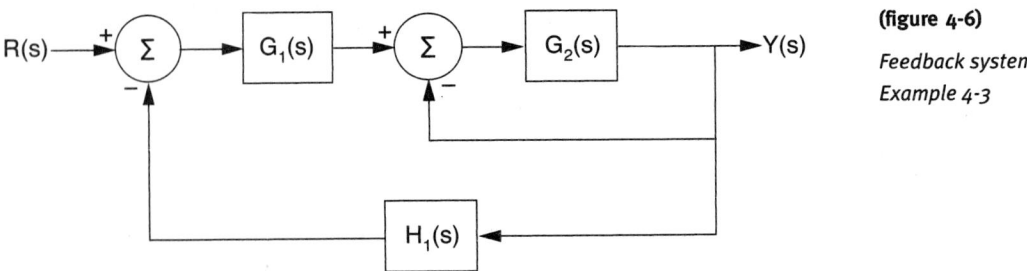

(figure 4-6)

Feedback system for Example 4-3

Example 4-3 (continued)

Solution

For this example, it is best to start at the output and reduce the subsystem there. The subsystem at the output is Form 1. As the Form 1 equation shows, this part of the system can be replaced by a single block with its transfer function.

$$\frac{G_2(s)}{1 + G_2(s)}$$

The block diagram then is simplified as in Figure 4-7.

The system can now be reduced to the second form by multiplying $G_2(s)$ by $G_1(s)$ and removing the $G_1(s)$ block as shown in Figure 4-8.

The transfer function of this system is easily found using Equation 4.3 for the transfer function.

$$E(s) = R(s) - Y(s)H_1(s)$$

$$Y(s) = \frac{G_1(s)G_2(s)}{1 + G_2(s)}R(s) - \frac{G_1(s)G_2(s)}{1 + G_2(s)}Y(s)H_1(s)$$

$$Y(s)\left[1 + \frac{G_1(s)G_2(s)}{1 + G_2(s)}H_1(s)\right] = \frac{G_1(s)G_2(s)}{1 + G_2(s)}R(s)$$

$$\frac{Y(s)}{R(s)} = \frac{\dfrac{G_1(s)G_2(s)}{1 + G_2(s)}}{1 + \dfrac{G_1(s)G_2(s)}{1 + G_2(s)}H(s)} = \frac{G_1(s)G_2(s)}{1 + G_2(s) + G_1(s)G_2(s)H(s)}$$

(figure 4-7)

Output block reduction for Example 4-3

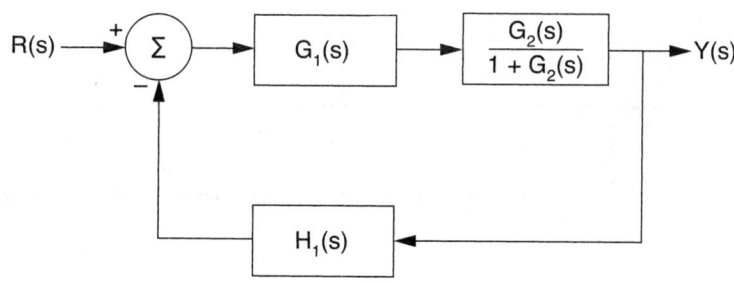

(figure 4-8)

Reduced block diagram for Example 4-3

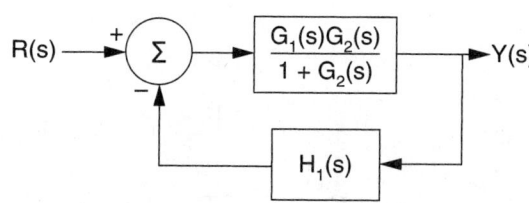

This transfer function could have been written by inspection. Since all feedback paths are from the output, with no independent internal loops, all that is required to write the transfer function is to count the number of feedback loops and add the necessary functions to the denominator of the transfer function. The procedure is as follows:

1. Count the number of feedback loops and note the number of gain blocks contained in each loop. The number of loops indicates the number of terms, excluding the one that is in the denominator of the transfer function.

2. In the numerator of the transfer function, write the forward path gain, the product, and any sums of all of the gains in the forward path of the system.

3. Write the denominator with a 1, plus the values in each of the feedback loops. In each term, include only the gains and feedback in the loop being considered. Do not add any of the other feedbacks in any other loop.

If there is an independent feedback loop within the block diagram, it must be reduced to its fundamental form. Then the method may be used if all remaining feedback comes from the output.

In the previous example, there are two feedback loops—one containing $G_1(s)$, $G_2(s)$, and $H_1(s)$ and the other containing only $G_2(s)$. The forward path is $G_1(s) \cdot G_2(s)$. The denominator will have feedback terms of 1 and 2: $1 + G_1(s) \cdot G_2(s)H_1(s) + G_2(s)$.

The next example also has two summing junctions, but an added gain at the input, which is summed at the output of the first gain block. This is called *feed forward gain*. The forward path will have $G_3(s)\left[G_1(s) + G_2(s)\right]$. Since all feedback loops start at the output of the system, the transfer function can be written by inspection, as in Example 4-4.

Example 4-4

There are two feedback loops: one containing all of the gains in the system and one with only a gain and feedback block. See Figure 4-9.

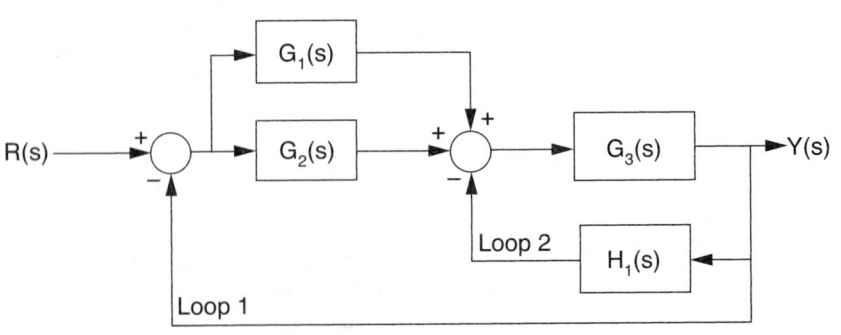

(figure 4-9)

Control system for Example 4-4

Example 4-4 (continued)

Solution

The transfer function is obtained by noting the gains of the system.

$$\mathrm{Forward\,Path\,Gain} = \big[G_1(s) + G_2(s)\big]G_3(s)$$
$$\mathrm{Loop\,1\,Gain} = \big[G_1(s) + G_2(s)\big]G_3(s)$$
$$\mathrm{Loop\,2\,Gain} = G_3(s)H_1(s)$$

The overall transfer function is

$$\frac{Y(s)}{R(s)} = \frac{\big[G_1(s) + G_2(s)\big]G_3(s)}{1 + \big[G_1(s) + G_2(s)\big]G_3(s) + G_3(s)H_1(s)}$$

Example 4-5

In this example, the generic G_X is used instead of $G_X(s)$. The G_2, G_3 blocks can be replaced with a single block, $G_1(G_2 + G_3)$; and the rest of the transfer function can be written by inspection. The system is shown in Figure 4-10.

Solution

The block diagram with the summed gains is shown in Figure 4-11.

(figure 4-10)

Control system for Example 4-5

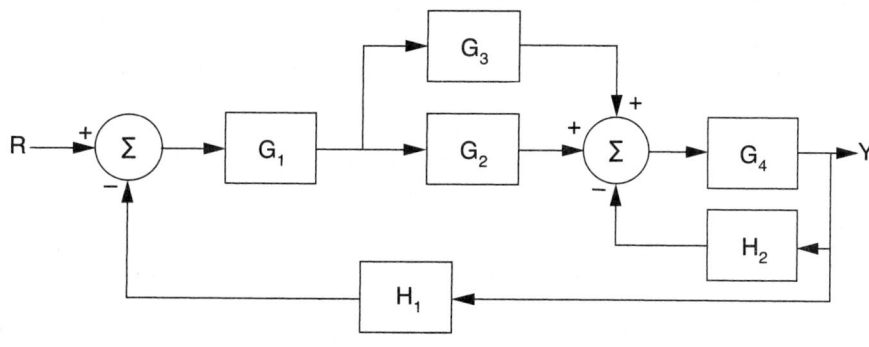

(figure 4-11)

First reduction of Example 4-5

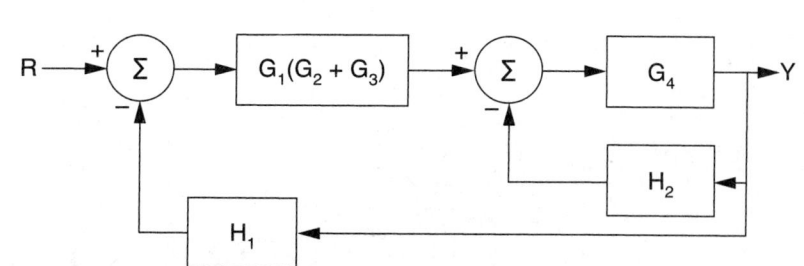

Example 4-5 (continued)

The system can now be reduced by inspection, and the transfer function is

$$\frac{Y}{R} = \frac{G_1 G_4 (G_2 + G_3)}{1 + G_1 G_4 (G_2 + G_3) H_1 + G_4 H_2}$$

Block Diagrams Showing a Disturbance Input

In some cases, the block diagram of a system will show a disturbance input at a specific point, usually an output where a load of some type is applied. The effect of the application of the load and disturbance can be calculated using the methods outlined previously and the superposition theorem to sum the load and disturbance.

Example 4-6

Consider the following system as shown in Figure 4-12:

Solution

Set the disturbance to zero and determine the transfer function without the disturbance. The transfer function of this system with the disturbance input, $D(s)$, set to zero is

$$\frac{Y(s)}{R(s)} = \frac{G_1(s)G_2(s)}{1 + G_1(s)G_2(s)H(s)}$$

To determine the effect of the disturbance on the system, allow the reference input $R(s)$ to equal zero. The block diagram can then be redrawn as shown in Figure 4-13.

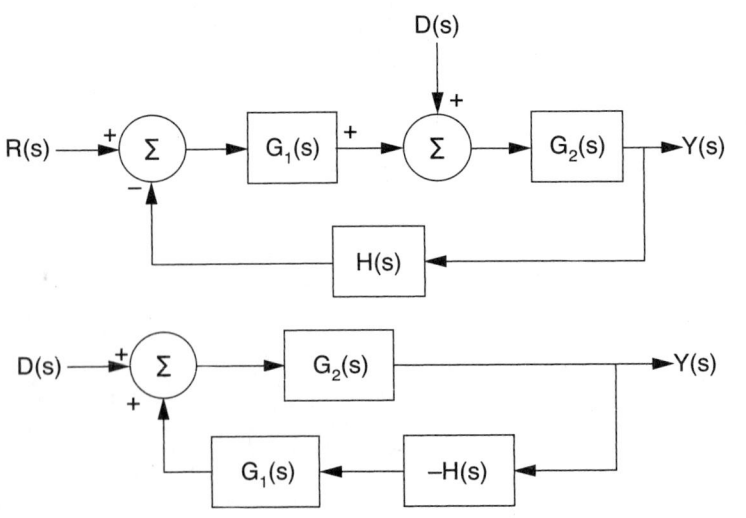

(figure 4-12)

Control system with disturbance input for Example 4-6

(figure 4-13)

Control system redrawn for disturbance input in Example 4-6

Example 4-6 (continued)

Notice that the feedback block has a negative sign; that preserves the polarity of the feedback. (The input summing junction could have been left in the path, and the negative sign would not have been needed on $H(s)$.) That produces the same denominator as in the transfer function with no disturbance. The transfer function for the system disturbance is

$$\frac{Y(s)}{R(s)} = \frac{G_2(s)}{1 + G_1(s)G_2(s)H(s)}$$

The total transfer function is the sum of the two transfer functions.

$$\frac{Y(s)}{R(s)} = \frac{G_1(s)G_2(s)}{1 + G_1(s)G_2(s)H(s)} + \frac{G_2(s)}{1 + G_1(s)G_2(s)H(s)}$$

Notice that the response to the disturbance is reduced by the gain factor G_1.

4.4 Block Diagram Identities

Some identities can be used to make block diagram reduction less tedious. You have already seen one in the form of the summed gain blocks in Example 4-5. Sometimes it is advantageous to shift summing junctions around to simplify the diagram. Other times two or more blocks can be combined simply by noting their configuration and replacing it with a single block. Some useful identities are shown in Table 4-1.

(table 4-1)

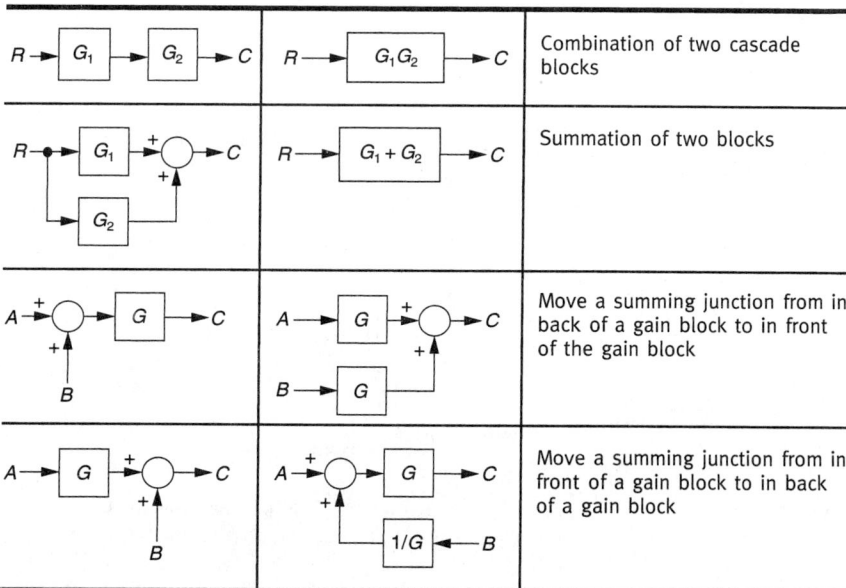

(table 4-1) *continued*

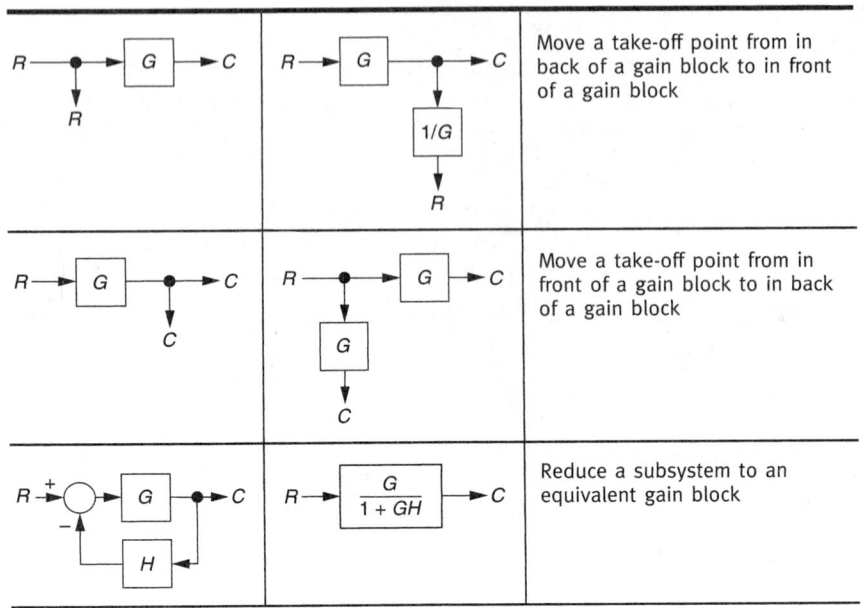

		Move a take-off point from in back of a gain block to in front of a gain block
		Move a take-off point from in front of a gain block to in back of a gain block
		Reduce a subsystem to an equivalent gain block

4.5 Block Diagrams and Transfer Functions of Control Systems

Example 4-7

Consider the following block diagram for a servo position control system as shown in Figure 4-14. The transfer function for the system is

$$\frac{\Omega(s)}{K(s)} = \frac{\dfrac{1.318 \cdot 10^6}{s^2 + 773s + 9.52 \cdot 10^5}}{1 + \dfrac{1.318 \cdot 10^6}{s^2 + 773s + 9.52 \cdot 10^5}} = \frac{1.318 \cdot 10^6}{s^2 + 773s + 2.27 \cdot 10^6}$$

$$\frac{\Theta(s)}{R(s)} = \frac{\dfrac{1.318 \cdot 10^6 K}{s(s^2 + 773s + 2.27 \cdot 10^6)}}{1 + \dfrac{1.318 \cdot 10^6 K}{s(s^2 + 773s + 2.27 \cdot 10^6)}}$$

$$= \frac{1.318 \cdot 10^6 K}{s^3 + 773s^2 + 2.27 \cdot 10^6 s + 1.318 \cdot 10^6 K}$$

Notice that the denominator of the transfer function contains all components of the numerator. The denominator is called the **characteristic equation** of the system. The characteristic equation is discussed in more detail in Chapter 5.

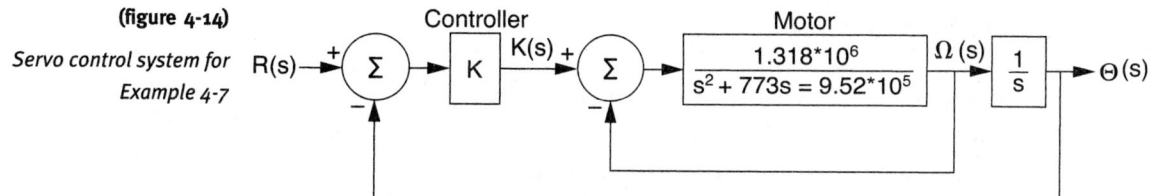

(figure 4-14)

Servo control system for Example 4-7

Example 4-8

Figure 4-15 shows a simple motor velocity control system. Determine the transfer function of the system.

Solution

The transfer function for the system is

$$\frac{\Omega(s)}{R(s)} = \frac{\dfrac{10K}{s^2 + 5s + 6}}{1 + \dfrac{10K}{s^2 + 5s + 6}} = \frac{10K}{s^2 + 5s + (6 + 10K)}$$

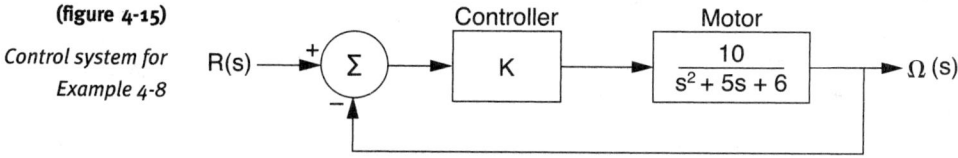

(figure 4-15)

Control system for Example 4-8

4.6 Signal Flow Graphs (SFGs)

The signal flow graph (SFG) can be considered a simplified block diagram. However, the mathematical rules of the flow graph are more restrictive than the block diagram. One type of signal flow graph is called the **phase variable form**. The phase variable form uses only integrators with a gain of 1 in the forward path. Links to the integrators may have any gain. A second form, which was introduced by Mason[1,2] in 1954, is a more general form in which each link in the graph is considered an amplifier or another component that may have any gain.

Fundamentals of the SFGs

SFGs are used to describe a control system transfer function. Generally, the graphs are simpler than a block diagram since only nodes, branches, and loops are used to show the configuration of the system.

Phase Variable Form

Several rules apply to the phase variable form of an SFG.

1. The fundamental building block consists of two outer nodes: one is an input, the other is an output, and a summing junction is in between. The gain of the input node may be any value, and the gain at the output node is 1/s multiplied by the input value. Signals may flow only in the direction of the arrows.
2. This type of graph applies only to linear systems.
3. The equations used in the SFG must be algebraic. (Integrodifferential equations are first transformed to the s-plane.)
4. Feedback and feed-forward loops are used to complete the design.
5. There are as many forward path integrations as the largest exponent of s in the system.
6. The equations generated are simultaneous equations of the system, which are easily transferred into matrices.
7. It is assumed that all integrators in the system are of the summing type and may have more than one input. All inputs are summed algebraically according to their sign.

The fundamental building block of the phase variable flow graph is shown in Figure 4-16.

$$R \circ \xrightarrow{\quad 1 \quad} \circ \xrightarrow{\quad 1/s \quad} \circ Y$$

(figure 4-16)

Fundamental block for signal flow diagrams

The outer circles indicate nodes, which are variables; and the lines, which are connecting branches, indicate transfer relationships. The central node is a summing junction. The SFG of a system is produced in the following way. Consider the transfer function of a general second-order system.

$$\frac{Y(s)}{R(s)} = \frac{a_1 s + a_0}{s^2 + b_1 s + b_0} \tag{4.4}$$

First, divide all terms by the highest power of s.

$$\frac{Y(s)}{R(s)} = \frac{\dfrac{a_1}{s} + \dfrac{a_0}{s^2}}{1 + \dfrac{b_1}{s} + \dfrac{b_0}{s^2}} \tag{4.5}$$

Since the largest exponent of s is 2, this SFG has two integrations. The two integrators and a node for the output $Y(s)$ will be drawn, which constitutes the forward path of the system as shown in Figure 4-17.

(figure 4-17)

Two integrators in forward path

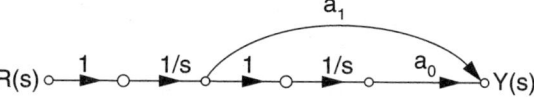

For this system, two feedback paths are located in the denominator and two feedforward paths are indicated in the numerator. The feed-forward paths are from the second integrator output to $Y(s)$ and from the output of the first integrator to $Y(s)$ and are positive values. Those components are drawn as shown in Figure 4-18.

(figure 4-18)

SFG with forward paths added

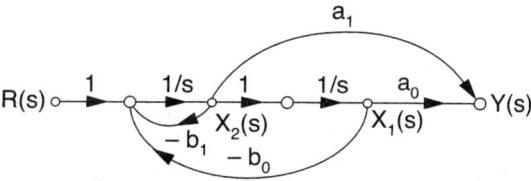

To finish the diagram, the feedback paths are entered as negative values.

The output nodes of the integrators are labeled from *right* to *left* with an arbitrary constant such as $X_1(s)$ as in Figure 4-19.

(figure 4-19)

Completed SFG

The values for $X_1(s)$ and $X_2(s)$ are readily found from the input to each node; that is, the input to $X_1(s)$ is $X_2(s) \cdot 1/s$. That forms the basis for the simultaneous equations of the system.

$$
\begin{aligned}
X_1(s) &= \frac{1}{s} X_2(s) \\
X_2(s) &= \frac{1}{s}[-b_0 X_1(s) - b_1 X_2(s) + R(s)] \\
Y(s) &= a_0 X_1(s) + a_1 X_2(s)
\end{aligned}
\tag{4.6}
$$

You form the simultaneous equations into matrices by taking s to the left-hand side of the equations and converting the equations to the time domain by using the dot (derivative of time) notation for the derivatives formed.

$$
\begin{aligned}
sX_1(s) &= X_2(s) \\
sX_2(s) &= -b_0 X_1(s) - b_1 X_2(s) + R(s)
\end{aligned}
\tag{4.7}
$$

$$\dot{x}_1 = \dot{x}_2$$
$$\dot{x}_2 = -b_0 \dot{x}_1 - b_1 \dot{x}_2 + r(t) \qquad (4.8)$$
$$y(t) = a_0 x_1 + a_1 x_2$$

$Y(s)$ remains the same as in Figure 4-19. From the equations, you can see that the matrices required are one (2×2) matrix and a column matrix for $r(t)$. It is now easy to form the matrices from the derivatives and coefficients of the differential equation. The subscripts on the left side of the equations give the matrix row number, and the subscripts on the right side of the equations give the column number of the matrix. $sX_1(s)$ indicates Row 1 of the matrix, and $X_2(s)$ indicates Column 2 of the matrix. The derived matrices are as follows:

$$\begin{bmatrix} \dot{x}_1 \\ \dot{x}_2 \end{bmatrix} = \begin{bmatrix} 0 & 1 \\ -b_0 & -b_0 \end{bmatrix} \begin{bmatrix} x_1 \\ x_2 \end{bmatrix} + \begin{bmatrix} 0 \\ 1 \end{bmatrix} r(t)$$

$$(4.9)$$

$$y(t) = \begin{bmatrix} a_0 & a_1 \end{bmatrix} \begin{bmatrix} x_1 \\ x_2 \end{bmatrix}$$

There are two types of systems to consider when using this model: the pole-zero model and the all-pole model.

Pole-Zero Model

The method for this type of model is the same as that shown above. Consider the system whose transfer function is

$$\frac{Y(s)}{R(s)} = \frac{4s^2 + 6s + 10}{s^3 + 6s^2 + 11s + 10} = \frac{\dfrac{4}{s} + \dfrac{6}{s^2} + \dfrac{10}{s^3}}{1 + \dfrac{6}{s} + \dfrac{11}{s^2} + \dfrac{10}{s^3}}$$

Since that is a third-order system, there will be three integrators in the forward path, three feed-forward paths, and three feedback paths. The SFG is shown in Figure 4-20.

(figure 4-20)

SFG of pole-zero control system model

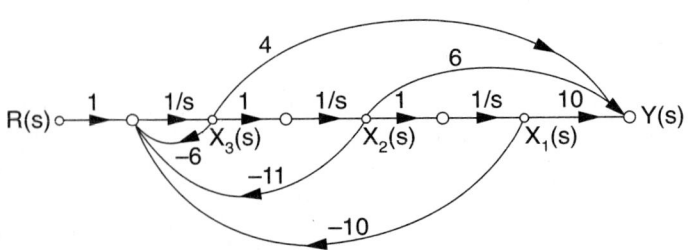

The state model is generated from the following equations:

$$\frac{Y(s)}{R(s)} = \frac{4s^2 + 6s + 10}{s^3 + 6s^2 + 11s + 10} = \frac{\dfrac{4}{s} + \dfrac{6}{s^2} + \dfrac{10}{s^3}}{1 + \dfrac{6}{s} + \dfrac{11}{s^2} + \dfrac{10}{s^3}}$$

$$sX_1(s) = X_2(s)$$
$$sX_2(s) = X_3(s)$$
$$sX_3(s) = -10X_1(s) - 11X_2(s) - 6X_3(s) + R(s)$$

$$\dot{x}_1 = x_2(t)$$
$$\dot{x}_2 = x_3(t)$$
$$\dot{x}_3 = -10x_1(t) - 11x_2(t) - 6x_3(t) + r(t)$$
$$y(t) = 10x_1(t) + 6x_2(t) + 4x_3(t)$$

The matrices of the system are

$$\begin{bmatrix} x_1 \\ x_2 \\ x_3 \end{bmatrix} = \begin{bmatrix} 0 & 1 & 0 \\ 0 & 0 & 1 \\ -10 & -11 & -6 \end{bmatrix} \begin{bmatrix} x_1 \\ x_2 \\ x_3 \end{bmatrix} + \begin{bmatrix} 0 \\ 0 \\ 10 \end{bmatrix} r(t)$$

$$y(t) = \begin{bmatrix} 10 & 6 & 4 \end{bmatrix} \begin{bmatrix} x_1 \\ x_2 \\ x_3 \end{bmatrix}$$

All-Pole Model

For this system, there are no zeros; so the SFG will have no feed-forward components. Consider the system described by

$$\frac{Y(s)}{R(s)} = \frac{20}{s^3 + 10s^2 + 25s + 20}$$

$$\frac{Y(s)}{R(s)} = \frac{\dfrac{20}{s^3}}{1 + \dfrac{10}{s} + \dfrac{25}{s^2} + \dfrac{20}{s^3}}$$

Once again, three integrators are required. Notice that only one path connects from the output of the third integrator to the output of the system, $Y(s)$. Its value is the numerator value of the transfer function. The SFG is shown in Figure 4-21.

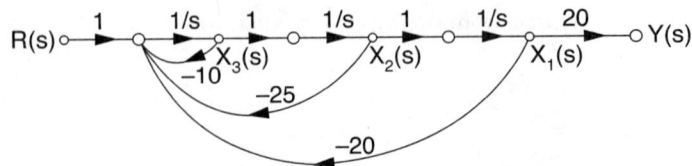

(figure 4-21)

SFG of all-pole model

The matrices are formed in a similar manner except that $y(t)$ has two zero values.

$$X_1(s) = \frac{X_2(s)}{s}$$

$$X_2(s) = \frac{X_3(s)}{s}$$

$$X_3(s) = \frac{1}{s}\left(-20X_1(s) - 25X_2(s) - 10X_3(s) + R(s)\right)$$

$$Y(s) = X_1(s)$$

$$sX_1(s) = X_2(s)$$

$$sX_2(s) = X_3(s)$$

$$sX_3(s) = -20X_1(s) - 25X_2(s) - 10X_3(s) + R(s)$$

$$\begin{bmatrix} x_1 \\ x_2 \\ x_3 \end{bmatrix} = \begin{bmatrix} 0 & 1 & 0 \\ 0 & 0 & 1 \\ -20 & -25 & -10 \end{bmatrix}\begin{bmatrix} x_1 \\ x_2 \\ x_3 \end{bmatrix} + \begin{bmatrix} 0 \\ 0 \\ 20 \end{bmatrix}r(t)$$

$$y(t) = \begin{bmatrix} 0 & 0 & 1 \end{bmatrix}\begin{bmatrix} x_1 \\ x_2 \\ x_3 \end{bmatrix}$$

Mason's Gain Formula and the SFG

The SFG by Mason provides simultaneous algebraic equations of any system. From a block diagram, an analogous SFG, which is an identical graphical structure, can be produced. Some of the rules for the phase variable form also apply to the general form of SFG, but there are differences. Those differences are presented in the following section.

Definition of SFG Terms

Input node. A node that has branches only leaving the node.

Loop. A path that starts on a node and ends on the same node, passing through no node more than once.

Loop gain. The path gain of a loop.

Nontouching loops. Any set of loops that do not share a common node.

Output node. A node that has branches only entering the node; if the node does not qualify as an output node, it can be made to qualify through the attachment of a unity gain branch labeled with the same node name.

Path. A set of branches that pass through nodes in the same direction; when the path passes through the nodes only once, it is called a forward path.

Path gain. The product of all of the gains in any given path.

SFG Algebra

The fundamental component in the SFG is shown in Figure 4-22. Notice that there is no central node as a summing integrator. The output of this component is $Y = GR$.

(figure 4-22)

Basic block for Mason's
SFG form

An example of a simple feedback system block diagram and its SFG equivalent are shown in Figure 4-23. You can see the similarity as well as the simplicity of the SFG compared to a block diagram.

(figure 4-23)

Block diagram with
Mason's SFG equivalent

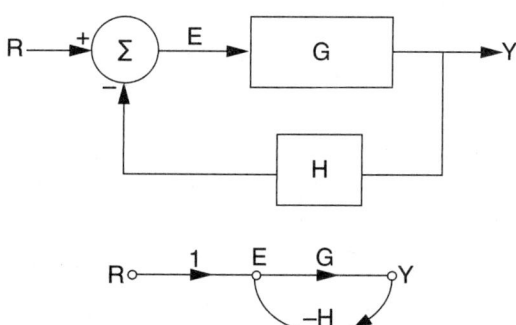

It is not necessary to show a summing junction. It is understood that the value of a node is equal to the algebraic sum of all signals entering the node. That is different from the SFG of the phase variable diagram. Subtraction is indicated by the placement of a minus sign on the loop connecting to a node. Input nodes are nodes with only outgoing branches; output nodes have only incoming branches. That graph provides a transfer function for any one (or a linear combination) of the dependent variables, X. A gain formula for any of the variables is formed as X/R_i, where R_i must be an input node and X is the variable of interest.

Figure 4-23 shows a path from the output back to an input node. That path is called a *loop*. To qualify as a loop, the connecting branches must start and end at the same node without entering any node more than once, another difference from the phase variable model, where the summing junction receives all feedback branches. The loop gain is the product of all of the branch values in the loop. Nontouching loops are loops that have no common nodes. A forward path is a way from R_i to another node that does not enter any node more than once. The path gain is the product of all branches in the path.

The numerator of Mason's gain formula accounts for all forward paths in the system, and the denominator accounts for all loops in the system. Mason's formula is

$$N = \frac{y_{out}}{x_{in}} = \frac{\sum\limits_{k=1}^{N} N_k \Delta_k}{\Delta} \qquad (4.10)$$

Where

x_{in} = input node variable

y_{out} = output node variable

N = gain between x_{in} and y_{out}

M = total number of forward paths between x_{in} and y_{out}

N_k = gain of the kth forward path between x_{in} and y_{out}

$\Delta = 1 - \sum\limits_{i} X_{i1} + \sum\limits_{j} X_{j2} - \sum\limits_{k} X_{k3} + \cdots$

Where

$\sum\limits_{i} X_{i1}$ = the sum of all single loop gains

$\sum\limits_{j} X_{j2}$ = the sum of the gain products of all combinations of two nontouching single loops

$\sum\limits_{k} X_{k3}$ = the sum of the gain products of all combinations of three nontouching single loops.

If there are sections of four or more nontouching single loops, the summations continue on as shown.

Δ is the system determinant. An example will help clarify the use of the formula.

Example 4-9

Solution

Consider the SFG shown in Figure 4-24.

The transfer function of this system can be determined using the methods of block diagram reduction shown earlier. The transfer function is

$$\frac{Y}{R} = \frac{G_1(G_2 + G_3)}{1 + G_1 H_1 + G_1 G_2 H_2}$$

Loop 1 is $\Delta_1 = L_1 = -G_1 H_1$, and Loop 2 is $\Delta_2 = L_2 = -G_1 G_2 H_2$. The sum of the two loop gains is $(-G_1 H_1 - G_1 G_2 H_2)$. Since there are no nontouching loops, the $\sum X_{i,j,k}$ is zero. And

$$\Delta = 1 - \left(-G_1 H_1 - G_1 G_2 H_2\right) = 1 + G_1 H_1 + G_1 G_2 H_2$$

Δ is the denominator of the transfer function.

The numerator of the transfer function is $G_1(G_2 + G_3)$; and since the forward path touches all of the loops, the value of Δ_k reduces to unity. The final result is

$$\frac{Y}{R} = \frac{G_1\left(G_2 + G_3\right)}{1 + G_1 H_1 + G_1 G_2 H_2}$$

That agrees with Figure 4-24.

(figure 4-24)

Control system for Example 4-9

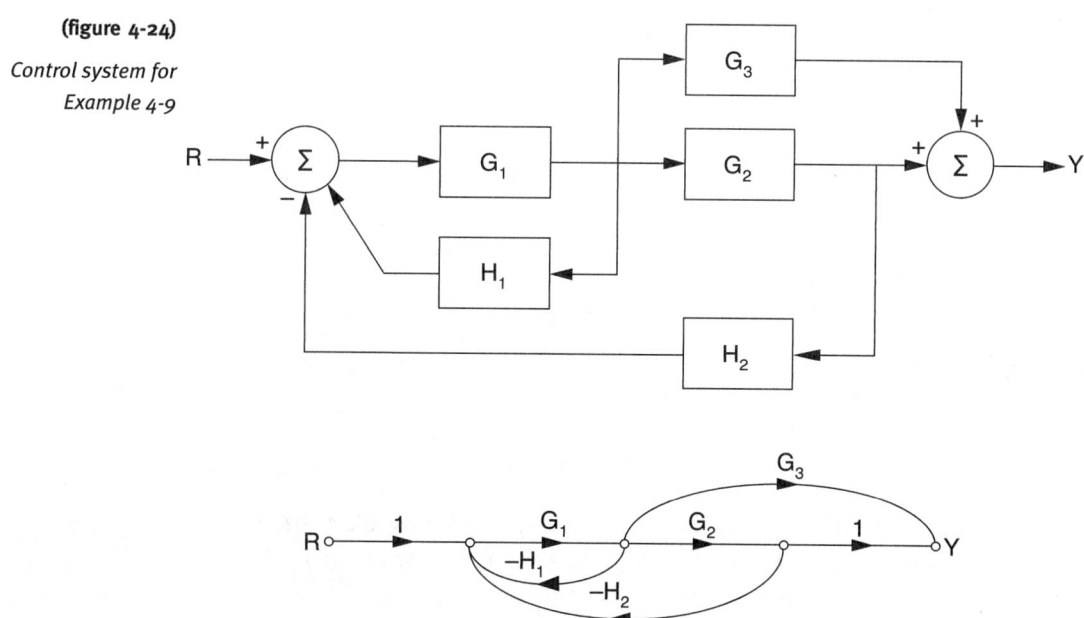

4.7 Use of MATLAB and Simulink for Transfer Functions and Simulation

MATLAB becomes more useful in this chapter because of its ability to operate on the required matrixes and to input a block diagram of a system. **Simulink** is used to input the block diagram. The response of the system is obtained by simulation of the block diagram.

Use of MATLAB

In **MATLAB**, transfer functions are usually given by row vectors describing the numerator and denominator of the transfer function. Many designs require consideration of the position of the poles and zeros of the system to achieve a specified response to an input.

MATLAB can be used to generate the closed-loop transfer function, using the built-in feedback and cloop functions. Consider the system shown in Figure 4-5. The transfer function can be generated in the following way:

ng = 3; dg = [1 4 3];	%Define the numerator and denominator.
h = 0.5;	%Define the feedback factor.
[ng,dg] = feedback(ng, dg, h, −1);	%Start defining the transfer function.
[n,d] = cloop(ng, dg, −1)	%Define the closed-loop transfer function.

The output of the **MATLAB** function is

n =

    0   0   3

d =

    1   4   4.5

From that, the closed-loop transfer function is

$$\frac{Y(s)}{R(s)} = \frac{3}{s^2 + 4s + 4.5}$$

For a given transfer function such as that of Figure 4-20, it is possible to separate the transfer function into its poles, zeros, and gain, using the *tf 2zp* function in **MATLAB**. To perform the process, the following method is used:

n = [4 6 10];   %Numerator vector of transfer function

d = [1 6 11 10];   %Denominator vector of transfer function

[z,p,k] = tf2zp(n,d)   %Get zeros, poles, and gain from transfer function.

The output of **MATLAB** for that function is

z =

$\qquad$ −0.7500 + 1.3919i

$\qquad$ −0.7500 − 1.3919i

p =

$\qquad$ −3.7963

$\qquad$ −1.1018 + 1.1917i

$\qquad$ −1.1018 − 1.1917i

k =

$\qquad$ 4

Similarly, given the poles, zeros, and gain of a system, the transfer function can be generated by using the *zp2tf* function and the following method:

k = 5;   %Gain of the system

z = −3;   %Zero of the system

p = [−3 + j*3, −3 − j*3];   %Poles of the system

[n, d] = zp2tf(z, p, k)   %Transform to transfer function form.

The output of the transformation is

n =

$\qquad$ 0   5   15

d =

$\qquad$ 1   6   18

The transfer function of the system is

$$\frac{Y(s)}{R(s)} = \frac{5(s + 3)}{s^2 + 6s + 18}$$

The response of the system to a step input is easily found using the *step(n,d)* function.

step(n, d)

The response is shown in Figure 4-25.

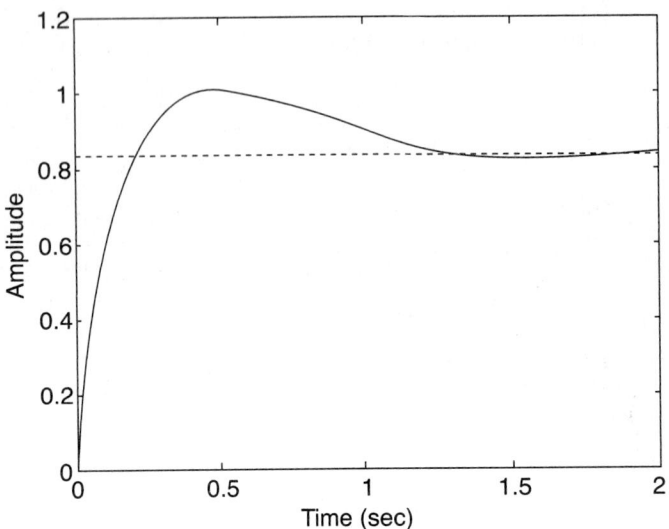

(figure 4-25)

MATLAB output of control system

Taking a slightly more complex transfer function, consider the system with the following parameters: $G_1(s) = 10$, $G_2(s) = (s + 5)/s(s + 2)$, $H_1(s) = s/(s + 4)$, and $H_2(s) = 1$. That problem is broken into two parts: the definition of the system at the output and its transfer function, then the overall transfer function of the total system.

ng1 = 1; dg1 = 1;   %Define G_1.

ng2 = [1 5]; dg2 = [1 2 0];   %Define G_2.

nh1 = [1 0]; dh1 = [1 4];   %Define H_1.

[nx1, dx1] = cloop(ng2, dg2, –1);   %Define transfer function of G_2.

[n, d] = feedback(nx1, dx1, nh1, dh1, –1)   %Define overall transfer function.

The output of MATLAB is

$n =$

    0   1   9   20

$d =$

    1   8   22   20

You can see the response of the system by using the command

step(n, d)

Use of Simulink

Additional to the command-line environment, **MATLAB** has a graphical interface that can be used to describe a system and simulate the system response, using blocks supplied by the program. The GUI provided is called **Simulink**. To invoke **Simulink**, type *simulink* while in the **MATLAB** window. The **Simulink** library browser will open. Otherwise, use the Launch Pad tab at the bottom of the **MATLAB** window to display the library browser. The library contains all of the blocks for drawing a block diagram (for example, transfer function block, gain block, integrator, differentiator, summing junction, and step input signal generator). The blocks are arranged into sublibraries according to the function they perform.

Double-clicking on any library opens the sublibrary and shows the various block components available in that library. The library browser is shown in Figure 4-26.

(figure 4-26)

*Toolbox for **Simulink***

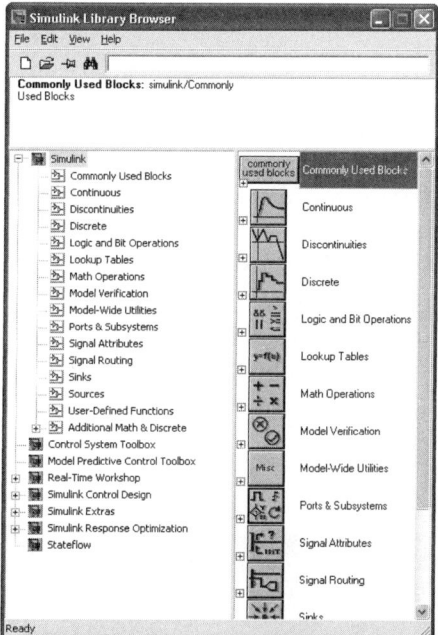

To start a new **Simulink** diagram, click the File menu in the browser and select New, then Model. Doing so opens a blank window for adding the necessary components. The example from the previous **MATLAB** simulation will be used to illustrate the use of **Simulink**. For this example, you will need only the Source, Sink, Math, and Linear libraries.

1. Open the Math Operations library. Copy two Sum blocks and a Gain block to the window by clicking and dragging each block to the working window. Double-click each Sum block to open a dialog box. Change one of the input signs of the summing block to (−) in the box labeled List of Signs.

2. From the Linear library, select a Zero-Pole block and a Transfer Fcn block to the window. Double-click the Zero-Pole block and enter values for the poles and zeros of the system. Zero and pole values are entered as values, not as a vector. For this system, the zero values are [−5], the poles are [0,−2], and the gain is 1.0. When you accept the values, the function in the block will change to reflect the values you entered. Double-click the Transfer Function block and change the numerator value to [1 0] and the denominator value to [1 4]. Since the feedback signal travels from right to left in the system, use the Format pull-down menu and select Flip Horizontal or press Ctrl + I to change the orientation of the block. Arrange the blocks as shown in Figure 4-27.

3. To connect the blocks, you can use either the (>) pointing out on the block or the (>) pointing in to the block. Click and drag from one to the other. When you release the mouse button, the wire is set between the blocks.

4. Now select a Step Input block from the Source library and a Scope block from the Sink library. Place them at the input and output of the block diagram. Then connect the source to the Sum block and the scope to the output of the Zero-Pole block. To set up the step input, double-click the step input. The dialog box gives three options: (1) the step time, which determines when the step changes value from zero (in this case, it should be set to 0); (2) the initial value of the step (in this case, also 0); and (3) the final value of the step (in this case, 1). Click on the Simulation pull-down menu and select Simulation parameters. In the Simulation Time, Stop Time textbox, change the time to 5 seconds and click Apply.

5. To simulate the system, click on Simulate. Then double-click the Scope and enlarge the graph to full screen. Right-click in the screen and select Autoscale. The output graph will change to full screen for the display.

The block diagram is shown in Figure 4-27; the output is shown in Figure 4-28.

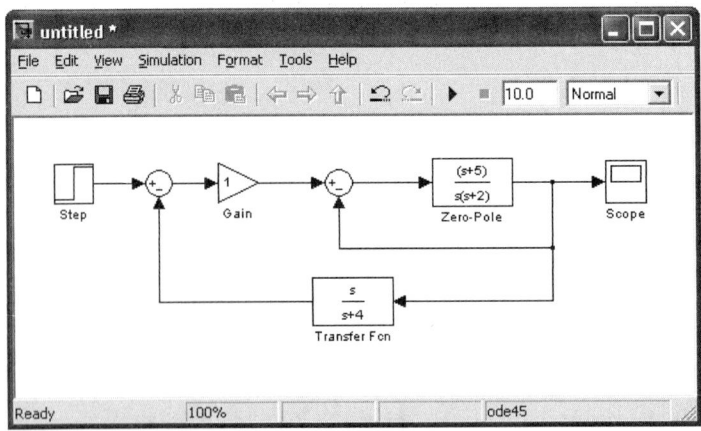

(figure 4-27)

Block diagram for simulation

(figure 4-28)

Simulink scope output of simulation

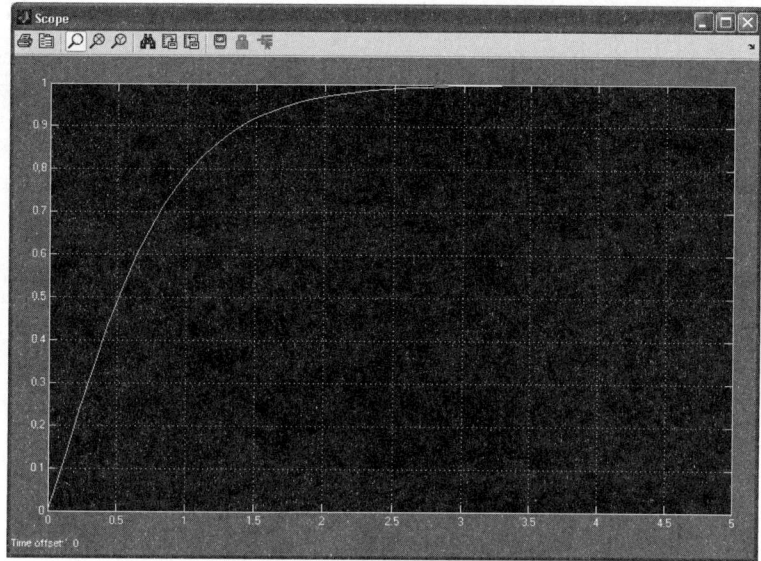

If you simulated the system from the command-line environment, you will see that that simulation produced the same output.

MATLAB and **Simulink** can also produce state-space functions for systems. That topic will be taken up in a later chapter.

Summary

➤ This chapter has introduced the concepts of block diagrams and the extraction of the transfer function of a system from the block diagram.

➤ The signal flow graph in both the phase variable and SFG form is used to provide the simultaneous equations of a system.

➤ The simultaneous equations are easily converted to state matrices.

➤ Mason's gain equation is used to provide the system transfer function for more complex systems.

➤ A number of examples have been given for each type of analysis.

➤ **MATLAB** is used to produce the system transfer function and the response graph of a system.

➤ **Simulink** is used to enter the system in block diagram form and provide the system response graph.

➤ **MATLAB** M-files are used to provide shortcut methods for producing transfer functions and response graphs of control systems in both the continuous and discrete time systems.

1. S. J. Mason, "Feedback Theory—Some Properties of Signal Flow Graphs," Proc. IRE, Vol. 41, No. 9, pp. 1144–1156, Sept. 1954.
2. S. J. Mason, "Feedback Theory—Further Properties of Signal Flow Graphs," Proc. IRE, Vol. 44, 1956.

Problems

Block Diagrams

P4.1   For the following block diagrams, determine the transfer function.

a.

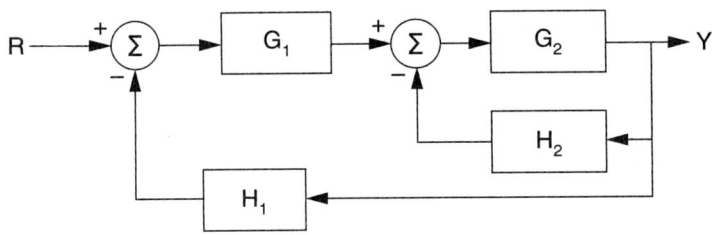

(figure P4-1a)

Control system for Part a

b.

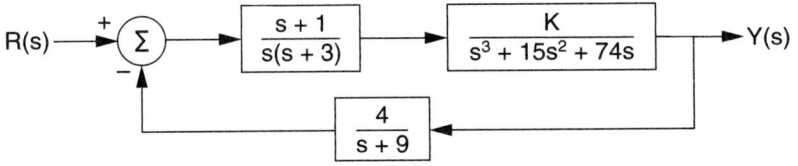

(figure P4-1b)

Control system for Part b

c.

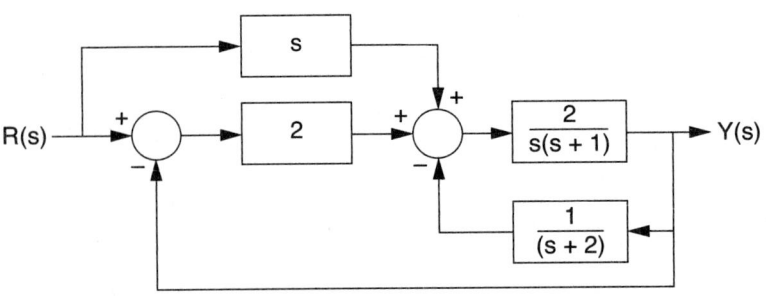

(figure P4-1c)

Control system for Part c

P4.2   Given the state variables as follows

$$\begin{bmatrix} x_1 \\ x_2 \\ x_3 \end{bmatrix} = \begin{bmatrix} 0 & 1 & 0 \\ 0 & 0 & 1 \\ -9 & -6 & -3 \end{bmatrix} \begin{bmatrix} x_1 \\ x_2 \\ x_3 \end{bmatrix} + \begin{bmatrix} 0 \\ 0 \\ 1 \end{bmatrix} r(t)$$

$$y(t) = \begin{bmatrix} 3 & 1 & 0 \end{bmatrix} \begin{bmatrix} x_1 \\ x_2 \\ x_3 \end{bmatrix}$$

Show the minimum phase diagram. Then write the s-plane transfer function as a ratio of polynomials.

P4.3   For the following transfer functions, draw a minimum phase diagram and write the state matrices of the systems.

a. $P(s) = \dfrac{8}{s^2 + 4s + 8}$

b. $P(s) = \dfrac{2s + 30}{s^2 + 24s + 30}$

c. $P(s) = \dfrac{s^2 + 4s + 3}{s^3 + 6s^2 + 11s + 25}$

d. $P(s) = \dfrac{8}{s^3 + 6s^2 + 24s + 16}$

P4.4   For the following minimum phase diagrams, write the Laplace transfer function.

a.

(figure P4-4a)

Minimum phase diagram for Part a

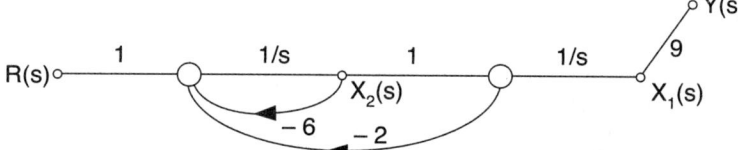

b.

(figure P4-4b)

Minimum phase diagram for Part b

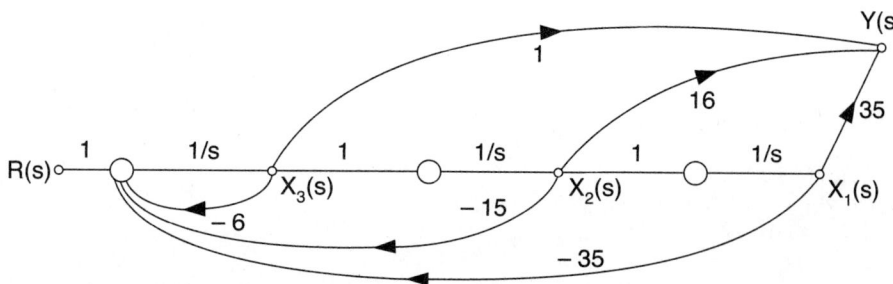

P4.5   Consider the state model below. Find $Y(s)/R(s)$ and express the result as a ratio of polynomials.

$$\begin{bmatrix} x_{13} \\ x_2 \\ x_3 \\ x_4 \end{bmatrix} = \begin{bmatrix} 0 & 1 & 0 & 0 \\ 0 & 0 & 1 & 0 \\ 0 & 0 & 0 & 1 \\ -40 & -120 & -60 & -9 \end{bmatrix} \begin{bmatrix} x_1 \\ x_2 \\ x_3 \\ x_4 \end{bmatrix} + \begin{bmatrix} 0 \\ 0 \\ 0 \\ 1 \end{bmatrix}$$

$$y(t) = \begin{bmatrix} 120 & 40 & 0 & 0 \end{bmatrix} \begin{bmatrix} x_1 \\ x_2 \\ x_3 \\ x_4 \end{bmatrix}$$

P4.6   Consider the block diagram of Figure P4-6. Determine the transfer function, $Y(s)/R(s)$, first by moving the extreme right summing junction to the left of G_2, then by moving the center-summing junction to the right of G_3. Show that both methods yield the same transfer function.

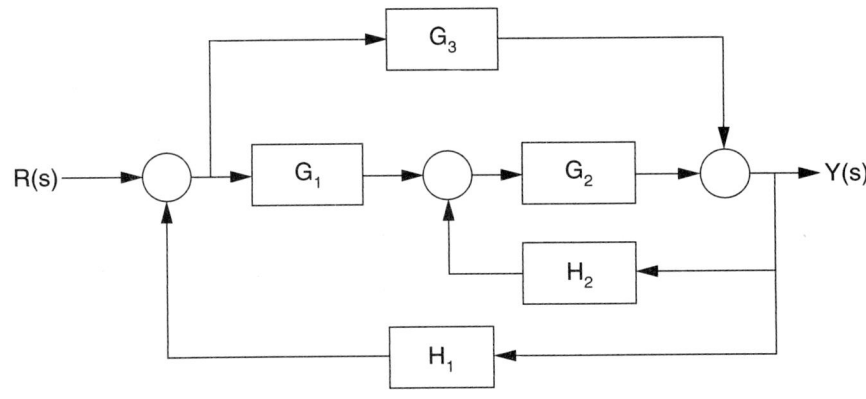

(figure P4-6)

Block diagram for Problem P4.6

P4.7   Convert the following block diagram to an SFG. Then use Mason's formula to determine the transfer function.

(figure P4-7)

Block diagram for Problem P4.7

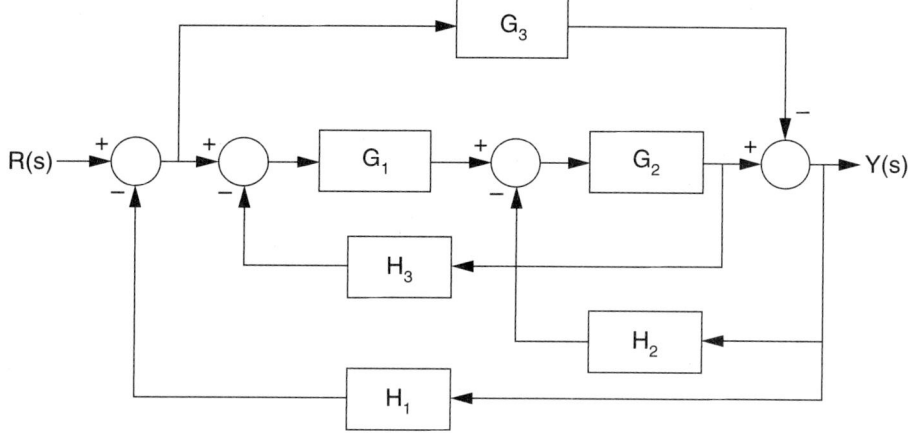

P4.8   Use Mason's gain formula to determine the transfer function of the following system. Then produce the state matrices of the system.

(figure P4-8)

Block diagram for Problem P4.8

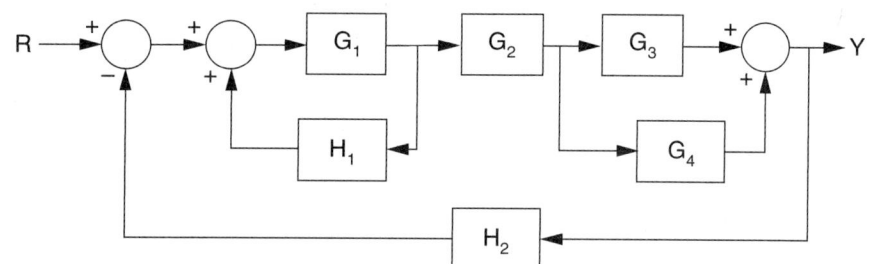

P4.9   Use any method to produce the transfer function $Y(s)/R(s)$ for the following block diagram.

(figure P4-9)

Block diagram for Problem P4.9

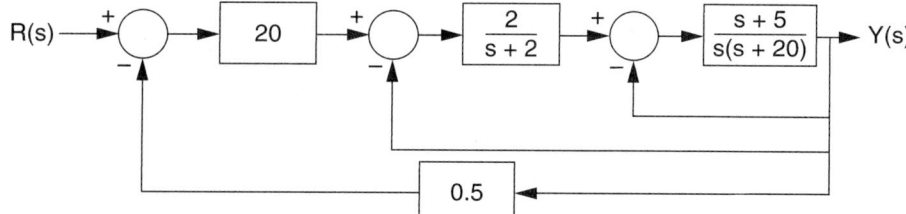

P4.10 Determine the output response of the system shown in Figure P4-10. $D(s)$ is a disturbance that will affect the output.

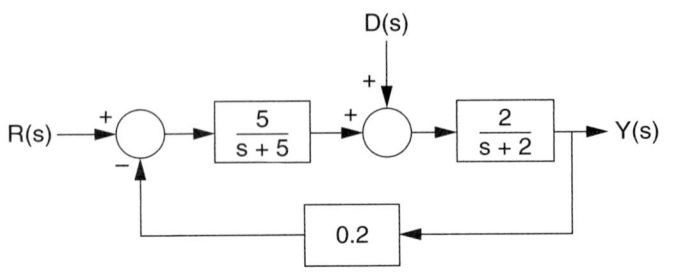

(figure P4-10)

*Block diagram for
Problem P4.10*

MATLAB and Simulink Problems

M4.1 For Problem 4.2, use **MATLAB** to convert the state equations to the transfer function, using the *tf* and *ss* commands. (See Appendix A for use of *ss* command.)

M4.2 For Problem P4.1c, use **Simulink** to provide an output simulation of the system.

M4.3 For Problem P4.9, use **Simulink** to provide the output of the system for a step input. Then remove the feedback block and resimulate the system. What is the effect of adding the feedback block to the system? What would be the effect of changing the feedback block to a value of 2.0?

M4.4 For Problem P4.10, use **Simulink** to provide the output response to a ramp input of $tu(t)$ with a disturbance impulse of amplitude 1 and duration of one second, occurring at three seconds. Use the signal builder in the Sources library to provide the impulse.

M4.5 For the following state matrices, use **MATLAB** to determine the transfer function of the system. Then produce the response of the system to a step input.

$$\begin{bmatrix} \dot{x}_1 \\ \dot{x}_2 \\ \dot{x}_3 \end{bmatrix} = \begin{bmatrix} 0 & 1 & 0 \\ 0 & 0 & 1 \\ -5 & -8 & -5 \end{bmatrix} \begin{bmatrix} x_1 \\ x_2 \\ x_3 \end{bmatrix} + \begin{bmatrix} 0 \\ 0 \\ 1 \end{bmatrix} u(t)$$

$$y(t) = \begin{bmatrix} 5 & 2 & 0 \end{bmatrix} \begin{bmatrix} x_1 \\ x_2 \\ x_3 \end{bmatrix}$$

Chapter 5 | Fundamentals of Stability

Chapter Objectives

After completing this chapter, you should be able to:

➤ State the conditions for asymptotic stability as an absolute requirement in any control system.

➤ Show that there are several methods to determine stability and that each method is important to ultimately determine if a system is unconditionally stable.

➤ Show that the characteristic equation of a system represents the closed-loop transfer function denominator.

➤ Show by example, that the pole positions in the s-plane determine if a control system is stable or unstable.

➤ Construct the Routh table and show the methods of continuing the table if a zero occurs in the first column of the table or a row of all zeros occurs.

➤ Demonstrate that the Routh table is an important approach to determining the stability of a system using the characteristic equation.

➤ Determine the limits of stable gain using the Routh table.

➤ Determine the frequency of oscillation at the limit of stable gain using the Routh table.

➤ Explain the concept of asymptotic stability.

➤ Use **MATLAB** to determine the roots of a characteristic equation.

5.1 Introduction

The total response of a system is the sum of its natural and forced responses. If the natural response increases without limit or oscillates, it quickly overcomes the forced response, control of the system is then lost. The characteristic equation of the system contains the poles and zeros of the system, and right-half plane (RHP) poles are an indication of instability.

With modern computer aids, the factoring of higher-order characteristic equation polynomials is simple; and the number of roots in the right-half of the s-plane can be determined easily using a calculator or program such as **MATLAB**. While knowing the number of roots in the RHP gives some insight into the stability of a system, the roots do not automatically determine the stability of the system for all conditions of the system.

The following methods have been developed to determine the asymptotic stability of a system:
- ➤ **s-Domain Methods**
  - ➤ The Routh table
  - ➤ The root locus
- ➤ **Frequency Domain Methods**
  - ➤ Bode's method
  - ➤ Nyquist's criteria
  - ➤ Nichols method

This chapter will investigate the use of the Routh table and its role in the determination of stability of systems. In some cases, the Routh table provides necessary and sufficient information to say that a system is unconditionally stable.

5.2 Definition of Stability

In all systems, three possibilities of stability (more precisely, asymptotic stability) or instability exist; but only one of the conditions is acceptable for any system.

1. The system may rise from zero exponentially to some final value and remain at that value or decay back to zero. This type of system is said to be asymptotically stable.
2. The system may rise exponentially without limit until the amplifiers in the system saturate and no further increase is possible; or the system may rise exponentially, then oscillate at the maximum and minimum values possible. This type of system is unstable.
3. The system may rise exponentially to some value and oscillate between two values, never increasing or decreasing. Such a system is said to be marginally stable, or a bounded oscillator.

The only acceptable response is number one.

5.3 Stability and Pole Locations

The poles of a transfer function's characteristic equation determine the stability of a control system. The zeros of a control system affect the system response to an input, but do not have a direct role in the stability of the system.

s-Plane

The variable s is the complex frequency operator. It has two components, a real value and an imaginary value.

$$s = \sigma + j\omega \qquad (5.1)$$

The axes of the s-plane are σ and $j\omega$. Poles and zeros are roots of the closed-loop transfer function and are plotted within the s plane. All poles should be in the left half, but often one or more of the poles are in the RHP. Zeros, which may be in either half of the plane, have an effect on the response of the system being investigated, but not on the stability of the system.

The s-plane is divided into two halves. The left-half plane is considered to be the stable side of the plane because poles on this side indicate exponentially decaying quantities (e^{-at}). The RHP is considered to be the unstable side of the plane because poles in this half represent exponentially increasing (e^{at}) quantities. The s-plane is shown in Figure 5-1.

(figure 5-1)

s-plane

As you can see, it is a simple Cartesian coordinate chart.

Fundamental Pole-Zero Representations in the s-Plane

Several fundamental pole-zero placements result in responses that can occur in a design. Each contributes to the overall response of the system, and several of the types may be contained in a single design. Knowledge of the type of time domain response produced by each of the forms provides insight into how a system will perform from examination of the pole-zero positions on the chart. The forms are:

➤ A single pole at the origin, representing a unit step.

➤ A pole on the $-\sigma$ axis, representing an exponential decay to zero.

➤ Two imaginary axis poles, representing a bounded sine wave oscillation.

➤ Two poles and one zero on the imaginary axis, representing a cosine wave oscillation.

➤ Two complex poles on the $-\sigma$ axis, representing an exponentially decaying sinewave.

➤ Two complex poles and a zero on the $-\sigma$ axis, representing an exponentially decaying cosine wave.

The graphical representations are shown in Figure 5-2.

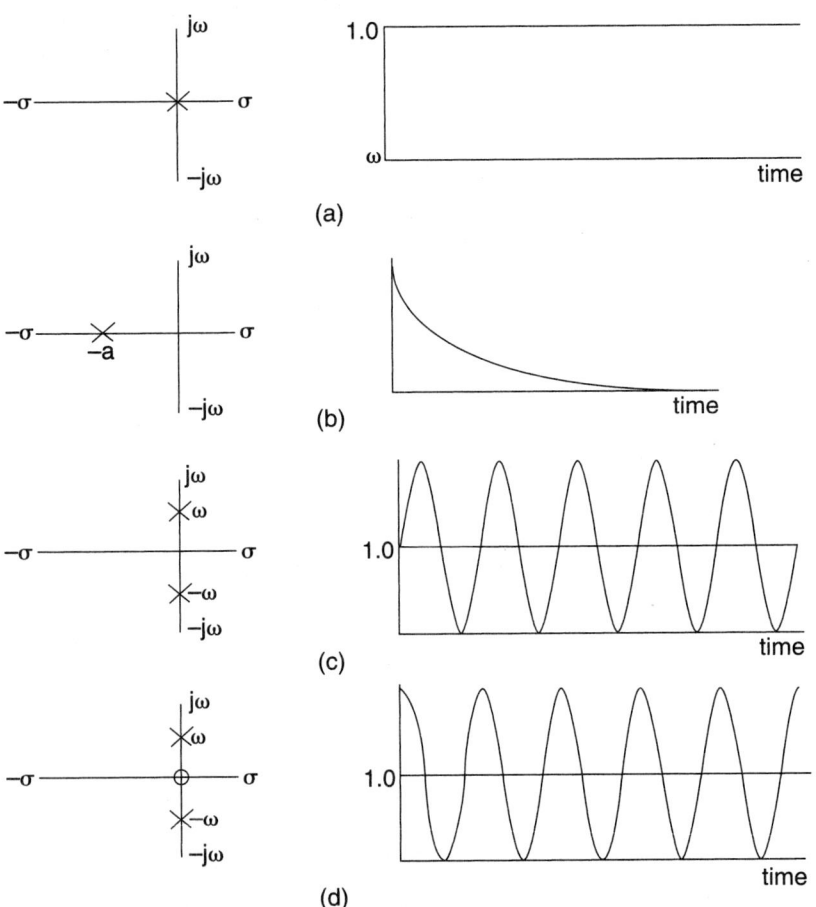

(figure 5-2)

Pole-zero representations in the s-plane

(figure 5-2) *continued*

Pole-zero representations in the s-plane

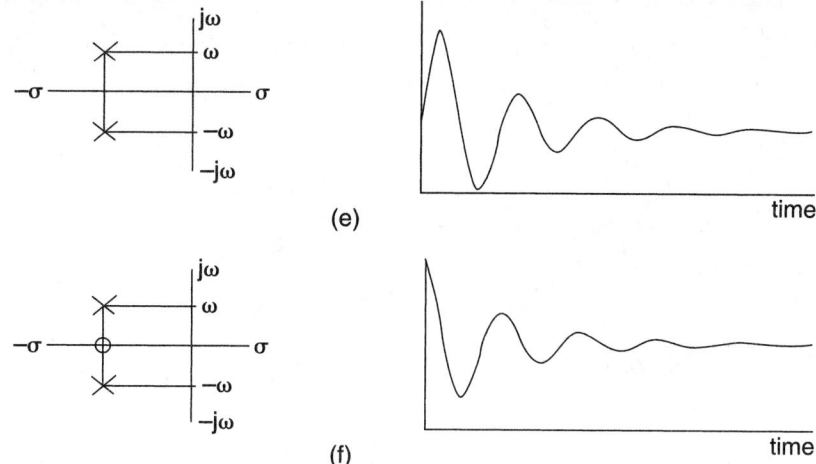

(e)

(f)

5.4 Routh Table as a Predictor of Stability

Stability is an absolute requirement in any control system. In the process of design, instability may not immediately be apparent. For this reason, methods to test the stability of a system have been devised. The Routh table is one such method and the methods of using the Routh table are illustrated here.

Characteristic Equation of a System

The characteristic equation of a system is defined as:

> *The denominator of the transfer function derived from a block diagram or an SFG, set equal to zero.*

In all cases, the response to an excitation signal is determined completely from the characteristic equation. If any coefficient in the characteristic equation is negative, the system has poles in the RHP and is unstable. If values of gain, K, can be determined for the system, the system may be stable over some or all of its range. To determine the characteristic equation the closed-loop transfer function is used. An example of forming the characteristic equation follows.

Example 5-1

Consider the following block diagram and its transfer function, as shown in Figure 5-3.

$$\frac{Y(s)}{R(s)} = \frac{\dfrac{18}{(s + 1)(s^2 + 4s + 3)}}{1 + \dfrac{18}{(s + 1)(s^2 + 4s + 3)} * 0.2}$$

$$\frac{Y(s)}{R(s)} = \frac{18}{s^3 + 5s^2 + 7s + 6.6}$$

Factor the characteristic equation and determine if all of the roots are in the left-half of the s-plane.

Solution

The characteristic equation for this system is

$$s^3 + 5s^2 + 7s + 6.6 = 0$$

The roots of the characteristic equation are:

$$s_1 = -3.5525$$
$$s_2 = -0.7237 + j1.155$$
$$s_3 = -0.7237 - j1.155$$

The factorization shows that all the roots are in the left-half s-plane and that the system appears to be asymptotically stable.

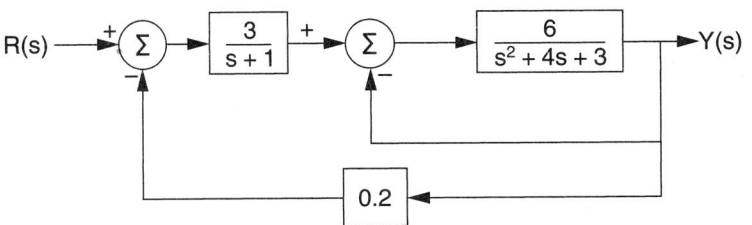

R(s) ──→ Σ ──→ 3/(s+1) ──→ Σ ──→ 6/(s²+4s+3) ──→ Y(s), feedback 0.2

(figure 5-3)

Control system for Example 5-1

A characteristic equation can contain all positive values, but may still have roots in the left-half of the s-plane or roots on the $j\omega$ axis. Consider the following characteristic equation:

$$s^4 + s^3 + 3s^2 + 3s + 2 = 0$$

The roots of that equation are $s_{1,2} = 0.13 \pm j1.59$, $s_{3,4} = -0.63 \pm 0.62$. Obviously, there are two roots in the right-half of the s-plane and, by definition, the system is not stable. A simple method to determine the number, not the value, of roots in the right-half s-plane is to use a method devised by Edward J. Routh.[1]

Form of the Routh Table

A technique for determining the stability of a continuous system was established by A. Hurwitz, in the 1800s. The criterion came to be called the Hurwitz criterion. The technique defined a determinant formed from the coefficients of the characteristic equation of the transfer function. The technique was cumbersome for higher-order equations and was modified by another scientist, E. Routh, into a tabular form. The total criterion came to be known as the Routh-Hurwitz criterion.

The tabular form of the determinant makes determining the number of roots (but not their values) in the RHP, if any, a simple task. In general, all coefficients of s must be present and nonzero. If any term is missing, the system is unstable and the table to be formed will have a zero in one of its starting rows. The form of the table is obtained as follows:

Consider the general characteristic equation of a system:

$$F(s) = a_n s^n + a_{n-1} s^{n-1} + a_{n-2} s^{n-2} + \cdots + a_1 s + a_0 = 0 \qquad (5.2)$$

The equation can be split into two rows of coefficients, one even and one odd. Doing so yields the coefficient field

s^n	a_n	a_{n-2}	a_{n-4}	$a_{n-6}\ldots$
s^{n-1}	a_{n-1}	a_{n-3}	a_{n-5}	$a_{n-7}\ldots$

In all cases, the highest exponent of s should have a coefficient of 1 for convenience. If the highest power of s has a coefficient greater or less than 1, divide the equation through by the coefficient of the highest power of s. The general form of a table is

s^7	a_n	a_{n-2}	a_{n-4}	a_{n-6}
s^6	a_{n-1}	a_{n-3}	a_{n-5}	a_{n-7}
s^5	b_1	b_2	b_3	0
s^4	c_1	c_2	c_3	0
s^3	d_1	d_2	0	0
s^2	e_1	e_2	0	0
s^1	f_1	0	0	0
s^0	g_1	0	0	0

Where

$$b_1 = \frac{a_{n-1} \cdot a_{n-2} - a_n \cdot a_{n-3}}{a_{n-1}} \quad c_1 = \frac{b_1 \cdot a_{n-3} - a_{n-1} \cdot b_2}{b_1}$$

$$b_2 = \frac{a_{n-1} \cdot a_{n-4} - a_n \cdot a_{n-5}}{a_{n-1}} \quad c_2 = \frac{b_1 \cdot a_{n-5} - a_{n-1} \cdot b_3}{b_1}$$

$$b_3 = \frac{a_{n-1} \cdot a_{n-6} - a_n \cdot a_{n-7}}{a_{n-1}} \quad c_3 = \frac{b_1 \cdot a_{n-7} - a_{n-1} \cdot 0}{b_1}$$

All rows below the s^{n-1} row are considered to be auxiliary equations of the characteristic equation and are used in some of the calculations.

Rules for Constructing the Routh Table

The rules for constructing the Routh table are as follows:
1. Decompose the characteristic equation into two rows of odd/even or even/odd components. These are the first two rows of the table.
2. Using the calculations shown, construct the remaining rows of the table.
3. The last coefficient of the highest even power row repeats as the last coefficient of all even power rows.
4. When the table is complete, all coefficients in the first column must be positive and greater than zero for the system to be stable.
5. If any of the coefficients of the first column is negative, the number of roots in the RHP is equal to the number of sign changes in the first column. This indicates that the system is unstable. Negative values anywhere else in the table do not indicate instability.
6. If a zero occurs in the first column only and all other coefficients are present, there may be roots on the $j\omega$ axis or in the RHP.
7. If a row of all zeros occurs, there are roots of equal magnitude but opposite sign, roots that are on the $j\omega$ axis, or complex roots that are diametrically opposed around the origin.
8. The Routh table can be used to determine the limits of stable gain for a system.

Routh Table Examples

There are several possibilities for the conditions of the Routh table. The table may terminate naturally, the table may terminate with a zero in the first column, or the table may terminate in a row of all zeros. These conditions will be taken up next.

Naturally Terminating Routh Tables

This section will investigate Routh tables that terminate naturally, with no first-column zeros or rows of zeros.

Example 5-2

The characteristic equation of a system is

$$3s^6 + 9s^5 + 12s^4 + 21s^3 + 66s^2 + 99s + 150 = 0$$

Use the Routh table to determine if the system is stable.

Solution

Dividing through by 3 yields

$$s^6 + 3s^5 + 4s^4 + 7s^3 + 22s^2 + 33s + 50 = 0$$

Rearranging the characteristic equation into even and odd power rows yields

s^6	1	4	22	50
s^5	3	7	33	

In all cases, the number of terms in the characteristic equation is the highest power of s plus 1; in this case, seven terms. To determine stability, the rest of the coefficients of the table are formed using the method shown above. Since there are seven coefficients, continue in the following way:

s^6	1	4	22	50
s^5	3	7	33	0
s^4	$\dfrac{3 \cdot 4 - 1 \cdot 7}{3} = \dfrac{5}{3}$	$\dfrac{3 \cdot 22 - 1 \cdot 33}{3} = 11$ $\quad \dfrac{3 \cdot 50 - 1 \cdot 0}{3} = 50$		0
s^3	$\dfrac{1.667 \cdot 7 - 3 \cdot 10}{1.667} = -11$	$\dfrac{1.667 \cdot 33 - 3 \cdot 50}{1.667} = -57$	0	0
s^2	$\dfrac{-11 \cdot 10 - 1.667 \cdot (-57)}{-11} = 1.364$	$\dfrac{-11 \cdot 50 - 1.667 \cdot 0}{-11} = 50$	0	0
s^1	$\dfrac{1.364 \cdot (-57) - (-11 \cdot 50)}{1.364} = 346.23$	0	0	0
s^0	$\dfrac{346.22 \cdot 50 - 1.364 \cdot 0}{346.23} = 50$			

The table shows a negative value in the s_3 row. Therefore there are two sign changes in the first column. The first sign change occurs when going from the s^4 row to the s^3 row; the second sign change, from the s^3 row to the s^2 row. There are two roots in the RHP, and the system is unstable.

Example 5-3

The characteristic equation of a system is

$$s^3 + 6s^2 + 11s + 12 = 0$$

Determine if the system is stable.

Example 5-3 (continued)

Solution

The Routh table is:

$$
\begin{array}{c c c}
s^3 & 1 & 11 \\
s^2 & 6 & 12 \\
s^1 & 9 & 0 \\
s^0 & 12 & 0 \\
\end{array}
\qquad s^1 = \dfrac{6 \cdot 11 - 1 \cdot 12}{6} = 9
$$

Since there are no sign changes in the first column and the table does not terminate prematurely, the system is stable.

Example 5-4

The characteristic equation of a system is

$$
s^5 - 10s^4 + 17s^3 + 64s^2 - 108s - 144 = 0
$$

The negative signs in the characteristic equation show the system to be unstable. Factoring the equation provides the following root values: $s_1 = 6$, $s_2 = 4$, $s_3 = -2$, $s_4 = 3$, $s_5 = -1$. Clearly, there are three roots in the RHP. The Routh table should corroborate that fact.

Solution

$$
\begin{array}{c c c c}
s^5 & 1 & 17 & -108 \\
s^4 & -10 & 64 & -144 \\
s^3 & 23.4 & 122.4 & 0 \\
s^2 & 116.9 & -144 & 0 \\
s^1 & 93.57 & 0 & 0 \\
s^0 & -144 & 0 & 0 \\
\end{array}
$$

There are two sign changes from the first row to the third row. The last value in the table is also negative, giving three sign changes and, therefore, three roots in the RHP. The system is unstable as indicated by factoring the characteristic equation.

When a Routh table terminates prematurely, methods are needed to continue the table so all possibilities can be examined. The methods used when a zero is in the first column or when a row of all zeros occurs are discussed in the following examples.

Routh Tables Terminating with a Zero in the First Column

When a Routh table terminates prematurely, some method must be provided to continue the table to determine if the system is unstable, conditionally stable or stable. The following examples show how to continue the table under the condition of premature terminaton.

Example 5-5

Consider the characteristic equation

$$s^4 + 10s^3 + s^2 + 10s + 20 = 0$$

Determine if there are any sign changes in the first column of the Routh table.

Solution

The start of the Routh table is

s^4	1	1	20
s^3	10	10	
s^2	0		

That table terminates prematurely with a zero in the s^2 row.
To continue the table, the following method is used:

1. Replace the zero with a small positive number ε. The value of ε is not defined, but it is extremely small.
2. Make the calculation for the next term, using ε in place of the zero. The real number in the s^1 power calculation is insignificant compared to the number divided by ε. The remaining value may be positive or negative and becomes the next coefficient.
3. Continue the table as normal.

s^2	$0 \to \varepsilon$	20
s^1	$\dfrac{10 \cdot \varepsilon - 200}{\varepsilon} \approx -\dfrac{200}{\varepsilon}$	
s^0	20	

There is one negative value in the first column, indicating two sign changes and two roots on the right hand s-plane. The system is unstable. One thing to be careful of is that the ε method may produce erroneous tabulations if the characteristic equation has pure imaginary roots.

Routh Tables Terminating with a Row of All Zeros

Another factor causing premature termination of the table is a row of all zeros occurring in the table. When that occurs, there are three possible causes:

1. The equation has two real roots of equal but opposite sign.

2. The equation has one or more pairs of imaginary roots.

3. The equation has complex-conjugate roots with symmetry about the origin of the s-plane.

The method to handle the situation uses the auxiliary equations of the Routh table.

To continue the table, an auxiliary equation is formed from the row above the row of all zeros: $(A(s) = 0)$. The row in which the zeros occur will be an odd-ordered row; therefore, the auxiliary equation will be formed from an even-ordered row. This will always be the case. The roots of the auxiliary equation will also be roots of the characteristic equation.

Example 5-6

The characteristic equation of a system is

$$s^5 + 4s^4 + 8s^3 + 8s^2 + 7s + 4 = 0$$

Determine if the system is stable.

Solution

The Routh table is

s^5	1	8	7
s^4	4	8	4
s^3	6	6	
s^2	4	4	
s^1	0	0	

Note that the row in which the zeros occur is odd-ordered. The auxiliary equation will be formed from the s^2 row. The auxiliary equation is formed as follows.

Set up the auxiliary equation in descending powers of s, using the even powers of the row; in this case, s^2 and a constant term. The auxiliary equation is

$$A(s) = 4s^2 + 4$$

Take the derivative of the auxiliary equation yielding $dA(s)/ds = 0$.

$$\frac{dA(s)}{ds} = 8s$$

Example 5-6 (continued)

Replace the row of zeros with the derivative constants.

s^5	1	8	7
s^4	4	8	4
s^3	6	6	
s^2	4	4	
s^1	8		
s^0	4		

When a row of all zeros occurs, the row above the row of all zeros contains roots of the characteristic equation. In this case, the system is marginally stable. Solving the auxiliary equation for its roots, you find that there are two roots at $\pm j$. The roots are located on the $j\omega$ axis of the s-plane. The imaginary roots are the cause of the row of all zeros.

The calculations made here for the Routh table are instructive, but do not suggest any solution for improving the stability of the system. By investigating the system further and determining the conditions that cause instability, you would be able to use those conditions to set limits for the stable gain of the system.

Routh Tables Containing Gain Factors

Routh tables with fixed characteristic equations are used to determine absolute stability, but suggest no way to determine if the system has gain factors that can be changed to produce a stable system from an unstable system. When a block diagram with a variable gain factor K is used to set up the Routh table, the limits of stable gain can be calculated.

Routh Table with Simple Gain Factor

The real value of the Routh table lies in the ability to use it to calculate the limits of stable gain of any system.

The limit of stable gain can be defined as the value of a gain parameter, K, which causes the characteristic equation to have at least one pair of roots on the $j\omega$ axis.

In all cases, the gains used must be positive. Negative gains are ignored. An example of how the Routh table is used to determine gain limits is shown in Example 5-7.

Example 5-7

From the following characteristic equation, determine the range of stable gain.

$$s^4 + 6s^3 + 11s^2 + 6s + K = 0$$

Solution

The Routh table is:

s^4	1	11	K
s^3	6	6	
s^2	10	K	
s^1	$\dfrac{60 - 6K}{10}$		
s^0	K		

To determine the limits of stable gain, again use the auxiliary equation. In this case, start with the s^1 row, $A(s) = 0$.

$$A(s) = \frac{60 - 6K}{10} = 0$$

That row shows that the value 10 is superfluous to the equation and that $60 - 6K = 0$ is all that you must deal with.

$$60 - 6K = 0, \qquad 6K = 60, \qquad K = 10$$

By examining the s^0 row (constant row), you can immediately see that if $K = 0$, the system is marginally stable. Therefore, the values $K = 10$ and $K = 0$ will place a root on the $j\omega$ axis; and you can conclude that, for stability, $0 \le K \le 10$.

Routh Table with more than One Coefficient Gain Factor.

When there is more than one coefficient of a characteristic equation with a gain factor, the gain function can become quadratic or higher-ordered. Example 5-8 shows such a condition.

Example 5-8

For this case, a quadratic gain will be considered. The characteristic equation is

$$s^4 + 6s^3 + 18s^2 + (K - 12)s + K = 0$$

Determine the range of stable gain.

Example 5-8 (continued)

Solution

The Routh table is

$$
\begin{array}{cccc}
s^4 & 1 & 18 & K \\
s^3 & 6 & (K-12) & \\
s^2 & \dfrac{120-K}{6} & K & \\
s^1 & \dfrac{-K^2+96K-1440}{120-K} & & \\
s^0 & K & &
\end{array}
$$

By using the auxiliary equation on the s^1 row, you find $A(s)$ to be

$$A(s) = \frac{-K^2 + 96K - 1440}{120 - K} = 0$$

$$K_1 = 77.4 \qquad K_2 = 18.6$$

Only positive real roots can be used for the limits of gain since only positive gain is allowed. In this case, the roots are $K_1 = 77.39$ and $K_2 = 18.61$. Since both gains are positive, the region of stable gain for this system lies between those two values, $18.61 < K < 77.4$. It should be noted that even though the s^0 row predicts that gain must be greater than zero, if the gain were 1.0, for instance, the value in the s^1 row and first column would be –11.3, clearly indicating that the system is unstable.

The value of K is not always positive in the equations developed, nor is it impossible for the gain equation to have complex roots. Since the requirement is that all gains be positive for a system, if one of the gains is positive and the other negative, only the positive gain applies.

Frequency of Oscillation at the Maximum Gain

Each of the rows of the Routh table is an auxiliary equation. As such, it is possible to use the equations to determine parameters of the system when the gain is set to the maximum or minimum limits predicted by the Routh table. Example 5-8 shows that there are two gains that cause oscillation. That means there are two different oscillation frequencies. Using the gains established in this table and the second-order row, which contains $s^2 + \omega^2$, you find

For $K = 77.39$

$$A_1(s) = \frac{120 - K_1}{6}s^2 + K_1 = 0$$

$$A_1(s) = \frac{120 - 77.4}{6}s^2 + 77.4$$

$$7.1s^2 + 77.4 = 0 \qquad s^2 = -\frac{77.4}{7.1} = -10.9$$

$$s = \pm j3.3 \, \text{rad/sec}$$

For $K = 18.61$

$$A_1(s) = \frac{120 - K_2}{6}s^2 + K_2 = 0$$

$$A_1(s) = \frac{120 - 18.6}{6}s^2 + 18.6 = 0$$

$$16.9s^2 + 18.6 = 0 \qquad s^2 = -\frac{18.6}{16.9} = -1.1$$

$$s = \pm j1.05 \, \text{rad/sec}$$

Example 5-9

For the following system, determine the range of stable gain and the oscillation frequency of the system at the maximum gain value. See Figure 5-4.

Solution

The transfer function of the system is

$$\frac{Y(s)}{R(s)} = \frac{K(s + 2)}{s^4 + 12s^3 + 44s^2 + (48 + K)s + 2K}$$

From which you derive the Routh table, using the denominator of the transfer function.

s^4	1	44	$2K$
s^3	12	$(48 + K)$	
s^2	$\dfrac{480 - K}{12}$	$2K$	
s^1	$\dfrac{-K^2 + 144K + 23040}{480 - K}$		$K_1 = 240, \quad K_2 = -96$
s^0	$2K$		

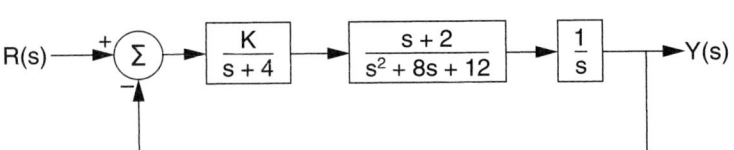

(figure 5-4)

Control system for Example 5-9

R(s) $\longrightarrow$ $+$ Σ $\dfrac{K}{s+4}$ $\dfrac{s+2}{s^2+8s+12}$ $\dfrac{1}{s}$ $\longrightarrow$ Y(s)

Example 5-9 (continued)

Factoring of the quadratic provides two gains, one positive and one negative. The negative gain is ignored since the requirement is that all gains be positive. The two gains are $K_1 = 240$ and $K_2 = -96$. The table shows that the system is stable for $0 < K < 240$, and only one oscillation frequency needs to be found. Using the maximum gain, the oscillation frequency is

$$A(s) = \frac{480 - K}{12} s^2 + 2K = 0$$

$$A(s) = \frac{480 - 240}{12} s^2 + 2 * 240 = 0$$

$$20s^2 + 480 = 0$$

$$s^2 = -\frac{480}{20} = -24$$

$$s = \pm j4.9 \text{ rad/sec}$$

Example 5-10

For the following closed-loop characteristic equation, use the Routh table to determine the limit of stable gain and the frequency of oscillation.

$$s^5 + 27s^4 + 281s^3 + 1413s^2 + (3438 + 4K)s + 4K = 0$$

Solution

The Routh table is

s^5	1	281	$(3438 + 4K)$
s^4	27	1413	$4K$
s^3	228.67	$3438 - 3.85K$	
s^2	$1007 + 0.46K$	$4K$	
s^1	$\dfrac{3.46 \cdot 10^6 + 1381K - 1.73K^2}{1007 + 0.46K}$		
s^0	$4K$		

The gains are found as usual by factoring the quadratic gain in the s^1 row. The gains are

$$A(s) = \frac{3.46 \cdot 10^6 + 1381K - 1.73K^2}{1007 + 0.46K}$$

$$K_1 = 1868.6 \qquad K_2 = -1070.3$$

Example 5-10 (continued)

The frequency of oscillation at the maximum gain is

$$A(s) = (1007 - 0.46K)s^2 + 4K = 0$$

$$A(s) = 147.4s^2 + 7474.4 = 0$$

$$s^2 = -\frac{7474.4}{147.4} = -50.7$$

$$s = \pm j7.12 \text{ rad/sec}$$

The Routh table is restricted to characteristic equations that are algebraic, with real coefficients. If any of the coefficients are not algebraic, such as e^{-at}, $\sin \omega t$, etc., the criterion cannot be applied.

5.5 Use of MATLAB for the Roots of a Characteristic Equation

Given a characteristic equation of the type

$$a_0s^n + a_1s^{n-1} + a_2s^{n-2} + \cdots + a_{n-1}s + a_n = 0 \qquad (5.3)$$

The equation can be decomposed into its factored form. The factored form contains only the roots of the equation.

$$(s - p_1)(s - p_2)(s - p_3) \ldots (s - p_n) = 0$$

MATLAB can be used to find the roots of a polynomial through use of the command *roots* in the following way.

roots([1 6 11 6])   % Polynomial to be factored

The output of **MATLAB** is

ans =

    −3.0000

    −2.0000

    −1.0000

An example of a system with a characteristic equation that has all positive real coefficients but has two roots in the RHP is

$$s^4 + 7s^3 + 7s^2 + s + 2 = 0$$

roots([1 7 7 1 2])   % Polynomial to be factored

ans =

    −5.8158

    −1.3007

    0.0583 + 0.5109i

    0.0583 − 0.5109i

It is clear that this system is unstable, yet the characteristic equation meets the basic requirements of all positive signs. The Routh table would have a zero in the first column.

Summary

➤ The most important characteristic of any system is its stability.

➤ The Routh table as a predictor of whether a system is stable provides a necessary insight into system stability.

➤ Methods have been developed for occasions when a zero occurs in the first column or a row of all zeros occurs in the Routh table.

➤ The method to determine the limits of stable gain of a system have also been established. Knowledge of the limits of stable gain aids in the construction of a root-locus plot of the migration of the roots of the characteristic equation as gain changes. Further, the frequency of oscillation of the system is determined at the limit of stable gain.

➤ The Routh table gives several points on the root locus, which help in the construction of the paths of root migration.

➤ MATLAB is used to factor the characteristic equation into its roots and to indicate if the system is stable or unstable (right-hand plane roots).

1.   Routh, E. J., *A Treatise on the Stability of a Given State of Motion*. London: Macmillan, 1877. Rpt. in *Stability of Motion*, Ed. A. T. Fuller. London: Taylor & Francis, 1975.

Problems

P5.1   Use the Routh test to determine if the following characteristic equations provide a stable, marginally stable, or unstable system.

a. $s^4 + 3s^3 + 15s^2 + 25s + 50 = 0$
b. $s^5 + 14s^4 + 35s^3 + s^2 + 20s + 15 = 0$
c. $s^3 + 6s^2 + 11s + 20 = 0$
d. $s^4 + 7s^3 + 19s^2 + 63s + 90 = 0$
e. $s^4 + 6s^3 + 5s^2 + 48s + 86 = 0$

P5.2   Use the Routh test to determine the range of stable gain for the following systems.

a. $s^4 + 3s^3 + 12s^2 + (K - 16)s + K = 0$
b. $s^3 + 3Ks^2 + (4 + K)s + 4 = 0$
c. $s^4 + Ks^3 + 2s^2 + (4 - K)s + 1 = 0$
d. $s^3 + (K + 4)s^2 + 3Ks + 15 = 0$

P5.3   For the system shown, determine the transfer function, the range of stable gain, and the radian velocity at the limit of stable gain.

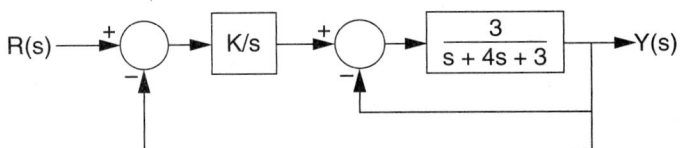

(figure P5-3)

Control system for Problem P5.3

P5.4   For the following system, determine the transfer function in the s-plane. Then use the Routh test to determine if the system is stable.

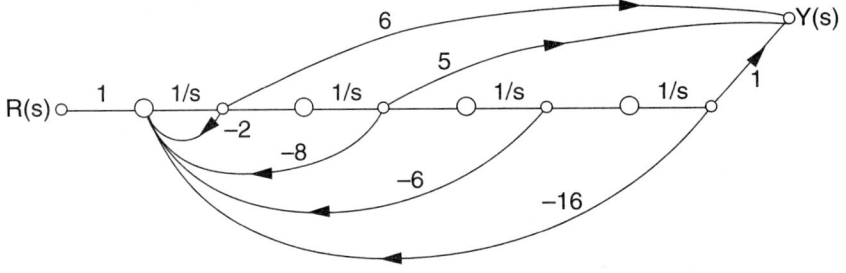

(figure P5-4)

SFG for Problem P5.4

P5.5   Determine the closed-loop transfer function of the following system and use the Routh test to determine the limits of stable gain.

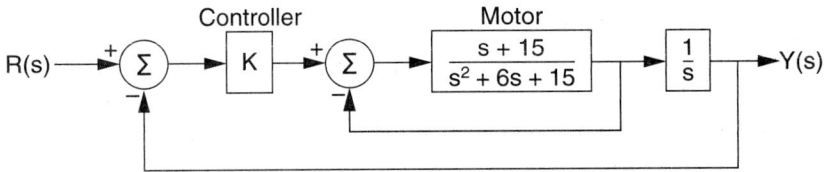

(figure P5-5)

Control system for Problem P5.5

P5.6   Given the following state matrices, determine the s-plane characteristic equation and determine the limits of stable gain, using the Routh test.

$$\begin{bmatrix} x_1 \\ x_2 \\ x_3 \\ x_4 \end{bmatrix} = \begin{bmatrix} 0 & 1 & 0 & 0 \\ 0 & 0 & 1 & 0 \\ 0 & 0 & 0 & 1 \\ -10K & -10K & -18 & -6 \end{bmatrix} \begin{bmatrix} x_1 \\ x_2 \\ x_3 \\ x_4 \end{bmatrix} + \begin{bmatrix} 0 \\ 0 \\ 0 \\ 1 \end{bmatrix} r(t)$$

P5.7   Use the Routh test to determine if the following system is asymptotically stable.

$$P(s) = \frac{5s^2 + 15s + 10}{s^5 + 21s^4 + 151s^3 + 471s^2 + 640s + 300}$$

MATLAB Problems

M5.1   For the following system, use **MATLAB** to generate the closed-loop transfer function, using the technique shown in Chapter 4. Then use the Routh test to determine if the system is asymptotically stable.

(figure M5-1)

Control system for Problem M5.1

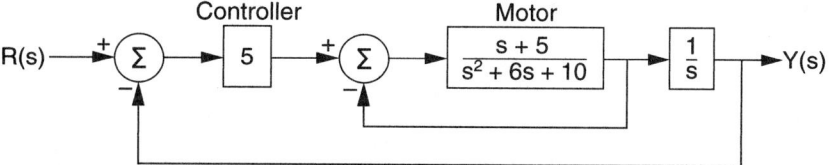

M5.2   Use **MATLAB** to find the roots of the characteristic equation in Problem P5.7

M5.3   Use **MATLAB** to generate the closed-loop transfer function for Example 5-9, using two gains, $K_1 = 50$ and $K_2 = 100$. Then determine the roots of the characteristic equation. Explain what is happening to the roots as the gain changes.

Chapter 6

Feedback and System Concepts

Chapter Objectives

After completing this chapter, you should be able to:

➤ Explain how feedback is used to alter the response of a system to an input stimulus.

➤ Explain what three signals are commonly used to evaluate the performance of a system.

➤ Explain what criteria influence the response of a system to a stimulus.

➤ Discuss the effect of system damping for overdamped, critically damped, or underdamped systems.

➤ Explain the factors that affect the risetime of a system.

➤ Explain what factors determine the peak time of a system.

➤ Explain what is meant by the settling time of a system.

➤ Explain how the peak amplitude of a system is determined.

➤ Explain how system error is affected by system configuration.

➤ Explain how system type numbers are derived based on input signal and error specification.

➤ Explain the effect of increasing the system order.

➤ Explain what is meant by the sensitivity of a system to changes in a system parameter.

➤ Use **MATLAB, Simulink,** and the **LTI Viewer** to produce specific response characteristics in the system being designed.

6.1 Introduction

The use of feedback in a system allows you to alter the response of a system to meet the required specifications of the system. Feedback has an effect on the stability of a system and its steady-state performance, rejection of disturbances, and sensitivity of the system to component variations. Feedback also affects the frequency response of a system and can be used to increase or decrease the phase and gain margin of a system.

All systems except a first-order system can be separated into linear and quadratic terms. Linear terms cause an exponential response, rising to a final value that is fixed in terms of its time constant. Second-order terms are more interesting in that their response is dependent on the position of their roots (poles) in the s-plane.

One of the methods used to test any system is application of a unit step input and observation of the risetime, overshoot, and settling time of the system. Linear systems are independent of the level of their input functions; that is, the overshoot and settling time of the system are unaffected by changes in the amplitude of the input signal. In this chapter, all of the systems considered will be linear systems.

6.2 Common Signal Types Used in Control System Analysis

Three common signal types are used to test control systems, and each has its specific Laplace transform. The signals are as follows:

➤ Step input

The unit step has an amplitude of R units (the input amplitude). It is zero for all time less than zero and R for all time greater than or equal to zero.

$$r(t) = \begin{cases} R, & t \geq 0 \\ 0, & t < 0 \end{cases} \qquad\qquad R(s) = \frac{R}{s} \qquad\qquad (6.1)$$

➤ Ramp input

The ramp signal is a signal whose amplitude varies linearly with time.

$$r(t) = Rtu_s(t) \qquad\qquad R(s) = \frac{R}{s^2} \qquad\qquad (6.2)$$

➤ Parabolic input

This is the parabolic signal whose amplitude increases according to the square of the time the signal is applied.

$$r(t) = \frac{Rt^2 u_s(t)}{2} \qquad\qquad R(s) = \frac{R}{s^3} \qquad\qquad (6.3)$$

Each of the signals requires a specific type of system for following the input signal with either zero error or some specified error.

6.3 Transient Response Criteria

System response is determined by the placement of poles in the s-plane. In most systems, there will be dominant poles, which are the poles closest to the origin of the s-plane because they have the slowest decay time. Other poles may exist but will be further to the left of the dominant poles.

Single Pole System

A single pole system having the form

$$\frac{Y(s)}{R(s)} = \frac{K}{Ts + 1} = \frac{K / T}{s + 1 / T} \tag{6.4}$$

Has a single pole located at $-1/T$ in the s-plane, where T is the time constant of the system. The time domain response of this system to a unit step input is

$$y(t) = K(1 - e^{-t/T}) \tag{6.5}$$

The system is well defined in terms of its time constant. For time t equal to T, the system rises to 63 percent of its final value as any simple exponential system will. In five time constants, the system is considered to be at its final value. The exponential rise of the system is shown in Figure 6-1.

(figure 6-1)

First-order system, exponential response

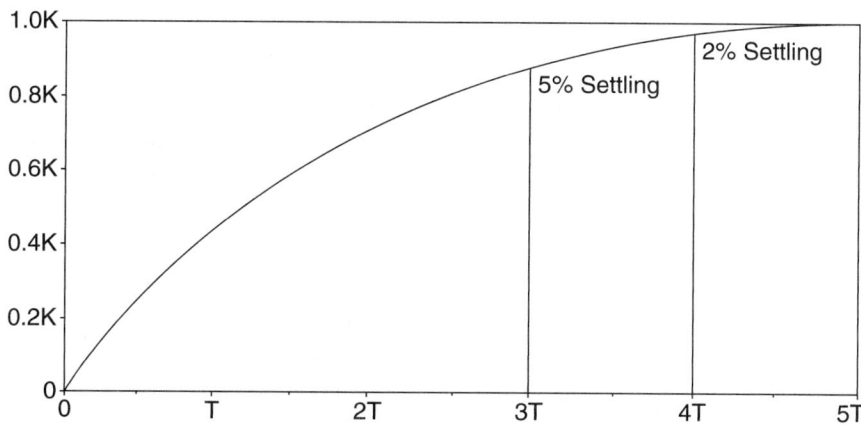

An important parameter of any system is the time required for the system to settle within a specified percentage of the final value (usually between 2 percent or 5 percent).

This value is the settling time of the system. For first-order systems, the 5 percent settling time is $2.99T \ (\approx 3T)$ and the 2 percent settling time is $3.912T$. Since a gain factor of K is included, the final value of the system will be the product of the gain, K, and the input. K is called the DC gain of the system.

Two Pole System

A more general form of the system with two dominant poles and no zeros is shown in Figure 6-2.

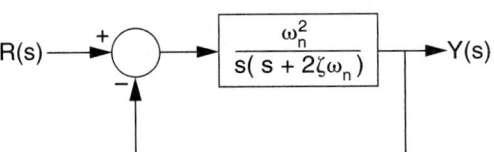

This system is called the prototype second-order system and has the transfer function

$$\frac{Y(s)}{R(s)} = \frac{\omega_n^2}{s^2 + 2\zeta\omega_n s + \omega_n^2} \tag{6.6}$$

In this system, the parameters ζ and ω_n are the damping ratio and the natural radian velocity of the system, respectively. In all cases, the roots must be in the left half of the s-plane for the system to be stable. Examining the discriminant of the quadratic set equal to zero shows three possible conditions for the equation:

1. The discriminant is greater than zero. In this case, the roots are real and unequal and lie on the σ axis. The roots are given by

$$s_{1,2} = -\zeta\omega_n \pm \omega_n\sqrt{\zeta^2 - 1} \tag{6.7}$$

   Note that the value of ζ is greater than 1 for these roots. ζ is the damping ratio of the system. Damping ratios greater than 1 indicate a slow system with long rise- and settling times. This type of system is said to be overdamped.

2. The discriminant is equal to zero. In this case, the roots are real and equal and lie on the σ axis. The roots are given by

$$s_{1,2} = -\zeta\omega_n \tag{6.8}$$

   The value of ζ in this case is unity and represents the fastest risetime of this type of system, with no overshoot. This type of system is said to be critically damped.

3. The discriminant is less than zero. In this case, the roots have a real and imaginary component. The roots are given by

$$s_{1,2} = -\zeta\omega_n \pm j\omega_n\sqrt{1 - \zeta^2} \tag{6.9}$$

In this case, the value of ζ is less than 1. Lower values of ζ cause the response of the system to overshoot the final value by some amount. As ζ approaches zero, the roots of the equation move toward the $j\omega$ axis. If ζ becomes zero, the roots lie on the $j\omega$ axis and the system is in oscillation. This system is said to be underdamped.

The three root conditions are shown in Figure 6-3.

(figure 6-3)

System pole positions in the s-plane

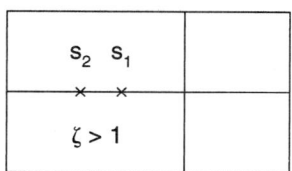

$$s_1 = -\zeta\omega_n + \omega_n\sqrt{\zeta^2-1}$$
$$s_2 = -\zeta\omega_n - \omega_n\sqrt{\zeta^2-1}$$

a.) Overdamped Systems.

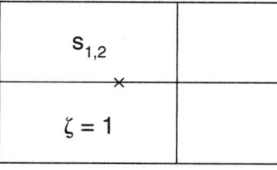

$$s_{1,2} = -\zeta\omega_n$$

b.) Critically Damped Systems.

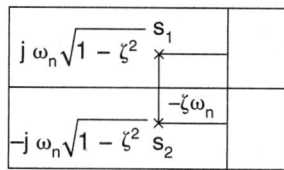

$$s_1 = -\zeta\omega_n + j\,\omega_n\sqrt{1-\zeta^2}$$
$$s_2 = -\zeta\omega_n - j\,\omega_n\sqrt{1-\zeta^2}$$

c.) Underdamped Systems.

The transfer function clearly shows that if s is allowed to approach zero, the final value of the system will be 1.

Overdamped Response of Two Pole Systems

In this case, the value of the poles of the system is real and different and both are located on the σ axis. Here you are dealing with two exponential responses that approach their final value at different rates. There is no possibility for overshoot since single first-order functions will not rise above the final value of the system.

For relatively large damping ratios $(\zeta > 1.4)$, the pole closest to the origin of the s-plane will become dominant; and the response of the system can be approximated by consideration of this pole. A simple example will clarify the concept. Given the second-order system $s^2 + 30s + 104$, the roots of the equation are $(s + 4)(s + 26)$ and produce two exponentials, e^{-4t} and e^{-26t}. The damping ratio is 1.471. It is clear that the pole at –26 decays at a rate greater than six times that of the pole at –4. Thus, the response can be considered to be controlled by the pole at –4.0 with little error.

Critical Damping of Two Pole Systems

When the damping ratio of a system is 1.0, both of the poles of the system occur at the same point on the σ axis and decay at the same rate. Given the following system with a unit step input

$$Y(s) = \frac{\omega_n^2}{s(s^2 + 2\omega_n s + \omega_n^2)} = \frac{\omega_n^2}{s(s + \omega_n)^2} \qquad (6.10)$$

The roots of the equation are $(s + \zeta\omega_n)^2$. The time domain transfer function of the system is

$$y(t) = 1 - (1 + \omega_n t)e^{-\omega_n t} \tag{6.11}$$

It is clear from Equation 6.8 that the response will never exceed 1.0.

Underdamped Response of Two Pole Systems

When the damping ratio of a pole system is less than 1, the poles of the system are placed at

$$s_{1,2} = -\zeta\omega_n \pm j\omega_n \sqrt{1 - \zeta^2} \tag{6.12}$$

When the poles are plotted in the s-plane, some characteristics of the system become apparent, as shown in Figure 6-4.

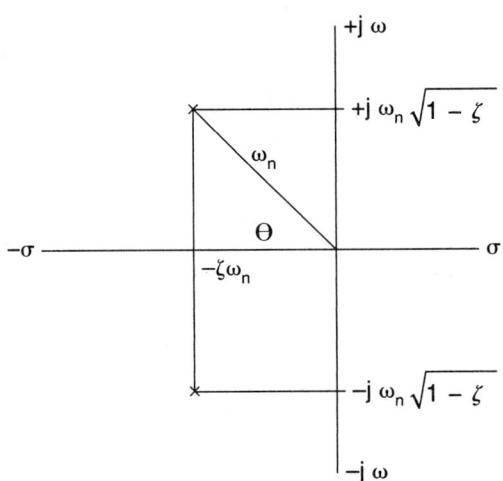

(figure 6-4)

Underdamped pole values

Using Pythagoras' theorem and using the cosine formula $\zeta = \cos^{-1}(\theta)$, you find that the vector from the origin to the poles is ω_n. Further, the imaginary value gives the damped radian velocity of the system.

Applying a unit step to the system produces

$$Y(s) = \frac{\omega_n^2}{s[(s + \zeta\omega_n)^2 + \omega_n^2(1 - \zeta^2)]} \tag{6.13}$$

For which the inverse Laplace transform is

$$y(t) = 1 - e^{-\zeta\omega_n t}\left(\cos\omega_n\sqrt{1-\zeta^2}\,t + \frac{\zeta}{\sqrt{1-\zeta^2}}\sin\omega_n\sqrt{1-\zeta^2}\,t\right) \qquad (6.14)$$

The transform clearly shows an oscillatory response due to the sine and cosine terms. Further examination of the equation indicates that as ζ decreases, the oscillations become more pronounced until, at last, if ζ becomes zero, the system will contain only the sine term and will oscillate at the natural frequency ω_n.

For second-order systems, the time constant is readily found from the fact that exponentially decaying systems decrease to 0.3679 of their final value in one time constant.

$$\ln(0.3679) = -\zeta\omega_n T$$
$$-1 = -\zeta\omega_n t \qquad (6.15)$$
$$t = \frac{1}{\zeta\omega_n}$$

A chart of the response of systems with damping ratios from zero to 2.0 is shown in Figure 6-5. Note that damping ratios greater than 0.7 produce little overshoot.

(figure 6-5)

Damping ratio curves

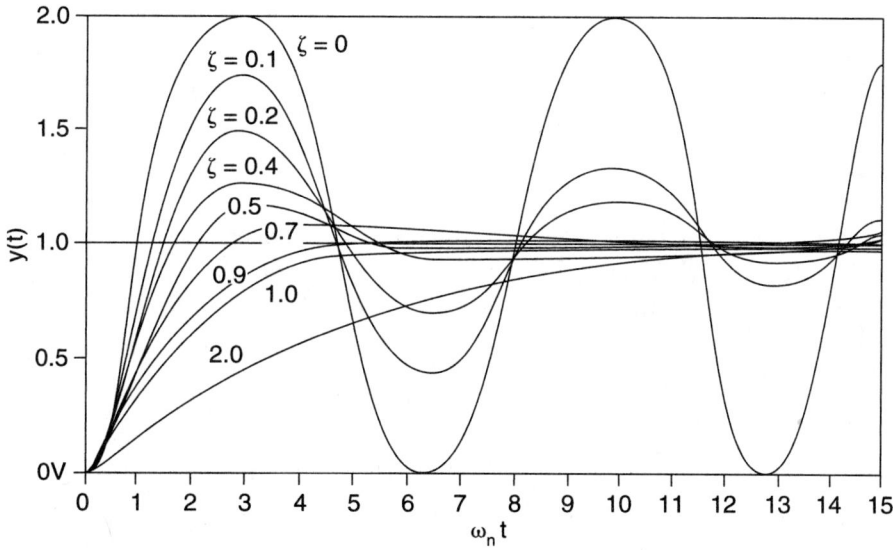

A number of other parameters can be obtained from Equation 6.14, such as the time for the first peak of the oscillation, the maximum overshoot in the system, and the approximate settling time for the system. The parameters are shown in Figure 6-6.

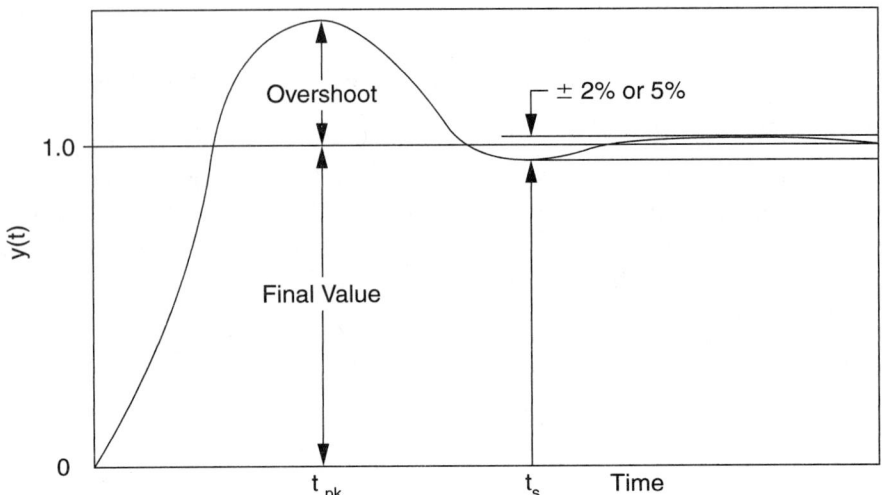

(figure 6-6)

Definitions of overshoot, final value, and settling time

6.4 Calculation of Risetime, Peak Time, Settling Time, and Peak Amplitude (M_P)

In all cases, systems will have specifications given. The system risetime and settling time are almost always given. In order for a system to operate properly, all the system specifications must be met and considered to be the minimum acceptable system.

Calculation of Risetime

Figure 6-5 shows that the oscillatory response is an exponentially decaying cosine wave. To obtain the risetime of the system, you need to consider the sine and cosine terms of Equation 6.14, replacing t with t_r. Since $e^{-\zeta\omega_n t_r}$ is not equal to zero, you obtain from Equation 6.14

$$\cos\omega_n\sqrt{1-\zeta^2}\,t_r + \frac{\zeta}{\sqrt{1-\zeta^2}}\sin\omega_n\sqrt{1-\zeta^2}\,t_r = 0 \qquad (6.16)$$

$$\frac{\zeta}{\sqrt{1-\zeta^2}}\sin\omega_m\sqrt{1-\zeta^2}\,t_r = -\cos\omega_n\sqrt{1-\zeta^2}\,t_r$$

$$\tan\omega_n\sqrt{1-\zeta^2}\,t_r = -\frac{\sqrt{1-\zeta^2}}{\zeta} \qquad (6.17)$$

From Figure 6-7 you can see that θ must lie between $\dfrac{\pi}{2}$ and π, from which the risetime from zero to 100 percent is calculated as follows:

$$\omega_n \sqrt{1 - \zeta^2}\, t_r = \tan^{-1}\left(\frac{\sqrt{1 - \zeta^2}}{\zeta}\right)$$

$$t_r = \frac{1}{\omega_n \sqrt{1 - \zeta^2}} \tan^{-1}\left(\frac{\sqrt{1 - \zeta^2}}{\zeta}\right)$$

and   (6.18)

$$\tan^{-1}\left(\frac{\sqrt{1 - \zeta^2}}{\zeta}\right) = \pi - \theta$$

$$t_r = \frac{\pi - \theta}{\omega_n \sqrt{1 - \zeta^2}}$$

The 10 percent to 90 percent risetime will be about 80 percent of this value. Keep in mind that these calculations are not exact; they are approximations. The actual risetime can be measured using the damping ratio chart, Figure 6-5.

Calculation of Peak Times

To obtain the time of occurrence of the peaks of an underdamped system, you need only consider the sine term in Equation 6.16. When the sine is zero, the cosine term will be at its peak. Thus, the peaks will occur every $n\pi$ radians. Thus, if you replace t with t_{pk} in the sine term

$$\sin\left(\omega_n \sqrt{1 - \zeta^2}\, t_{pk}\right) = 0 \tag{6.19}$$

From which

$$\omega_n \sqrt{1 - \zeta^2}\, t_{pk} = n\pi, \qquad n = 0,1,2,3,\ldots \tag{6.20}$$

And the peak times can be calculated from

$$t_{pk} = \frac{n\pi}{\omega_n \sqrt{1 - \zeta^2}} \tag{6.21}$$

The time of the first peak is usually most important and is at π radians.

The amplitude of the peaks is easily obtained using the envelope of the exponentially decaying peaks since the cosine term is unity at the peak times and the sine term is zero. Substituting the peak time into $y(t)$ yields

$$M_{pk} = 1 \pm \exp\left(-\frac{\pi\zeta}{\sqrt{1-\zeta^2}}\right) \tag{6.22}$$

Note that all odd-numbered peaks are added to the final value and that all even-numbered peaks are subtracted from the final value.

The maximum percent of overshoot in the system is simply 100 times the exponential term of the peak formulas. A short table of overshoot percentages is shown in Table 6-1, as well as its accompanying figure for $\zeta = 0.2$ to $\zeta = 1.0$.

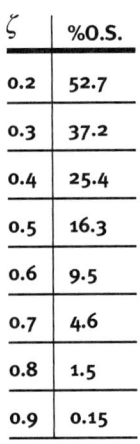

ζ	%O.S.
0.2	52.7
0.3	37.2
0.4	25.4
0.5	16.3
0.6	9.5
0.7	4.6
0.8	1.5
0.9	0.15

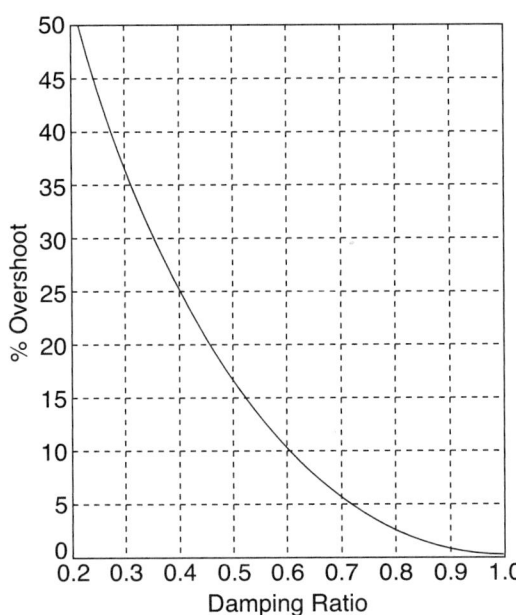

(table 6-1)

Calculation of Settling Time

Settling time is usually specified to be the time at which the response of a system enters within a specified band on either side of the final value and remains there. The bands are usually 2 percent or 5 percent. For example, if the desired settling is to be within 5 percent, the exponential term of Equation 6.14 can be used to determine when the output of the system is within 0.05 of the final value. The 2 percent settling time is arrived at similarly.

$$e^{-\zeta\omega_n T_s} = 0.05$$
$$-\zeta\omega_n T_s = \ln(0.05) \qquad\qquad 0 < \zeta < 0.69$$
$$T_s \approx \frac{3}{\zeta\omega_n} \; (5\%), \qquad\qquad T_s \approx \frac{4}{\zeta\omega_n} \; (2\%) \qquad\qquad (6.23)$$

Note that the response of the system for $\zeta < 0.69$ can enter the band of 1.05 and 0.95 from either above or below the final value. For damping ratios greater than 0.69, the response must approach the final value from below since there is only one peak whose overshoot will be less than 5 percent. For $\zeta > 0.69$, the settling time is approximately $4.5/\zeta\omega_n$. Remember that these formulas rely on the fact that the damping ratio is less than 1.0. The concepts are shown in Figures 6-7a and b.

(figure 6-7a)

Settling time for damping ratios ‹ 0.69

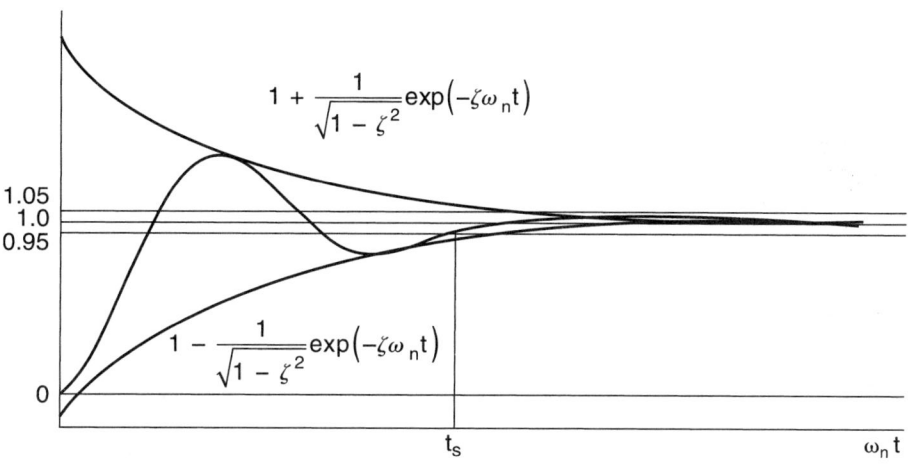

(figure 6-7b)

Settling time for damping ratios › 0.7 and ‹ 1.0

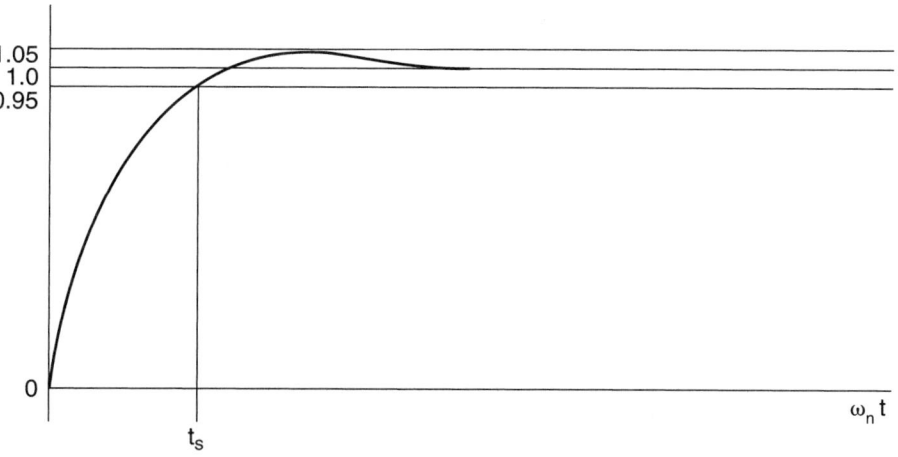

6.5 System Types and Error Functions

The goal in all systems is that when a reference input is applied, the output will have the exact value of the input in the steady state. That is, the output should have no error with regard to the input. In terms of the system

$$e(t) = r(t) - y(t) = 0 \tag{6.24}$$

Where $e(t)$ is the error of the system and the steady-state error is defined as

$$\underset{t \to \infty}{Lim}\ f(t) = \underset{s \to 0}{Lim}\ s \cdot F(s) \tag{6.25}$$

Systems are classified by type. Generally, the type number will not exceed two or three, but it can be greater. The system type determines what form of signal the system can follow with either a finite error or a zero error. The system type is given by the number of poles of the system at the origin of the s-plane.

Derivation of the Error Function

By examining a simple system, you can determine the error by simple block diagram reduction. Figure 6-8 and the following equations show the error for a unity gain system.

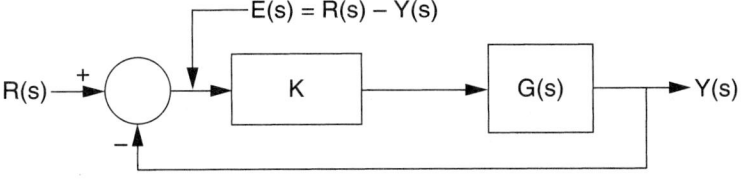

(figure 6-8)

Control system for error analysis

$$Y(s) = \frac{R(s) \cdot KG(s)}{1 + KG(s)} \tag{6.26}$$

$$E(s) = R(s) - \frac{R(s) \cdot KG(s)}{1 + KG(s)} = \frac{R(s) + R(s) \cdot KG(s) - R(s) \cdot KG(s)}{1 + KG(s)} \tag{6.27}$$

$$= \frac{R(s)}{1 + KG(s)}$$

Using the final value theorem allows you to calculate the value of the steady-state error of a system using a unit step function, R/s. Allowing $G(s)$ to be in the form of a factored polynomial such as

$$G(s) = \frac{(\tau_1 s + 1)(\tau_2 s + 1)\cdots}{s^n(\tau_3 s + 1)(\tau_4 s + 1)\cdots} \tag{6.28}$$

Where the exponent of s^n $(n = 0, 1, 2, ...)$ indicates the number of poles at the origin of the s-plane and also the system type number.

Type Zero System Error

Consider a system with no poles at the origin. $G(s)$ has $s^n = s^0 = 1$ and a step input. Analysis of the error function shows

$$E(s) = \lim_{s \to 0} s \cdot F(s) = \lim_{s \to 0} s \cdot \frac{R}{s[1 + KG(s)]} = \frac{R}{1 + KG(s)} \qquad (6.29)$$

$$E(s) = \lim_{s \to 0} \cdot \frac{K(\tau_1 s + 1)(\tau_2 s + 1)}{(\tau_3 s + 1)(\tau_4 s + 1)} = K = K_P \qquad (6.30)$$

$$e_{ss} = \frac{R}{1 + K_P} \qquad (6.31)$$

K_p is known as the position error constant.

If $R(s)$ is a ramp signal, R/s^2, for the same function, you find

$$E(s) = \lim_{s \to 0} s * F(s) = \lim_{s \to 0} s * \frac{R}{s^2[1 + KG(s)]} = \frac{R}{s[1 + KG(s)]}$$

$$\qquad (6.32)$$

$$= \frac{R}{0} = \infty$$

That indicates that the system cannot follow a ramp signal because the error will increase without limit. The error would be the same for a parabolic signal input. This type of system is a type zero system (s^0).

Example 6-1

Consider the following system:

$$G(s) = \frac{K(s + 1)}{(s + 1)(s + 5)}$$

Determine the error for each of the three signal types.

Solution

Step:

$$R(s) = \frac{R}{s}$$

$$e_{ss} = \lim_{s \to 0} s \cdot \frac{R}{s\left(1 + \dfrac{K(s + 1)}{s^2 + 6s + 5}\right)} = \frac{R}{1 + 0.2K}$$

Example 6-1 (continued)

Ramp:

$$R(s) = \frac{R}{s^2}$$

$$e_{ss} = \underset{s \to 0}{Lim}\ s \cdot \frac{R}{s^2\left(1 + \dfrac{K(s + 1)}{s^2 + 6s + 5}\right)} = \frac{R}{0} = \infty$$

Parabolic:

$$R(s) = \frac{R}{s^3}$$

$$e_{ss} = \underset{s \to 0}{Lim}\ s \cdot \frac{R}{s^3\left(1 + \dfrac{K(s + 1)}{s^2 + 6s + 5}\right)} = \frac{R}{0} = \infty$$

Type One System Error

A type one system has a single pole at the origin and can follow a step input with a zero error. Type one systems follow a ramp signal with a finite error, as follows:

$$G(s) = \frac{(\tau_1 s + 1)(\tau_2 s + 1)}{s(\tau_3 s + 1)(\tau_4 s + 1)} \tag{6.33}$$

$$e_{ss} = \underset{s \to 0}{Lim}\ s \cdot \frac{R}{s^2} \cdot \frac{1}{1 + \dfrac{K(\tau_1 s + 1)(\tau_2 s + 1)}{s(\tau_3 s + 1)(\tau_4 s + 1)}}$$

$$= \underset{s \to 0}{Lim} \cdot \frac{R}{s + \dfrac{sK(\tau_1 s + 1)(\tau_2 s + 1)}{s(\tau_3 s + 1)(\tau_4 s + 1)}} = \frac{R}{K_v} \tag{6.34}$$

Where K_v is known as the velocity error. That equation indicates that a type one system can follow a ramp signal with a finite error.

Example 6-2

Consider the following system:

$$KG(s) = \frac{K(s + 6)}{s(s + 1.2)(s + 3.6)}$$

Determine the error for each of the three signal types.

Example 6-2 (continued)

Solution

Step:

$$R(s) = \frac{R}{s}$$

$$e_{ss} = \underset{s \to 0}{Lim} \; s \cdot \frac{R}{s\left(\dfrac{s + 6}{s(s + 1.2)(s + 3.6)}\right)} = \frac{R}{\infty} = 0$$

Ramp:

$$R(s) = \frac{R}{s^2}$$

$$e_{ss} = \underset{s \to 0}{Lim} \; s \cdot \frac{R}{s^2\left(\dfrac{K(s + 6)}{s(s + 1.2)(s + 3.6)}\right)} = \underset{s \to 0}{Lim} \; \frac{R}{\dfrac{K(s + 6)}{(s + 1.2)(s + 3.6)}} = \frac{R}{1.389K}$$

Parabolic:

$$R(s) = \frac{R}{s^3}$$

$$e_{ss} = \underset{s \to 0}{Lim} \; s \cdot \frac{R}{s^3\left(\dfrac{K(s + 2)}{s(s + 3)(s + 5)}\right)} = \frac{R}{0} = \infty$$

Type Two System Error

A type two system will have two poles at the origin of the s-plane and will follow a step or ramp signal with zero error and a parabolic signal, R/s^3, with a finite error of R/K_a, which is called the acceleration error.

Example 6-3

Consider the following system:

$$KG(s) = \frac{K(s + 2)}{s^2(s + 3)(s + 5)}$$

Determine the error for each of the three signal types.

Example 6-3 (continued)

Solution

Step:

$$R(s) = \frac{R}{s}$$

$$e_{ss} = \lim_{s \to 0} s \cdot \frac{R}{s\left(\dfrac{K(s+2)}{s^2(s+3)(s+5)}\right)} = \frac{R}{\infty} = 0$$

Ramp:

$$R(s) = \frac{R}{s^2}$$

$$e_{ss} = \lim_{s \to 0} s \cdot \frac{R}{s^2\left(\dfrac{K(s+2)}{s^2(s+3)(s+5)}\right)} = \frac{R}{\infty} = 0$$

Parabolic:

$$R(s) = \frac{R}{s^3}$$

$$e_{ss} = \lim_{s \to 0} s \cdot \frac{R}{s^3\left(\dfrac{K(s+2)}{s^2(s+3)(s+5)}\right)} = \frac{7.5R}{K}$$

Table 6-2 shows the system types and their error functions.

System Type	Step	Ramp	Parabolic
0	$\dfrac{R}{1+K_P}$	∞	∞
1	0	$\dfrac{R}{K_v}$	∞
2	0	0	$\dfrac{R}{K_a}$

(table 6-2)

Example 6-4

Consider the system shown in Figure 6-9. The model is for a radar antenna that must track an object with a rate of 0.1 rad/sec and a maximum tracking error of 0.5° or less. When an object is tracked, the input signal is a ramp of $0.1tu_s(t)$. Further, the damping ratio is to be 0.7 or greater. Determine the value of K required to fulfill all of the specifications.

Solution

The first step is to convert the tracking error in degrees to radians. The error is 0.0087 rad. Since you have the error, you can calculate the value of K, using the error equation for a ramp input. Then you determine the range of K that fulfills the requirements.

$$e_{ss} = \frac{R}{K_v}$$

$$0.0087 = \operatorname*{Lim}_{s \to 0} s \cdot \frac{0.1}{s^2\left(\dfrac{20K}{s(s+30)}\right)} = \frac{0.15}{K}$$

$$K = 17.24$$

The damping ratio can be determined from the closed-loop transfer function of the system.

$$\frac{Y(s)}{R(s)} = \frac{20K}{s^2 + 30s + 20K} = \frac{345}{s^2 + 30s + 345}$$

$$\zeta = \frac{30}{2 \cdot \sqrt{345}} = 0.807$$

The minimum ζ is 0.7. The gain required to produce that value is found by setting $\zeta = 0.7$, calculating ω, then calculating K.

$$2\zeta\omega_n = 30, \qquad \omega_n = \frac{30}{2 \cdot 0.7} = 21.43 rad / sec, \qquad \omega_n^2 = 459.2$$

$$459.2 = 20K, \qquad K = \frac{459.2}{20} = 22.96 \approx 23$$

You now have the range of gains that satisfy the specifications. The normalized output of the system as a function of time is shown in Figure 6-10.

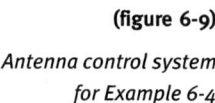

(figure 6-9)

Antenna control system for Example 6-4

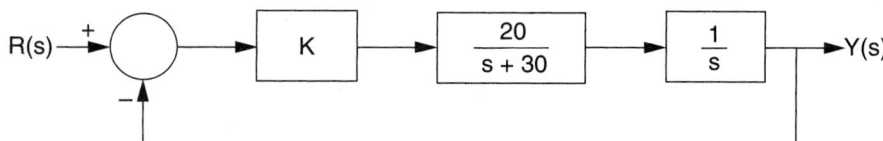

R(s) → + / − → K → $\dfrac{20}{s+30}$ → $\dfrac{1}{s}$ → Y(s)

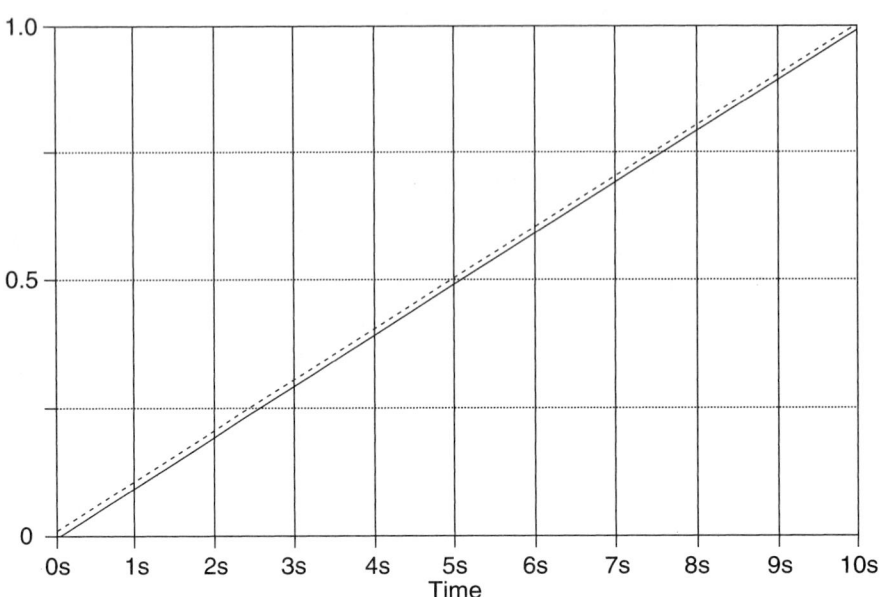

(figure 6-10)

Output of control system of Example 6-4

6.6 Effect of Increasing the System Order

In some cases, it may be desirable to increase the order of a system to achieve zero error and to improve steady-state performance of the system. However, care must be taken when doing so because instability may result. In the case of Example 6-4, if a cascade integrator is added to the forward path in the system, the system will become type two; but examination of the closed-loop transfer function shows that the system is unstable under all conditions.

$$\frac{Y(s)}{R(s)} = \frac{20Ks}{s^3 + 30s^2 + 20K}$$

Another method that produces a more stable system is the introduction of an integrator with a gain that is summed with the gain K, as shown in Figure 6-11.

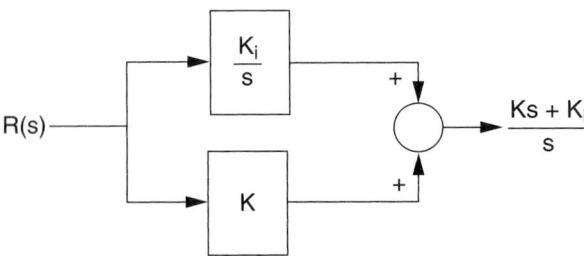

(figure 6-11)

Proper addition of a pole at the origin of the s-plane

When that is done, the system is still type two and is called a **PI,** or proportional-integral, system. Note that a new pole at the origin and a zero at K_i/K have been added to the system. This type of system will be considered in more detail in Chapter 9. The new forward path is then

$$\frac{Ks + K_i}{s} \cdot \frac{20}{s(s + 30)} = \frac{20Ks + 20K_i}{s^3 + 30s^2} \tag{6.35}$$

The Routh test can be used to determine the range of K that provides a stable system.

s^3	1	$20K$
s^2	30	$20K_i$
s^1	$20K - 0.67K_i$	$\dfrac{K_i}{K} = 30$
s^0	$20K_i$	

The test shows that as long as K_i is greater than zero and K is greater than $K_i/30$, the system is stable. However, the closer K is to $K_i/30$, the more oscillatory the system becomes. More clearly illustrating the effect of the gains, Figure 6-12 shows the step response of the system with K_i held constant at 3.0 and K varied from 1.0 to 10.0. Notice that as K increases, the risetime decreases, as does the overshoot of the system.

(figure 6-12)

Effect of increasing K on the PI system

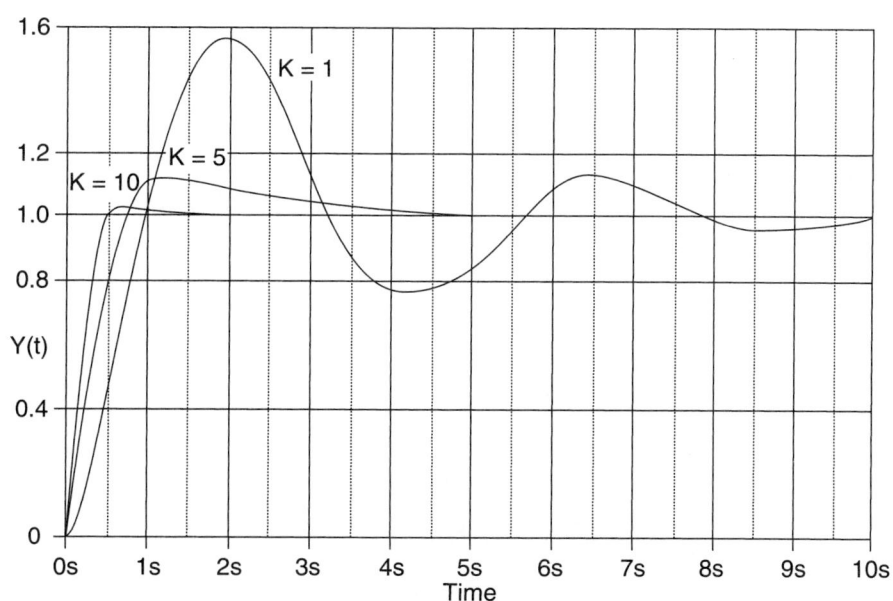

6.7 System Sensitivity to Parameter Changes

All systems are sensitive to variations in the parameters of the system. It is important to know how the overall system reacts when a change occurs in a specific parameter value so the design can be made to minimize the change in the output of the system. An appreciation for how the system reacts to a change in a parameter can be gained using the system shown in Figure 6-13.

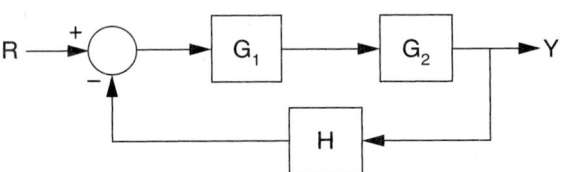

The transfer function of this system is

$$T = \frac{Y}{R} = \frac{G_1 G_2}{1 + G_1 G_2 H} \tag{6.36}$$

It should be clear from the equation that if $G_1 G_2 H \gg 1$, the transfer function is primarily sensitive to changes in H. The forward path gains have little effect on the output of the system. An analytical expression for the sensitivity of the transfer function to a change in a system parameter, a, is

$$S_a^T = \frac{\dfrac{\partial(T)}{T}}{\dfrac{\partial(a)}{a}} = \frac{\partial(T)}{\partial(a)} \cdot \left(\frac{a}{T}\right) \tag{6.37}$$

If the magnitude of the sensitivity is less than 1, the variation in T will be less than the normalized variation in the parameter a. When the magnitude of the sensitivity is 1.0, there is a one-to-one correspondence in the variation of the transfer function and the parameter. When the magnitude of the sensitivity is a negative real number, there is an inverse correspondence between the change in a and the change in T.

Applying the sensitivity function to G_1, the sensitivity of the system to changes in G_1 is given by

$$S_{G_1}^T = \frac{\partial T}{\partial G_1} \cdot \left(\frac{G_1}{T}\right) = \frac{G_2}{\left(1 + G_1 G_2 H\right)^2} \cdot \left(\frac{G_1}{\dfrac{G_1 G_2}{\left(1 + G_1 G_2 H\right)}}\right) = \frac{1}{1 + G_1 G_2 H} \tag{6.38}$$

Application of the sensitivity function to G_2 produces the same result. Application of the sensitivity function to H produces an inverse sensitivity function; and if $G_1 G_2 H \gg 1.0$, the corresponding sensitivity function is approximately -1.0.

$$S_H^T = \frac{\partial T}{\partial H} \cdot \frac{H}{T} = \frac{-G_1 G_2 H}{1 + G_1 G_2 H} \tag{6.39}$$

For an illustration of the use of the sensitivity function, examine the following example.

Example 6-5

For Figure 6-13, $G_1 = K$, $G_2 = 20/s(s + 30)$, and $H = 1$. The closed-loop transfer function is

$$\frac{Y(s)}{R(s)} = \frac{20K}{s^2 + 30s + 20K}$$

Solution

It is necessary to know the effect of changes in K on the system.

$$S_K^T = \frac{\partial T}{\partial K} \cdot \left(\frac{K}{T}\right) = \frac{(s^2 + 30s + 20K)20 - 20K \cdot 20}{(s^2 + 30s + 20K)^2} \cdot \frac{K}{\dfrac{20K}{s^2 + 30s + 20K}}$$

$$= \frac{20s^2 + 600s}{(s^2 + 30s + 20K)^2} \cdot \frac{s^2 + 30s + 20K}{20}$$

$$= \frac{s(s + 30)}{s^2 + 30s + 20K}$$

The final value theorem shows that the sensitivity of this system is zero for changes in the gain, K, for a unit step function.

6.8 Use of MATLAB LTI Viewer and Simulink for System Simulation

We already know that **MATLAB** can be used to produce the transfer function and graphic output of a system. In this chapter we will investigate the use of another two functions that are part of the control system toolbox, **Simulink** and the **LTI Viewer**. **Simulink** is a sub program that allows us to input a system in block diagram form, and the **LTI Viewer** allows us to see the graphical output of the block diagram. While **Simulink** has its own graphic output viewer, no data other than gross information is available in the **Simulink** viewer. The **LTI Viewer** solves this problem.

Use of the LTI Viewer in the MATLAB Workspace

MATLAB and **Simulink** can be used to show the time domain performance of a system. Better yet, they can be used to show the risetime, settling time, and peak time of a system with a control system toolbox device called the **LTI Viewer**. The **LTI Viewer** can be used with the standard workspace or with **Simulink**. The **LTI Viewer** can show up to six response curves at a time, as well as step, impulse, Bode, and other types of plots. To get an idea of the use of the **LTI Viewer**, look at Example 6-4. The usual numerator and denominator functions will be set up first; then they will be changed to a transfer function, using the *tf* function in **MATLAB**. That is a necessary step since a transfer or another type of function must exist in the workspace or in a file before the **LTI Viewer** can be used.

num = [20 60]; den = [1 30 20 60];	% numerator and denominator
	% functions
sys_tf = tf(num, den)	% convert to a transfer function
ltiview	% opens LTI Viewer with step
	% response plot

The *tf* command produces the transfer function using the row vectors you have entered. The *ltiview* command opens a blank window. Click on the File, then Import menu functions. A dialog box will open with the transfer function you have just constructed. Click on the transfer function; then click OK. (Double-clicking does not work.) The **LTI Viewer** now shows the step response of the system. Step response is the default plot type. Right-clicking in the **LTI Viewer** window causes a pop-up menu to appear. There are seven choices in the menu. The first is the plot type, and there are nine plot types from which to choose. For this exercise, you are interested only in the step response.

From the pop-up menu select *Characteristics,* then select *Peak Response,* then repeat the process for Settling Time and Rise Time. Blue dots appear at the appropriate points on the curve, and horizontal and vertical lines mark the points at which the measurements were taken. Moving the mouse pointer to any of the dots causes a small box containing the measurement values to appear. When you move off the dot, the box will disappear unless you left-click on the dot. You turn on the grid for the system by clicking *Grid On.* Similarly, the output range can be normalized, although it is not necessary for this example. The Properties function calls a dialog box that has several options. The only one of importance for this example is the Characteristics tab. The default settling time percentage is 5 percent. You can change it by typing in 2% and pressing the Enter key. The lines on the graph will change position to reflect the changes you made.

Figure 6-14 shows the response with the labels.

(figure 6-14)

*LTI Viewer—system
response to step input*

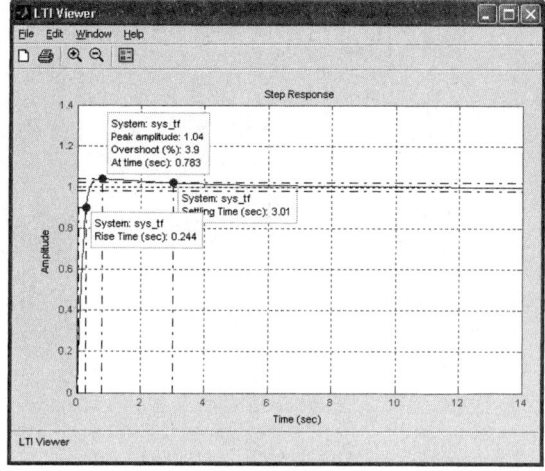

Use of the LTI Viewer with Simulink

The **LTI Viewer** can also be used with **Simulink**. The procedure is slightly different from when it is invoked from the workspace. The first step is to construct the block diagram as shown in Figure 6-15.

(figure 6-15)

*Simulink system for
Example 6-4*

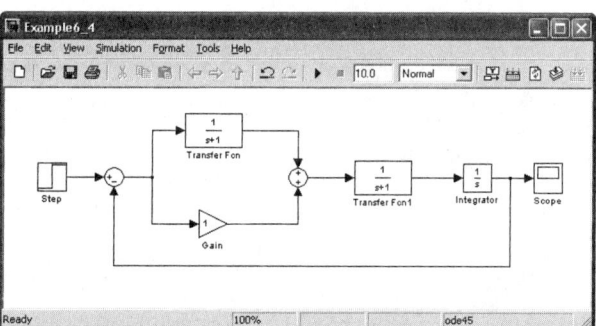

Notice that space has been left between the first summing junction and the output scope, a second scope has been added to the diagram, the additional space is for the viewer components, and the second scope is used to show the error signal of the system. We will open the **LTI Viewer** with two responses; the output and the error signal, you can have several responses at the same time in the viewer.

First, from the tools menu in **Simulink**, click on ***Control Design*** then select ***Linear Analysis***. This causes the viewer window to appear, with another dialog box, the "Controls and Estimation Manager" shown in Figure 6-16.

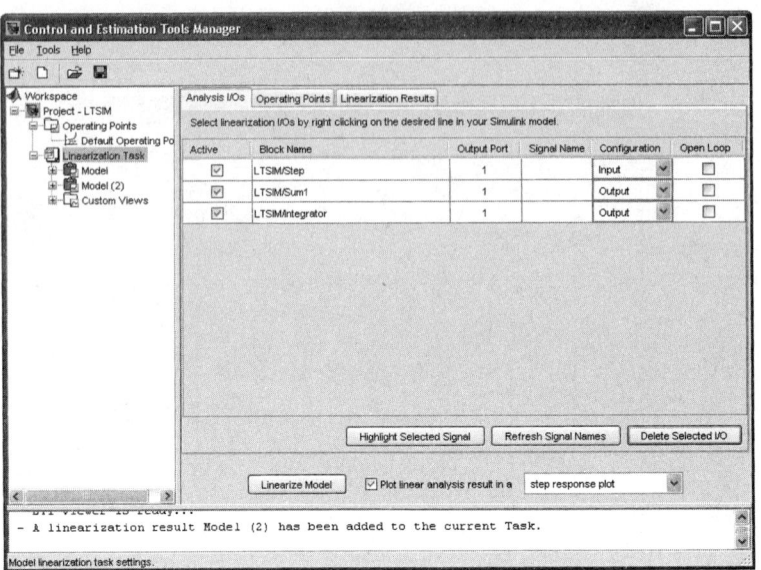

(figure 6-16)

Input-output points for Figure 6-15

To select input and output points for graphing the outputs of the simulation, right click on a line and then select "Linearization Port" from the menu and then "Input or Output."

Place the input point between the step input and summing junction, and the two output points on the lines to the scopes. Figure 6-17 shows the points in place.

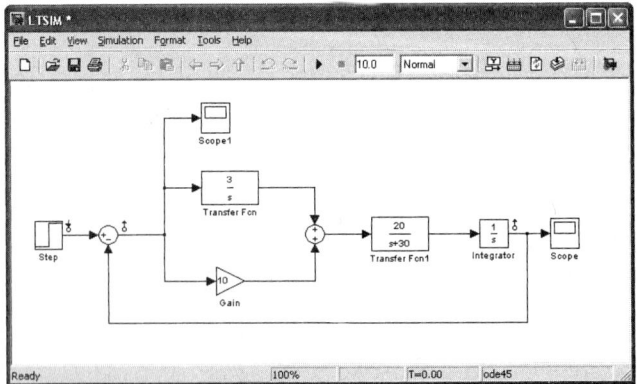

(figure 6-17)

Simulink system with input-output points installed

From the manager screen, select "Linearize Model," this opens the viewer window with both responses. To display the risetime, settling time and peak values right click in the windows as in the workspace method. The response is shown in Figure 6-18.

(figure 6-18)

LTI Viewer response of Example 6-4

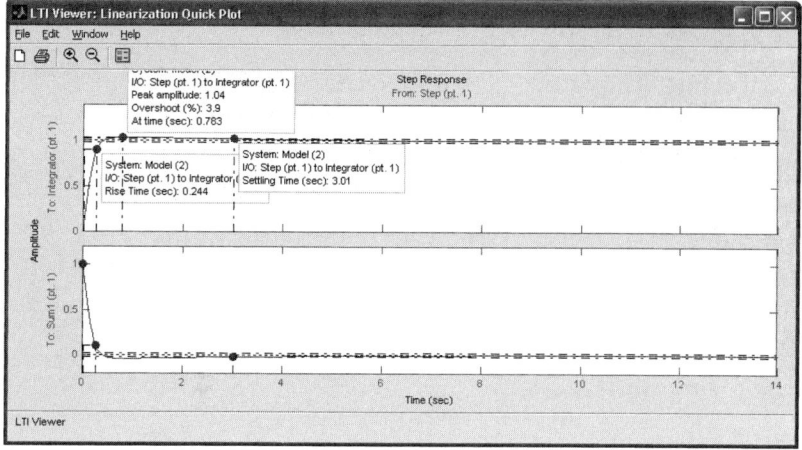

(figure 6-18)

LTI Viewer response of Example 6-4

Summary

➤ The second-order prototype system allows us to determine system risetimes, peak responses, and damping of control systems.

➤ The information to be gained from an examination of second-order systems is the concept of damping and radian velocity and how the position of the poles of the system affects the response of the system.

➤ Calculations for the maximum overshoot of the system, the risetime of the system, the peak times, and the settling time of the system give insight into whether or not the system will respond to the system specifications correctly.

➤ The ability of a system to follow a specific type of input signal and the requirements for the system to have a zero or small error depend on the system type.

➤ The error of systems is dependent on the system type number.

➤ Type zero systems can follow only a unit step input, and then with a finite error which is dependent on the system gain. The error for a ramp or parabolic input is infinite.

➤ Type one systems can follow a unit step input with zero error, and a ramp input with a finite error. The error for a parabolic input is infinite.

➤ Type two systems can follow both a unit step and ramp input with zero error. The error for a parabolic input is finite.

➤ The effect of increasing the order of a system to improve performance of a system is to change the system error for a given input signal type.

➤ The use of the sensitivity function S^T_a can predict how a system responds to changes in specific system parameters such as the system gain.

➤ **MATLAB** and **Simulink** can be used to determine risetime, peak time, and settling time.

➤ **Simulink** is used to input the block diagram of a system for simulation.

➤ **MATLAB, Simulink,** and the **LTI Viewer** are used to show several different types of graphical output of a system.

Problems

P6.1 For the following open-loop transfer functions, determine the system type and the error of each transfer function for the unit step $u_s(t)$, ramp $tu_s(t)$, and parabolic $t^2u_s(t)$ input signals.

a. $G(s) = \dfrac{100}{(s + 10)(s + 15)}$

b. $G(s) = \dfrac{50}{s^2(s + 5)(s + 12)}$

c. $G(s) = \dfrac{K(s + 10)}{s(s + 5)(s + 10)}$

P6.2 The systems of Problem P6.1 are canonic form 1. Provide the closed-loop characteristic equation for each of the systems. Use a Routh table to determine the limits of stable gain if they exist.

P6.3 Consider the system shown below to determine the following:

a. The gain K for a damping ratio of 0.7.

b. The damping ratio and first peak time for $K = 20$.

c. The settling time (5%) for $K = 5$.

d. The time domain output for a unit step input for $K = 10$.

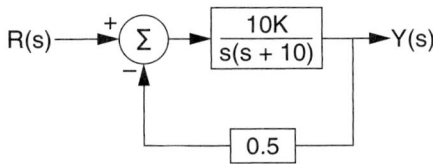

R(s) ⟶ + Σ ⟶ $\dfrac{10K}{s(s + 10)}$ ⟶ Y(s), feedback 0.5

(figure P6-3)

Control system for Problem P6.3

P6.4 The block diagram of a control system is shown below. Determine the system type number and error of the system for a unit ramp input signal.

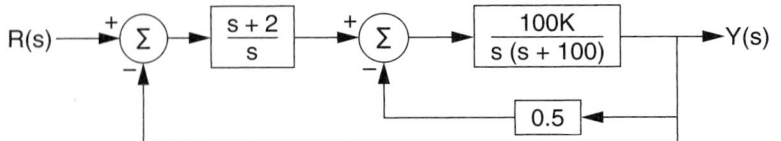

R(s) ⟶ + Σ ⟶ $\dfrac{s + 2}{s}$ ⟶ + Σ ⟶ $\dfrac{100K}{s(s + 100)}$ ⟶ Y(s), feedback 0.5

(figure P6-4)

Control system for Problem P6.4

P6.5 An all pole second-order system is to have a damping ratio of 0.5 and a settling time less than or equal to 0.5 second to 2 percent. Determine and plot the placement of the poles in the s-plane and the characteristic equation for the system.

P6.6   For the system shown, $K_2 = 10$. Determine the settling time, ζ, and the maximum peak value and time for the following K_1 gains:

$$K_1$$
200
100
75
50

(figure P6-6)

Control system for Problem P6.6

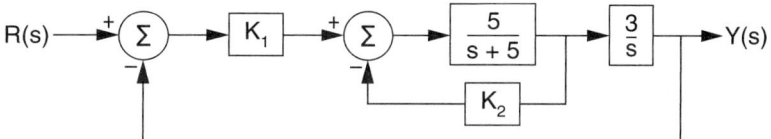

P6.7   For Problem P6.6, if K_1 is 50, determine a value of K_2 such that the maximum overshoot is 4.3 percent.

P6.8   For the system in Problem P6.6, $ts = 0.1$ second and maximum overshoot is 6 percent. Calculate $K1$ and $K2$ to produce the desired response. Check the design, using a computer program of your choice.

P6.9   For the system in Problem P6.6, determine the step, ramp, and parabolic error constants. Assume the system is stable.
$(r(t) = u_s(t), r(t) = tu_s(t), r(t) = t^2 u_s(t)/2)$

P6.10   For the unit step response shown, determine a second-order prototype transfer function of a unity gain system.

(figure P6-10)

Parameters for second-order prototype transfer function of Problem P6.10

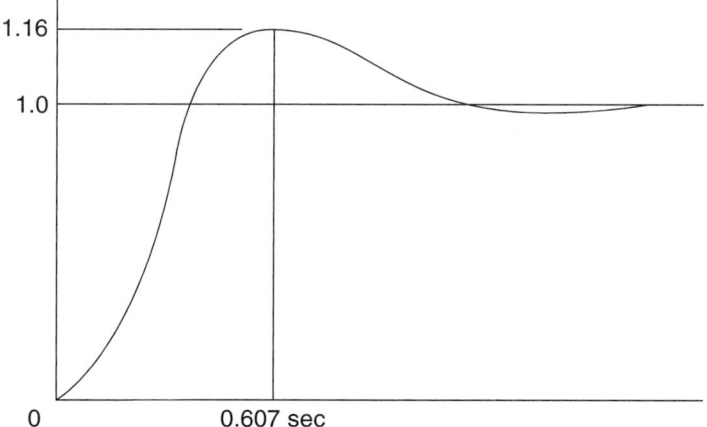

P6.11   For the following closed-loop transfer function, it is suggested that the poles located at –200 can be ignored. Assuming that is true, calculate the approximate risetime, percent of overshoot, peak time, and final value of the system.

$$\frac{Y(s)}{R(s)} = \frac{2 \cdot 10^6}{(s^2 + 6s + 25)(s + 200)^2}$$

P6.12   Considering the system shown below, determine the system type; the range of gain for asymptotic stability; and the unit step, ramp, and parabolic error constants of the system.

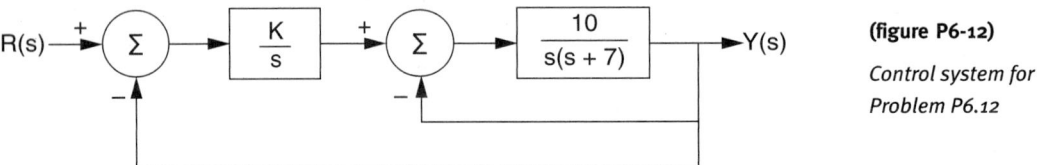

(figure P6-12)

Control system for Problem P6.12

P6.13   The block diagram of a control system is shown below. For the conditions shown, determine which provides the best rejection of a unit step disturbance.

a. $G_1(s) = 2$, $G_2(s) = \dfrac{10}{s + 4}$, $H = 0.25$

b. $G_1(s) = 10$, $G_2(s) = \dfrac{2}{s + 4}$, $H = 0.25$

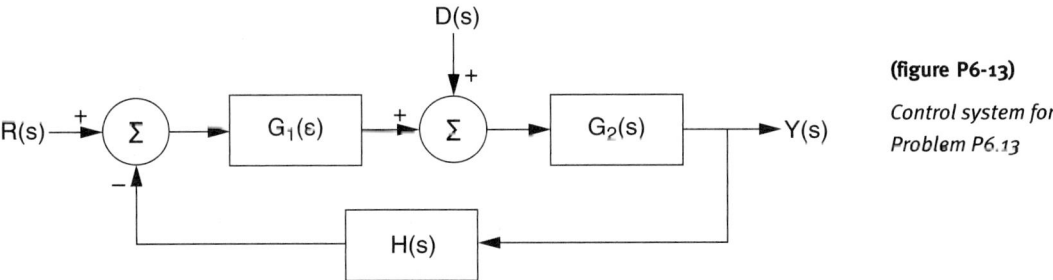

(figure P6-13)

Control system for Problem P6.13

P6.14   One control axis of shipboard radar has the block diagram shown below. The roll of the ship has a rate of $0.2tu_s(t)$.

a. Determine K for a damping ratio of 0.707.

b. Calculate the steady-state error produced by the roll of the ship.

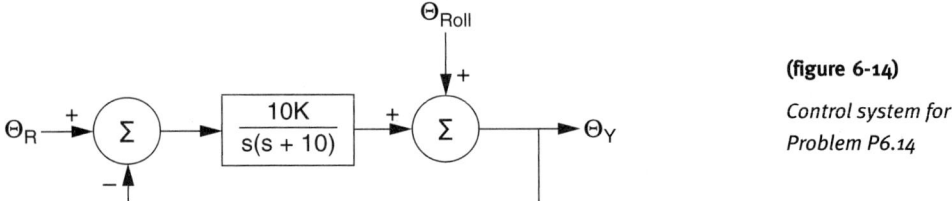

(figure 6-14)

Control system for Problem P6.14

P6.15   For the system shown below, determine the range of stable gain and, by trial and error method, determine a value of K that produces approximately 4 percent of overshoot. Determine the error constant for a unit ramp input.

(figure 6-15)

Control system for Problem P6.15

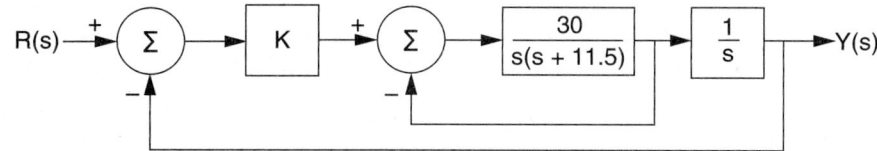

P6.16   The following is the closed-loop transfer function of a control system. Calculate a value of K that produces a damping ratio of 0.707. Calculate the 0 to 100 percent risetime and simulate the response of the system. Show the correlation of the calculation and the plotted risetime. Use whatever computer program is available.

$$\frac{Y(s)}{R(s)} = \frac{2K}{s^2 + 4s + 2K}$$

P6.17   Given the forward path transfer function

$$G(s) = \frac{10K}{s(s + 5)(s + 10)}$$

For a unity gain system, determine the value of K for which the system has sustained constant amplitude oscillation and determine the frequency of the oscillation.

P6.18   For the system shown below, calculate the sensitivity functions for both K_1 and K_2.

(figure P6-18)

Control system for sensitivity of K_1 and K_2 for Problem P6.18

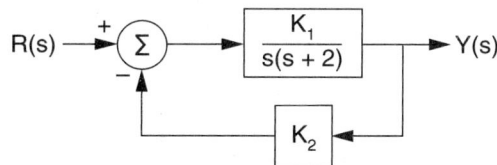

P6.19   For the system shown below, calculate the sensitivity function for K.

(figure 6-19)

Control system for Problem P6.19

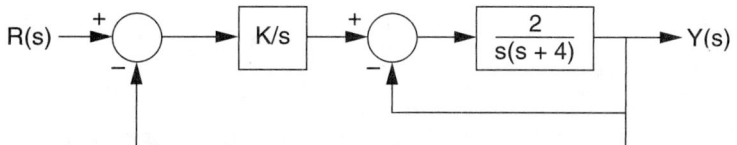

MATLAB Problems

M6.1   A position servo block diagram is shown below. Use **MATLAB** to show the output response to a step input when the gain is 5. Also show the risetime peak value, and settling time for the system, using the **LTI Viewer** for gains of 1, 5, and 10. Show all plots on the same screen. Comment on the effect of changing the gain on the risetimes and peak values.

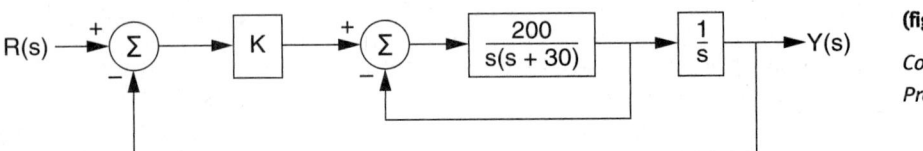

(figure M6-1)

Control system for Problem M6.1

M6.2   For the system shown in Problem P6.11, use the **LTI Viewer** to show that the poles at –200 can be ignored. Calculate both transfer functions and plot them on the same screen. To do that, import both transfer functions from the **LTI Browser**. Comment on the effect of the poles at –200.

M6.3   For Problem P6.15, use **Simulink** to show that the gain you calculated produces an approximate damping ratio of 0.7. If the input to the system is a constant $tu_s(t)$, plot the output on the same graph. You can use the **LTI Viewer** to determine the overshoot for a step input. What gain produces a damping ratio of unity?

M6.4   For Problem P6.11, use **MATLAB** to show that the response of the system is approximately the same for the full and the order reduced system.

M6.5   For Problem P6.17, use **MATLAB** to show that the value of K calculated for sustained bounded oscillation produces oscillation.

Chapter 7

Root-Locus Design

Chapter Objectives

After completing this chapter, you should be able to:

➤ Explain how the root locus of a system shows the change in pole position as a system gain is changed.

➤ Explain how the migration of the poles gives an indication of how the system response changes the damping of the system as a system gain is increased or decreased.

➤ Explain why the system gain for almost all systems must be positive.

➤ State the rules for construction of the root locus.

➤ Account for the qualities of a system that are obtained from the root locus.

➤ Show that the root locus is asymptotic to lines in the s-plane that are called asymptotes.

➤ To calculate the angles the asymptotes make with the real axis.

➤ Explain the concept of breakaway and break-in points on the σ axis and what they mean.

➤ Recover the open-loop transfer function from the characteristic equation.

➤ State the concept of the complementary root locus.

➤ Use **MATLAB** to formulate the open-loop transfer function and construct the root locus.

7.1 Introduction

The concept of root locus was first introduced by Evans.[1] Its purpose in design is to evaluate the effect on the stability of a system as some system parameter (usually a value of gain) is changed. Values such as the minimum damping ratio of the system, the gain at a specific pole, or the tendency toward oscillation can be predicted using the root locus.

The characteristic equation contains both the poles and zeros of a system obtained from the open-loop transfer function, but it is the poles that determine the stability of the system. For a system to be stable, all of the poles must be in the left-half plane or must migrate toward the left-half plane.

The influence of the poles of a system in system response cannot be understated. The poles give insight into how the system stability is influenced by changes in gain. The system may be stable for all values of gain that are greater than zero, it may be stable for values of gain greater or less than a specific value, or it may be stable for gains between two values.

The use of the root locus in conjunction with a Routh table provides a comprehensive understanding of how a system will react to changes in gain. The root locus gives insight into what the system damping ratio and radian velocity are at closed-loop poles and how the placement of zeros will affect the time response of a system.

7.2 Root-Locus Concepts, $K > 0$

First, all gains in a system are assumed to be greater than zero. The case for gains less than zero, the complementary root locus, will be taken up later in the chapter. The root locus is a plot of a control system's pole migration as some variable, usually a gain, is varied.

Magnitude and Angle Criterion

Consider a single-loop control system with a gain factor K whose closed-loop transfer function is given in Equation 7.1.

$$\frac{Y(s)}{R(s)} = \frac{KG(s)}{1 + KG(s)H(s)} \tag{7.1}$$

The denominator of the equation, set to zero, is the characteristic equation of the system.

$$1 + KG(s)H(s) = 0 \tag{7.2}$$

The characteristic equation is generally a quotient of factored polynomials containing all of the system's poles and zeros. When rewritten in that manner, the characteristic equation becomes

$$1 + \frac{K(s + z_1)(s + z_2)(s + z_3)\cdots}{(s + p_1)(s + p_2)(s + p_3)\cdots} = 0$$

$$\frac{K(s + z_1)(s + z_2)(s + z_3)\cdots}{(s + p_1)(s + p_2)(s + p_3)\cdots} = -1$$

(7.3)

When set equal to –1, the left-hand side of the equation is the open-loop transfer function of the system. The open-loop transfer function provides the poles and zeros of the system when $K = 0$.

The gain factor K is not constrained and may have any value between $-\infty \leq K \leq \infty$. However, for this discussion, only values of K greater than zero are considered. If the characteristic equation is rearranged to provide a value for K, the result is a negative value for K that can be rewritten as $|K|\angle 180°$, from which the magnitude and angle criterion is developed as follows:

$$|K| = \frac{(s + p_1)(s + p_2)(s + p_3)\cdots}{(s + z_1)(s + z_2)(s + z_3)\cdots} \angle 180°$$

(7.4)

Since negative feedback is being used, any angle that equates to 180° or an integer multiple of 180° is acceptable for the criteria, and the criteria can be rewritten as

$$\angle G(s)H(s) = (2k + 1)180°, \; where \; k = 0,1,2,\cdots$$

(7.5)

The significance of Equation 7.3 is that when plotting a root locus, it is important to know the roots of the closed-loop characteristic equation for numerous values of K since a change in K produces a change in the root values of the characteristic equation. The use of a calculator with a polynomial root finding program is useful for finding the roots of the characteristic equation; the program **MATLAB** can also be used to plot the root locus.

7.3 Rules for Constructing the Root Locus

In producing a root locus, several concepts and rules can make the construction easier:

1. **Start and end points.** The root locus starts on the open-loop poles and ends on the open-loop zeros of the transfer function. In all cases, the root locus is symmetrical about the real axis.

2. **Closed-loop segments.** The root loci exist to the left of an odd-numbered open-loop pole or finite zero on the σ axis and end on an open-loop zero or pole. If no poles are on the σ axis, the root locus starts on the open-loop poles and ends on the open-loop zeros, and there are as many segments of the root locus as the order of the characteristic equation.

3. **Range of gain.** The gain (K) is zero at the open-loop poles and infinite at the open-loop zeros.

4. **Breakaway and break-in points.** A breakaway point exists between two adjacent open-loop poles on the σ axis, and a break-in point exists between two adjacent open-loop zeros on the σ axis.

5. **Asymptotic properties.** Root loci are asymptotic to straight lines in the s-plane with angles given by M, the number of poles, and N, the number of zeros.

6. **Angles of departure and arrival at a pole or zero.** The angle of departure or arrival of the root locus at an open-loop pole or zero is given by the angle of the vectors from all open-loop poles and zeros to the open-loop pole or zero being investigated.

7. **Gain at a specific closed-loop pole.** The gain at a closed-loop pole is equal to the product of the vectors of the open-loop poles to the pole being calculated divided by the product of the open-loop zeros to the same pole.

System Qualities Obtained from the Root Locus

The Routh table provides us with certain limits based on the roots of the characteristic equation. There is, however, no way to determine the gain or damped radian velocity for a specific damping ratio, or whether or not the system is tending toward the right-half plane as one of the system gains in the system is increased. The root-locus provides us with this information.

System gain at a closed-loop pole

The system gain at any closed-loop pole can be obtained by drawing a vector from all open-loop poles and zeros to the closed-loop pole being investigated. The gain is equal to the product of all pole vectors to the closed-loop pole divided by the product of all zero vectors to the closed-loop pole. The concept is shown in Figure 7-1. The angle made by the vectors to the selected pole must be equal to 180° or a multiple thereof to fulfill both the angle and magnitude criteria.

$$|K| = \frac{\displaystyle\prod_{i=1}^{M}|(s + p_i)|}{\displaystyle\prod_{j=1}^{N}|(s + z_j)|} \tag{7.6}$$

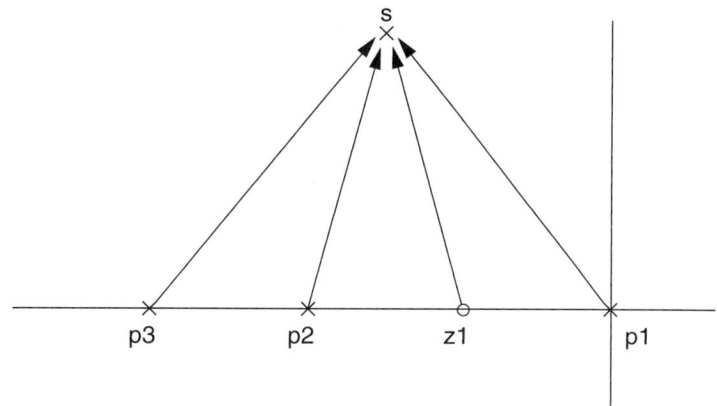

System damping ratio at a closed-loop pole

The angle θ, which is made by a vector from the origin of the s-plane to a closed-loop pole on the root locus, provides the damping ratio at the closed-loop pole. The cosine of the angle θ is the damping ratio.

$$\zeta = \cos^{-1}\theta \qquad (7.7)$$

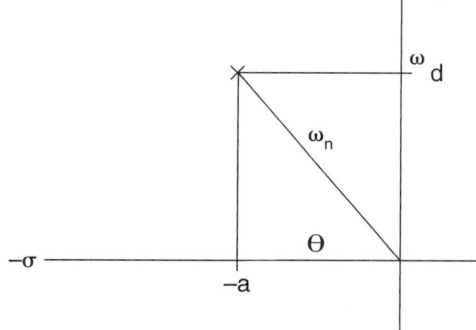

System damped radian velocity at a closed-loop pole

The damped radian velocity (ωd) differs from the natural radian velocity (ωn) in that damping is taken into account and the damped radian velocity is always less than the natural radian velocity. The damped radian velocity can be determined from the angle of the origin of the s-plane to the closed loop pole. The value is

$$\omega_d = \omega_n Sin\theta = \omega_n\sqrt{1 - \zeta^2} \qquad (7.8)$$

Use of the Rules to Start the Root Locus

To get started, a pole-zero chart of the open-loop transfer function is generated. Remember that when there are no finite open-loop zeros, there are as many open-loop zeros at infinity as there are open-loop poles and there will be as many segments of the root locus as the order of the system.

The following all-pole characteristic equation will be used to determine the quantities needed to plot the root locus:

$$G(s)H(s) = 1 + \frac{K}{s(s + 1)(s^2 + 4s + 13)}$$

A Routh table is useful for this system as the root locus crosses the $j\omega$ axis. For that to happen, the closed-loop characteristic equation is needed. The characteristic equation and Routh table are as follows:

$$s^4 + 5s^3 + 17s^2 + 13s + K = 0$$

s^4	1	17	K
s^3	5	13	
s^2	14.4	K	$s = \pm j1.612 \; rad/s$
s^1	$13 - 0.347K$		$0 < K < 37.44$
s^0	K		

The gain calculated from the table shows that the system is stable for gains less than a maximum of 37.44 $(0 < K < 37.44)$. Also, the root locus crosses the $j\omega$ axis at 1.612 rad/sec. From the Routh table, a point on the root locus has been found.

Rule 1 states that the root locus begins on the open-loop poles and ends on the open-loop zeros of the system. From this, if there are four open-loop poles and no finite open-loop zeros, portions of the root locus will start at the poles, 0 and −1 and −2 ± $j3$. As K is increased, the poles on the σ axis migrate toward each other and converge to a double root. There is a complete segment of the root locus on the σ axis. The poles break away from the σ axis at the double-pole value. The poles at −2 ± $j3$ will depart from the open-loop poles at an angle and end on the open-loop zeros at infinity.

Rule 2 states that the root locus will occur to the left of an odd-numbered pole or finite zero located on the σ axis. The chart in Figure 7-3 shows the pole segment of the root locus plotted from the poles at 0 and −1 according to Rule 2.

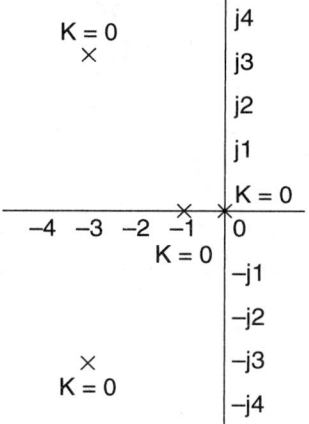

Rule 3 states that the gain at the open-loop poles is zero and the gain at the open-loop zeros is infinite.

Rule 4 states that if the gain is increased sufficiently, a double pole will occur between two adjacent open-loop poles. Increases in gain after that value will cause the roots to depart from the σ axis (breakaway).

Finding the breakaway/break-in points involves finding multiple roots of the characteristic equation where the characteristic equation equals zero. Complex conjugate roots can also be breakaway points when they fulfill the characteristic equation or when they are points on the root locus. Breakaway/break-in points occur at a double pole on the real axis when the poles are real. The method to determine the breakaway/break-in points uses the derivative of the gain equation of the system. If the gain is a quotient of poles and zeros, the derivative of a quotient rule is used.

$$\frac{d}{dx}\left(\frac{u}{v}\right) = \frac{v\dfrac{du}{dx} - u\dfrac{dv}{dx}}{v^2}$$

The gain equation for the system is determined using the open-loop transfer function. The roots of the resulting derivative are then found by factoring the third-order equation that results. The calculation and roots are shown in Figure 7-5.

$$K = s^4 + 5s^3 + 17s^2 + 13s$$

$$\frac{dK}{ds} = 4s^3 + 15s^2 + 34s + 13 = 0$$

Roots at $s_1 = -0.466$, $s_{2,3} = -1.642 \pm j2.067$

From the derivative, a real root is on the segment of the root locus on the σ axis and is a breakaway point. The complex roots are useful for plotting the root locus. One more point is now on the root locus.

Rule 5 states that the root locus will become asymptotic to lines in the s-plane with angles that are determined by the number of poles and finite zeros of the system. The number of asymptote lines is equal to the value of the number of open-loop poles minus the number of open-loop zeros. The lines in the s-plane will emanate from a centroid that is the arithmetic average of the pole-zero positions. The asymptote angles are

$$\Psi = \frac{(2k + 1)180°}{M - N} \qquad [k = 0,1,2,3,\cdots,(M - N - 1)] \qquad (7.9)$$

For that transfer function, $M - N = 4$ and the value of k is $(M - N - 1) = 3$. The angle calculations are

$$\Psi_1 = \frac{(2 * 0 + 1)180°}{4} = 45° \qquad \Psi_2 = \frac{(2 * 1 + 1)180°}{4} 135°$$

$$\Psi_3 = \frac{(2 * 2 + 1)180°}{4} = 225° = -135° \qquad \Psi_4 = \frac{(2 * 3 + 1)180°}{4} = 315° = -45°$$

The asymptote angles are $\pm 45°$ and $\pm 135°$. The centroid of asymptotes is

$$\sigma_c = \frac{\sum\limits_{i=1}^{M}(p_i) - \sum\limits_{i=1}^{N}(z_i)}{M - N} \qquad (7.10)$$

$$\sigma_c = \frac{0 - 1 - 2 - j3 - 2 + j3}{4} = -1.25$$

A short table of asymptote angles for $M - N$ up to five is shown in Table 7-1.

(table 7-1)

M - N	Asymptote Angles
0	No asymptotes
1	180°
2	±90°
3	±60°, 180°
4	±45°, ±135°
5	±36°, 180°, ±144°

Rule 6 gives the angle of departure of the root locus from a particular open-loop pole or the angle of arrival at an open-loop zero. A few rules are needed to calculate the angle of departure or arrival of the root locus at a pole or at zero.

First, the angle of departure or arrival must meet the angle criteria; second, angles from zeros to poles add and angles from pole to pole or zero to zero subtract. All angle measurements are made with respect to the positive real axis.

The formula for the angle of departure (arrival) is

$$-(\theta_1 + \theta_2 + \cdots) + (\phi_1 + \phi_2 + \cdots) = (2k + 1)180° \tag{7.11}$$

Where θ are the angles from pole to pole and ϕ is the angle from zero to pole. An example of the method is shown in Figure 7-4.

(figure 7-4)

Angle of departure at an open-loop pole

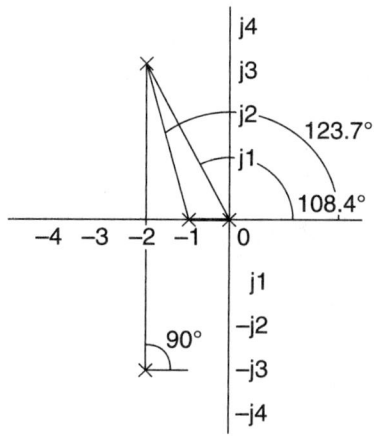

$$-(90° + 108.4° + 123.7) = (2k + 1)180°$$

For the transfer function being used, the departure from the $-2 \pm j3$ pole is

$$\theta_{depart} = 180° - 322.1° = -142.1°$$

In this case, k was set equal to -1; that provides an additive 180° for the total angle, which is $-142.1°$. If you had chosen either $k = 0$ or $k = 1$, you would have eventually arrived at the same angle, but by a more roundabout way.

The complete root locus is shown in Figure 7-5.

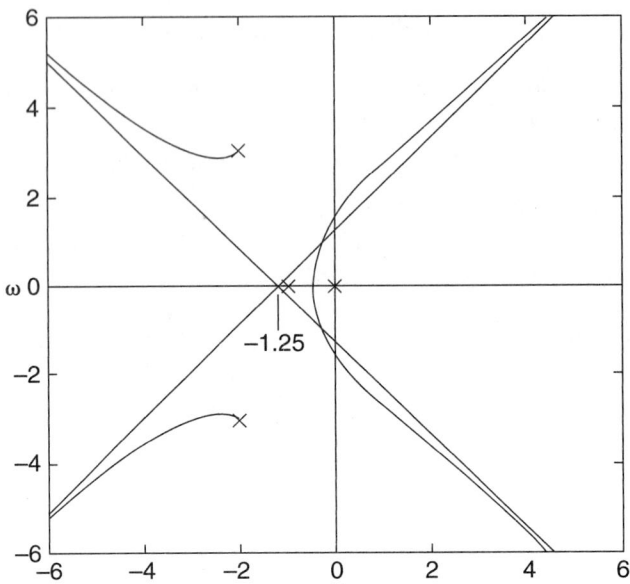

(figure 7-5)

Root locus for using the rules of the root locus

7.4 Second-Order Examples

The simplest root locus is a second-order system with no zeros. Example 7-1 shows such a locus.

Example 7-1

Consider the control system shown in Figure 7-6.

Solution

The transfer function of the system is

$$\frac{Y(s)}{R(s)} = \frac{\dfrac{2K}{s^2 + 8s + 15}}{1 + \dfrac{2K}{s^2 + 8s + 15}} = \frac{2K}{s^2 + 8s + (15 + 2K)}$$

If K is set to zero, the roots (open-loop poles) of the characteristic equation are −3 and −5. As K is increased from zero, the poles move toward each other; and when the gain is 1.0, a double pole occurs at −4. Increases in gain above 1.0 will cause the roots to plot to a real value of −4.0 plus an imaginary value. As the gain is increased to greater and greater values, the real part of the roots does not change. As the gain approaches infinity, the root locus will terminate on the two zeros located there. It appears that no value of gain will cause the system to become unstable; it will always settle to a final value since the root locus never migrates toward the right-half plane. Keep in mind, however, that due to mechanical constraints, the system may not be able to perform physically if the gain is raised too high. The root locus is shown in Figure 7-7.

(figure 7-6)

Control system for
Example 7-1

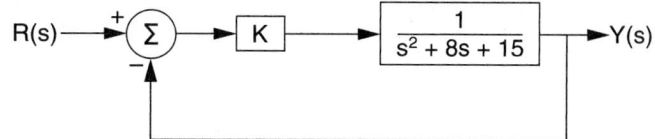

(figure 7-7)

Root locus for Example 7-1

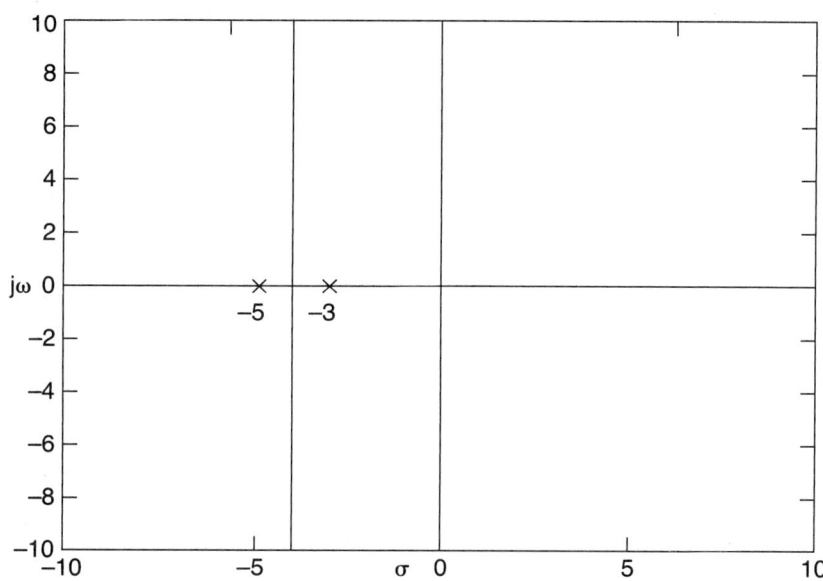

Example 7-2

For a second example, still second-order, consider the block diagram shown in Figure 7-8.

Solution

The transfer function of this system is

$$\frac{Y(s)}{R(s)} = \frac{\dfrac{K(s+4)}{s(s-3)}}{1 + \dfrac{K(s+4)}{s(s-3)}} = \frac{K(s+4)}{s^2 + (K-3)s + 4K}$$

The open-loop transfer function will provide the root locus. The poles are located at $s_1 = 0$ and $s_2 = +3.0$, and the single zero is at $s = -4.0$. Obviously, there is a pole in the right-half plane. A segment of the root locus exists to the left of the pole at +3 and extends to the pole at zero; therefore, a breakaway point exists between those values (Rule 4). The single zero also has a segment of the root locus to its left, ending on the zero at infinity. A break-in point will exist to the left of that value.

Example 7-2 (continued)

To determine the breakaway/break-in points, you must first determine the expression for K in the characteristic equation. The characteristic equation for this system is

$$1 + \frac{K(s + 4)}{s(s - 3)} = 0$$

The gain equation and the derivative and gains are

$$-K = \frac{s(s - 3)}{(s + 4)} = \frac{s^2 - 3s}{s + 4}$$

$$-\frac{dK}{ds} = \frac{(s + 4)(2s - 3) - (s^2 - 3s)}{(s + 4)^2} = \frac{2s^2 + 5s - 12 - s^2 + 3s}{(s + 4)^2} = \frac{s^2 + 8s - 12}{(s + 4)^2} = 0$$

$$s_1 = -9.292, \qquad s_2 = 1.292$$

This root locus is circular, and the asymptotes for the system are at 180° since the number of poles minus the number of zeros is one. The root locus is shown in Figure 7-9.

In general, second-order root loci will be straight lines as shown in Example 7-1 or circles as shown in Example 7-2. They also may be segments of circles if the poles of the open-loop transfer function are complex and a zero is included.

Higher-order systems generally have more interesting root-loci shapes and are not restricted as to the shape.

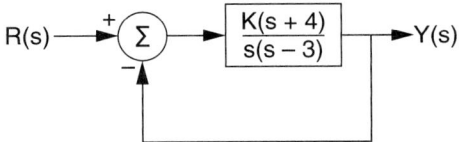

R(s) ——→ + Σ —→ $\dfrac{K(s + 4)}{s(s - 3)}$ ——→ Y(s)

(figure 7-8)

Control system for Example 7-2

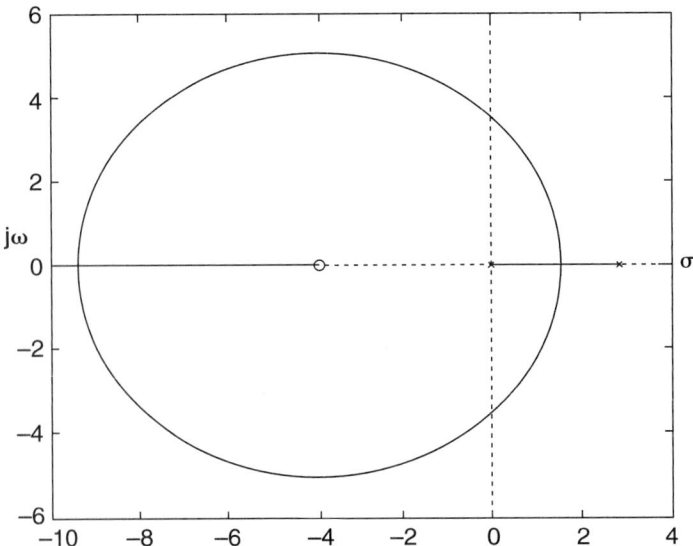

(figure 7-9)

Root locus for Example 7-2

7.5 Higher-Order Examples

When the system order is greater than two, the root loci take on more interesting shapes and become asymptotic to lines originating on and at an angle to the σ axis. The determination of where those lines originate and the angles were discussed earlier. The angles of the lines are determined by the number of poles, M, and the number of zeros, N.

Example 7-3

Consider the following control system characteristic equation:

$$1 + \frac{K}{s(s + 1)(s + 2)(s + 3)} = 0$$

Solution

By inspection, you can see that there are no zeros and four poles for this system. The open-loop poles are located at 0, –1, –2, and –3. The roots and segments are plotted in Figure 7-10.

The starting point of the asymptote lines, called the centroid $\left(\sigma_c\right)$ of asymptotes, is

$$\sigma_c = \frac{(-1) + (-2) + (-3) - (0)}{4} = \frac{-6}{4} = -1.5$$

The asymptotes originate at –1.5 on the σ axis.

The angle the asymptotes make with the real axis, using the asymptote angle formula, with $k = M - N - 1$ are $4 - 0 - 1 = 3$ and $k = 0, 1, 2, 3$.

$$\Psi_1 = \frac{(2 * 0 + 1)180°}{4} = 45°, \qquad \Psi_2 = \frac{(2 * 1 + 1)180°}{4} = 135°,$$

$$\Psi_3 = \frac{(2 * 2 + 1)180°}{4} = 225°, \qquad \Psi_4 = \frac{(2 * 3 + 1)180°}{4} = 315°$$

The angles are ±45° and ±135°.

Since all four poles lie on the σ axis, two segments of the root locus also lie on the σ axis. With no zeros, there will be two breakaway points. Further, since no break-in point exists, the root locus will curve toward the asymptotes. A Routh table will show four points on the root locus and the gain at those points. The characteristic equation for the Routh table is generated from the closed-loop transfer function.

s^4	1	11	K
s^3	6	6	
s^2	10	K	
s^1	$\dfrac{60 - 6K}{10}$		
s^0	K		

$$0 \le K \le 10$$

Example 7-3 (continued)
_____

Using the Routh table, you can determine that the maximum value of K for stability is 10 and that the crossing point of the root locus is at $\pm j1$ rad/sec. Those figures also apply to the segments of the root locus that do not cross the $j\omega$ axis. Therefore, there are four points on the root locus. Next, you must determine the breakaway points. First, determine the gain equation of the system and its derivative from the characteristic equation.

$$K = s^4 + 6s^3 + 11s^2 + 6s$$

The roots of that expression are

$$\frac{dK}{ds} = 4s^3 + 18s^2 + 22s + 6 = 0$$

Only two of the three roots lie on the root locus of the system; they are the breakaway points. The third root is not used since it is not on the open-loop root locus. The fact that the third root is the same value as the centroid of the asymptotes is coincidental. This method does not produce the centroid of asymptotes as a by-product. A pole-zero map and the closed-loop segments showing the asymptotes is provided in Figure 7-24.

$$s_1 = -0.382, s2 = -1.5, \ s3 = -2.618$$

The complete root locus is shown in Figure 7-11.

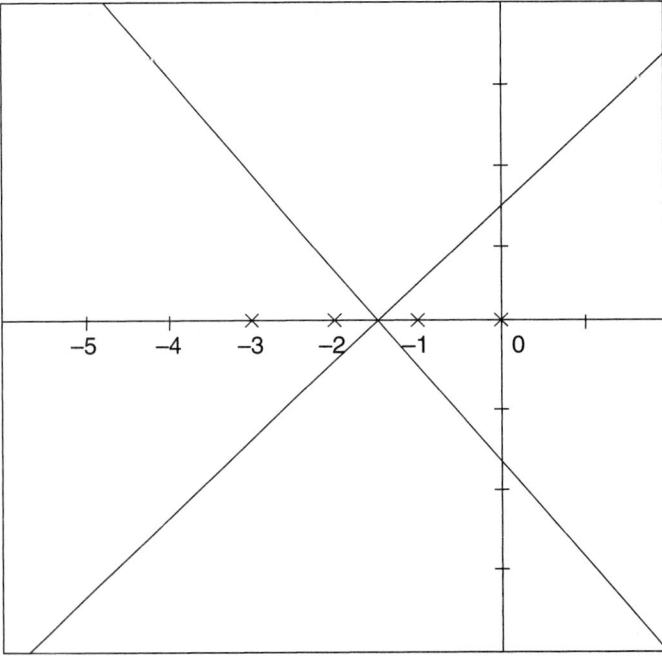

(figure 7-10)
Pole placements and asymptotes for Example 7-3

(figure 7-11)

Root locus for Example 7-3

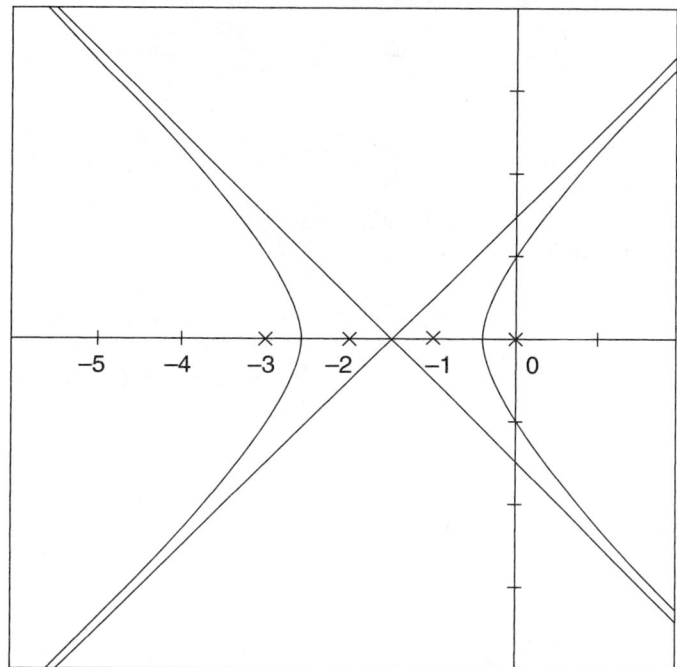

7.6 Open-Loop Poles and Zeros and the Characteristic Equation

Sometimes you are given the characteristic equation of a system. From the equation, you can determine the poles and zeros of the open-loop system. The next two examples will demonstrate the process.

Example 7-4

Consider the following characteristic equation with unknown gains:

$$s^4 + 12s^3 + (41 + K)s^2 + (30 + 5K)s + 6K = 0$$

Determine the open-loop characteristic equation of the system.

Solution

The method for recovering the poles and zeros of the system is not very difficult. The steps are as follows:

1. Separate the characteristic equation into separate terms—those containing K and those not containing K.

$$s^4 + 12s^3 + 41s^2 + Ks^2 + 30s + 5Ks + 6K = 0$$

Example 7-4 (continued)

2. Collect like terms.

$$s^4 + 12s^3 + 41s^2 + 30s + Ks^2 + 5Ks + 6K = 0$$

3. Remove any common multipliers and separate the equation into two parts—those not containing K and those containing K.

$$s(s^3 + 12s^2 + 41s + 30) + \left[K(s^2 + 5s + 6) \right] = 0$$

4. Divide both parts of the equation by the parts not containing K.

$$\frac{s(s^3 + 12s^2 + 41s + 30)}{s(s^3 + 12s^2 + 41s + 30)} + \frac{K(s^2 + 5s + 6)}{s(s^3 + 12s^2 + 41s + 30)} = 1 + \frac{K(s^2 + 5s + 6)}{s(s^3 + 12s^2 + 41s + 30)}$$

5. Factor the numerator and denominator for the poles and zeros of the open-loop transfer function.

$$1 + \frac{K(s^2 + 5s + 6)}{s(s^3 + 12s^2 + 41s + 30)} = 1 + \frac{K(s + 2)(s + 3)}{s(s + 1)(s + 5)(s + 6)}$$

6. Produce the pole-zero chart for the poles and zeros found. The poles and zeros for this function are as follows:

Poles	Zeros
$s_1 = 0$	$s_1 = -2$
$s_2 = -1$	$s_2 = -3$
$s_3 = -5$	
$s_4 = -6$	

The pole-zero plot of this system is shown in Figure 7-12.

Based on Rule 2 for root loci, there will be segments of the root locus on the σ axis between the poles at 0, –1 and –5, –6. There will also be a segment between the two zeros at –2, –3. Since there are four poles and two zeros, the asymptotes for this system are at $\pm 90°$. Rule 4 states that breakaway and break-in points exist between adjacent poles and zeros; therefore, there will be two breakaway points and one break-in point for this system. The centroid of asymptotes is found from

$$\sigma_c = \frac{(0 - 1 - 5 - 6) - (-2 - 3)}{4 - 2} = -\frac{7}{2} = -3.5$$

Example 7-4 (continued)

The root locus starts at 0 and −1 and moves together as K is increased. At some point, a double pole occurs and the root locus breaks away from the real axis and moves toward the zeros at −2, −3. You can infer from that that the root locus will not cross the $j\omega$ axis since the other poles must tend toward the asymptotes. A Routh table is not necessary for this plot.

The breakaway and break-in points are found from the gain equation.

$$K = \frac{s^4 + 12s^3 + 41s^2 + 30s}{s^2 + 5s + 6}$$

Differentiating that equation for the required expression for breakaway/break-in points gives

$$\frac{dK}{ds} = \frac{(s^2+5s+6)(4s^3+36s^2+82s+30)-(s^4+12s^3+41s^2+30s)(2s+5)}{(s^2+5s+6)^2} = 0$$

$$\left(4s^5+56s^4+286s^3+656s^2+642s+180\right)-\left(2s^{5+}+29s^4+142s^3+265s^2+150s=0\right.$$

$$2s^5 + 27s^4 + 144s^3 + 391s^2 + 492s + 180 = 0$$

The roots are

$$s_1 = -0.586, s_2 = -2.447, s_3 = -5.464, s_{4,5} = -2.501 \pm j2.286$$

To be a breakaway/break-in point, the root value must fall on a segment of the root locus. Therefore, the complex roots are extraneous and, therefore, are ignored. The three other roots all fall on a segment of the root locus; therefore, the roots s_1 and s_3 are breakaway points. s_2 is a break-in point. The root locus is shown in Figure 7-13.

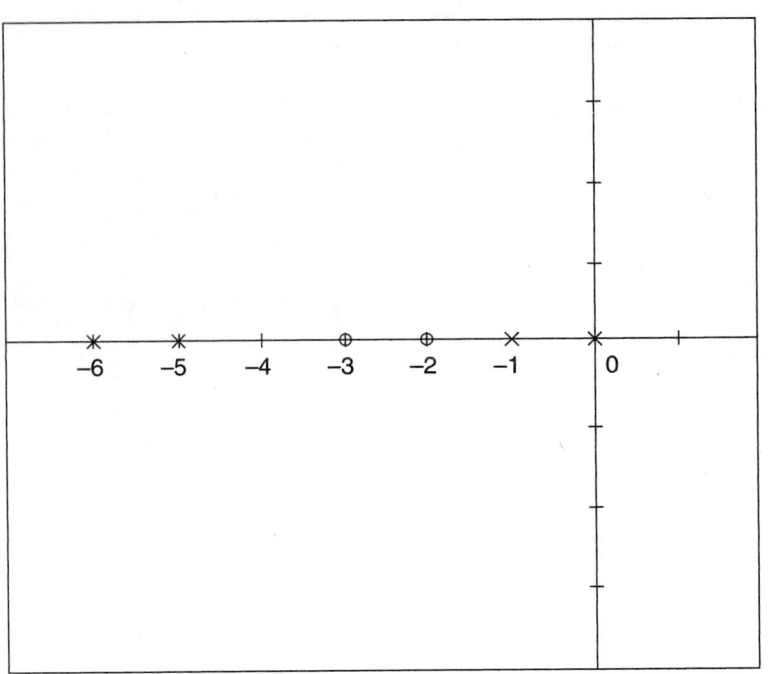

(figure 7-12)
*Pole-zero placement for
Example 7-4*

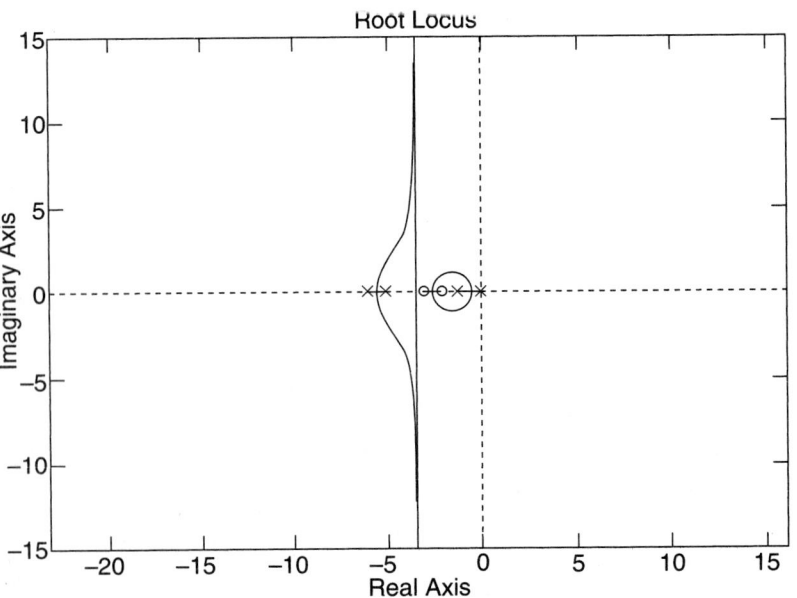

(figure 7-13)
Root locus for Example 7-4

Example 7-5

Consider the system shown in Figure 7-14, which is to be used in a positioning system. The motor exhibits a second-order transfer function that relates output shaft speed to armature voltage. Integrating the output shaft speed produces distance or position. One common method of compensation for this type of system adds a pole and zero, with an adjustable gain, to the system.

Solution

The zero location is left as a constant, and its position is to be decided based on the following criteria: the positioning system should exhibit high speed with zero overshoot, and the dominant poles should have unity damping ratio. (Unity damping ratio occurs at the breakaway point.)

The open-loop transfer function and the characteristic equation are as follows:

$$D(s)G(s) = \frac{K(s + a)}{s(s + 12)(s + 5)(s + 8)}$$

$$1 + \frac{K(s + a)}{s(s + 12)(s + 5)(s + 8)} = 0$$

$$s^4 + 25s^3 + 196s^2 + (480 + K)s + Ka = 0$$

A chart of only the poles is shown in Figure 7-15.

You need to make an assumption here for the breakaway point. Placing the zero too close to the origin causes the pole at zero to be dominant, which indicates a sluggish system. Placing the zero too far from the origin causes the system to exhibit an oscillatory response. Since the damping ratio is unity at the breakaway point, the breakaway will be arbitrarily specified at -2.5, halfway between the pole at the origin and the first pole at -5.0. The poles produce a second-order product, which is $(s + 2.5)(s + 2.5) = s^2 + 5s + 6.25$. Performing long division and equating the derived values produce the following parameters, which give the gain at the required zero position.

$$
\begin{array}{r}
s^2 + 20s + 89.8 \\
s^2 + 5s + 6.25 \overline{\smash{\big)}\, s^4 + 25s^3 + 196s^2 + (480 + K)s + Ka} \\
\underline{s^4 + 5s^3 + 6.25s^2} \\
20s^3 + 189.8s^2 + (480 + K)s \\
\underline{20s^3 + 100.0s^2 + 125s} \\
89.8s^2 + (355 + K)s + Ka \\
\underline{89.8s^2 + \quad 448.8s + 561} \\
(K - 93.8)s + (Ka - 561)
\end{array}
$$

$$K - 93.8 = 0, \quad K = 93.8$$

$$Ka - 561 = 0, \quad a = \frac{561}{93.8} = 5.98 \approx 6.0$$

The asymptote angles for the system are $\pm 60°$ and $180°$, from Table 7-1. The centroid of the asymptotes is at -7.667. The complete root locus showing the breakaway point at -2.5 and the asymptotes is provided in Figure 7-16.

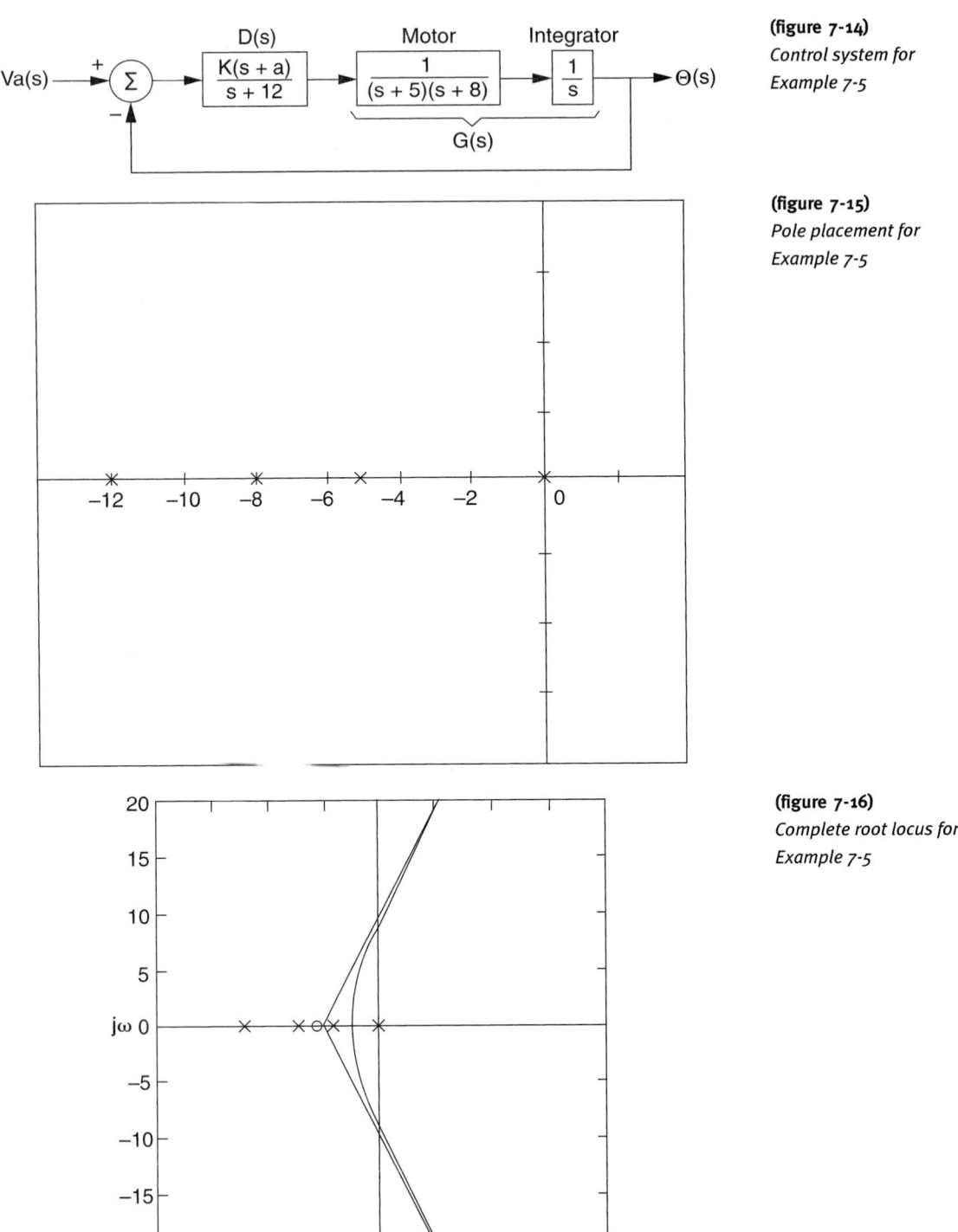

(figure 7-14)
Control system for Example 7-5

(figure 7-15)
Pole placement for Example 7-5

(figure 7-16)
Complete root locus for Example 7-5

The last example further shows the use of the methods.

Example 7-6

Consider the following characteristic equation:

$$s^4 + 11s^3 + 51s^2 + (91 + K)s + (50 + 3K) = 0$$

Determine the gains for asymptotic stability and plot the root locus of the system.

Solution

The required open-loop transfer function is found using the method outlined previously.

$$G(s)H(s) = \frac{K(s + 3)}{(s + 1)(s + 2)(s^2 + 8s + 25)}$$

The poles of the system are at −1, −2, −4 ± j3; and the zero is at −3. A Routh table will be of some use in determining the limits of stable gain and the radian velocity at the jω crossing point. The Routh table is as follows:

$$
\begin{array}{cccc}
s^4 & 1 & 51 & (50 + 3K) \\
s^3 & 11 & (91 + K) & \\
s^2 & \dfrac{470 - K}{11} & (50 + 3K) & \\
s^1 & \dfrac{36720 + 13K - K^2}{470 - K} & & \\
s^0 & (50 + 3K) & &
\end{array}
$$

The gains calculated by factoring the second-order gain term are 198.235 and −187.237. The usable gain is 198.235; the other gain, being negative, is discarded. The radian velocity at a gain of 198.235 is 5.108 rad/sec.

$$\left(\frac{470 - K}{11}\right)s^2 + (50 + 3K) = 0$$

$$\left(\frac{470 - 198.235}{11}\right)s^2 = -(50 + 3 * 198.235)$$

$$s^2 = -\frac{644.7}{24.71} = 26.1$$

$$s = \pm 5.108 \; rad/s$$

There are four poles and one zero; therefore, the asymptotes are ± 60° and 180°. The centroid of asymptotes is

$$\sigma_c = \frac{-(1 + 2 + 4 + 4) - (-3)}{3} = -\frac{8}{3} = -2.667$$

The complete root locus is shown in Figure 7-17.

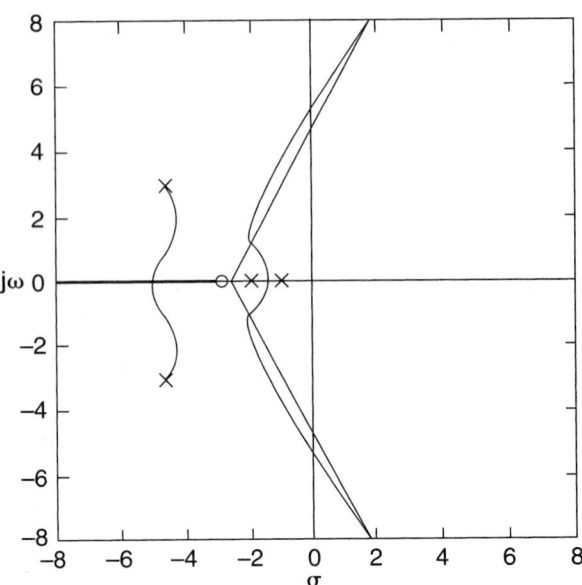

(figure 7-17)

Complete root locus for

Example 7-6

Effect of Adding Poles and Zeros to $G(s)H(s)$

Many compensation schemes require that poles or zeros be added to a system to improve system performance. Therefore, it is important to know the effect of adding poles or zeros on the root locus.

Generally, the effect of adding a pole to $G(s)H(s)$ is to move the root locus toward the right-half plane, due to the additional integration. Adding a left-half plane zero to $G(s)H(s)$, generally, is to move the root locus toward the left-half plane, due to the additional differentiation.

Consider a second-order system with no zeros. The asymptotes are at $\pm 90°$. If a pole is added to the system, the asymptotes change to $\pm 60°$ and $180°$ and the root locus bends to cross the $j\omega$ axis.

Now consider a second-order system with a zero. If the zero is greater than the poles, a break-in point is created and the root locus becomes circular, breaking away between the poles and breaking in between the zeros. If the zero is added between the poles, the root locus lies entirely on the σ axis.

The effects are shown in Figures 7-18a-d.

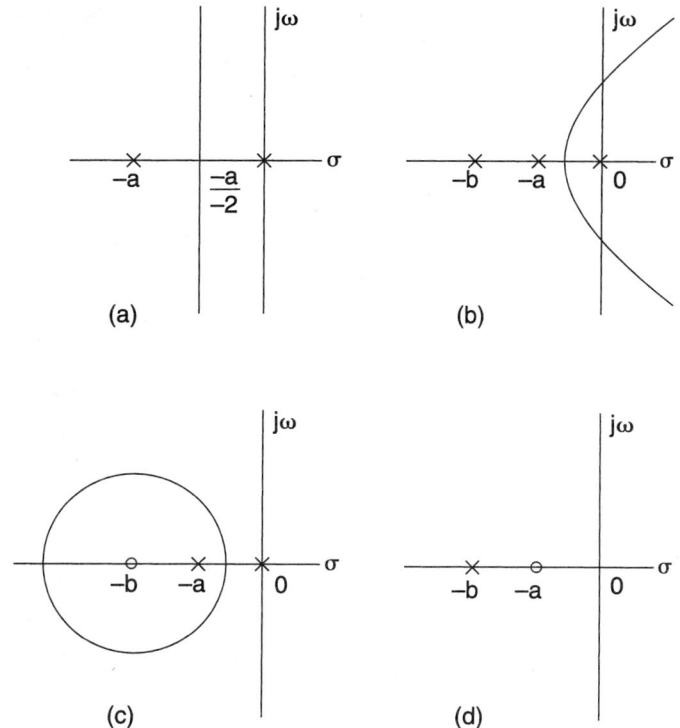

(figure 7-18)

Effects of adding poles and zeros to G(s)H(s)

Breakaway Points Not on the Real Axis

In some cases, breakaway points may occur that are complex if the points satisfy the characteristic equation. Consider the following example.

Example 7-7

Consider a system whose characteristic equation is

$$1 + KG(s)H(s) = 1 + \frac{K}{s(s + 8)(s^2 + 8s + 80)}$$

Determine any breakaway points of that system.

Solution

The roots of the transfer function are 0, −8, and −4 ± j8. The root locus is symmetrical about the point $s = -4$. The breakaway points at −4 ± j4.9 also are points on the root-locus and fulfill the angle criterion; they are breakaway points, as well. The root locus is shown in Figure 7-19.

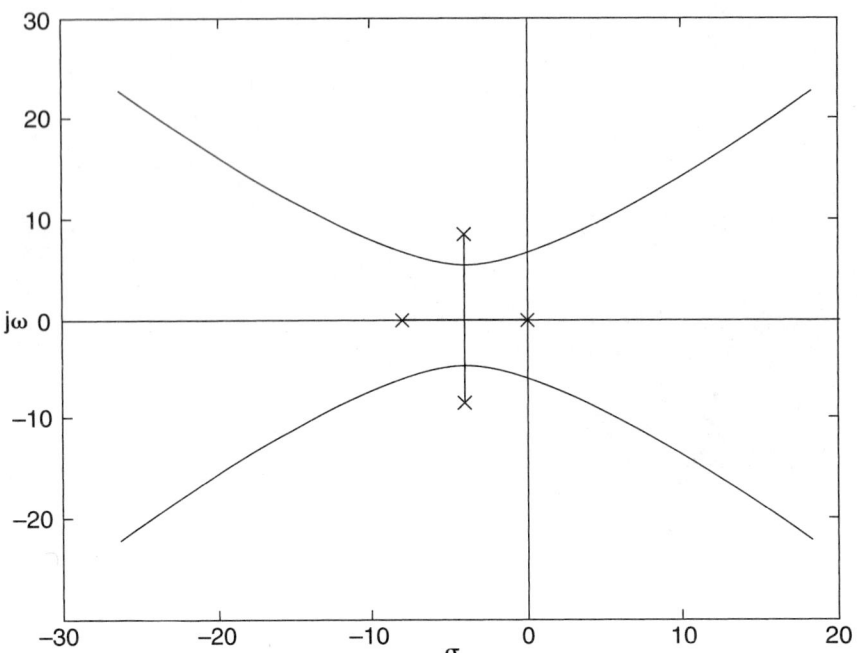

(figure 7-19)

Root locus for Example 7-7

7.7 Complementary Root Locus, *K* ‹ 0

What happens when K is less than zero? A root locus can still be generated for $-\infty < K \leq 0$; it is called the **complementary root locus,** or **CRL.** Some of the basic concepts must be changed to accommodate this type of locus. The changes are made to the angle criterion and to where the segments of the CRL appear on the s-plane.

The angle criterion must be modified such that the angle is an integral multiple of 360°, including 0°. The angles are given by

$$\Psi = \frac{360° \cdot k}{M - N} \qquad (k = 0,1,2,...) \qquad (7.12)$$

Any of the rules that depend on the angle criterion must also be modified to meet this requirement. Any point that is on the root locus must also fulfill these criteria.

The intersection of the asymptotes has a similar formula to that for the root locus.

$$\sigma_{CRL} = \frac{\sum_{i=1}^{M}(-p_i) - \sum_{j=1}^{N}(-z_j)}{M - N} \qquad (7.13)$$

Segments of the CRL on the real axis exist when there is an even number of poles or finite zeros to the right of a selected pole or zero on the σ axis. An example will help clarify the process.

Example 7-8

Consider a system with the open-loop transfer function

$$KG(s) = \frac{K(s - 4)}{s(s + 5)}$$

Examination of the characteristic equation shows the system to be unstable. However, if K is made negative, there is a stable region of gain. Substitution of $K = -1$ yields a double pole at -2; and if the gain is increased to -25, a double pole exists at $+10$. There is, then, a breakaway point at -2, between the pole at 0 and -5, and a break-in point at 10. The root locus for this system is circular, as shown in Figure 7-20.

From the transfer function of the system, you can see that when $K = -5$, the system root locus crosses the $j\omega$ axis at 4.47 rad/sec.

The angle criterion says that the summation of angles from zeros to poles minus the summation of angles from poles to poles must equal 0° or an integral multiple of 360°. In determining if the point $4 \pm j6$ is a point on the root locus, the angle from the zero minus the angles from the poles must equal $0 + k \cdot 360°$. The summation shown in Figure 7-21 clearly meets the criteria.

(figure 7-20)
Complementary root locus for Example 7-8

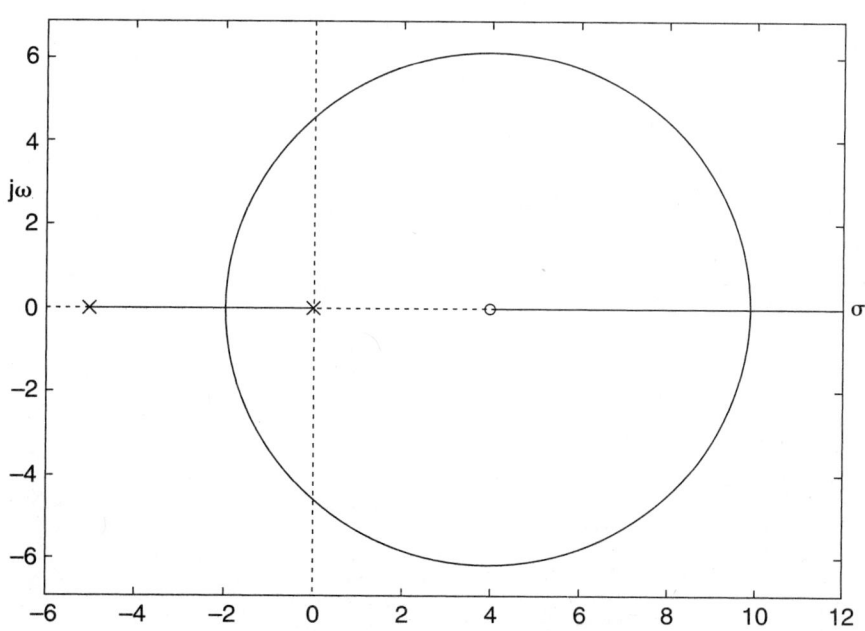

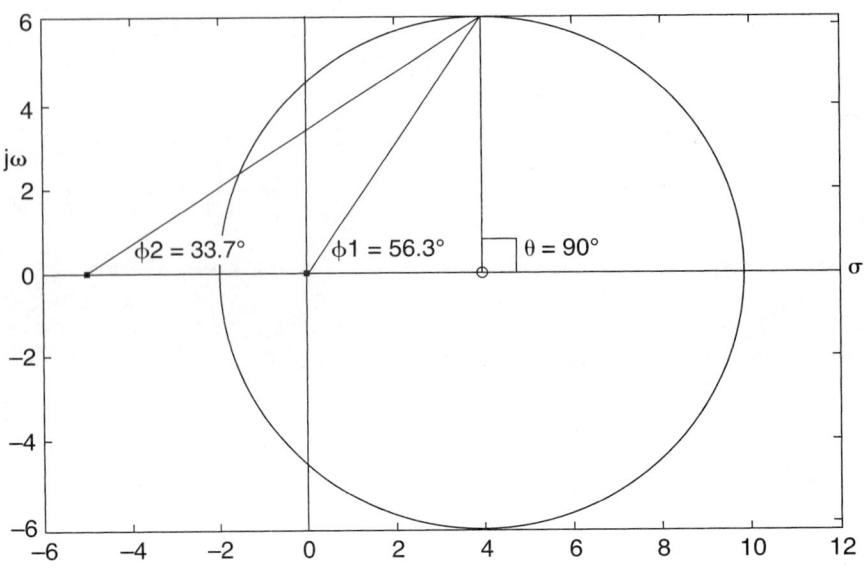

(figure 7-21)
Angle criteria for Example 7-8

7.8 Use of MATLAB for the Root Locus

MATLAB is useful for generating the root loci of systems, but another tool provided by The MathWorks, Inc., is the SISO Design Tool.

Consider the system of Example 7-4. A root locus can be made by typing the open-loop transfer function into **MATLAB** as follows:

sys = tf([1],[1 6 11 6 0]);   % Enter the open-loop transfer function.

rlocus(sys)   % Plot the root locus of the system.

sgrid   % Print lines of constant ζ and ω on the chart.

That produces a simple root locus for the system shown in Figure 7-22.

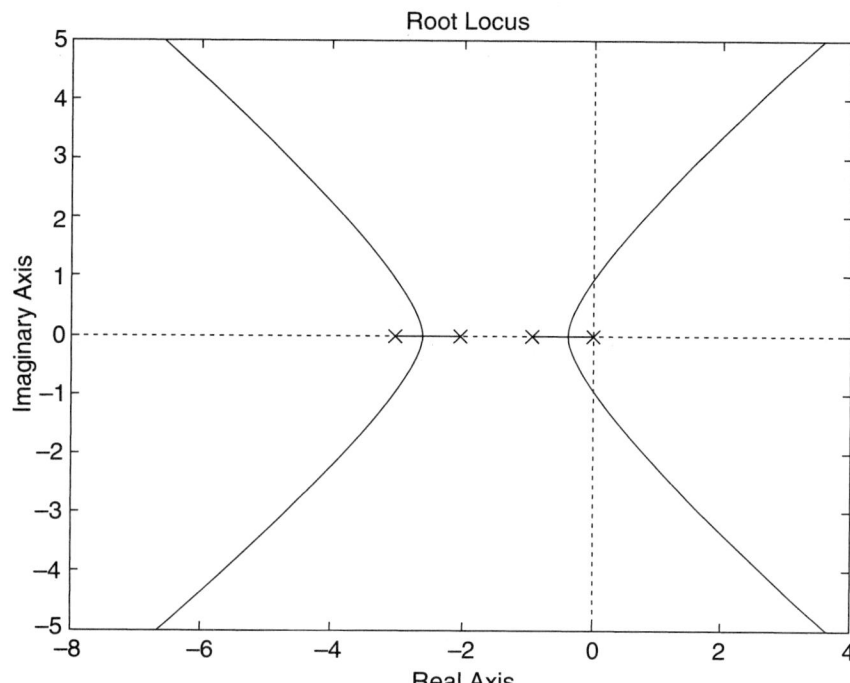

You can determine closed-loop pole positions, gain, damping ratio, and radian frequency by clicking anywhere on the root locus. You also can make the root locus plot display lines of constant ζ and ω by adding a command to the **MATLAB** workspace; the command is ***sgrid***, which places the lines on the grid. The root locus and the grid structure are shown in Figure 7-23.

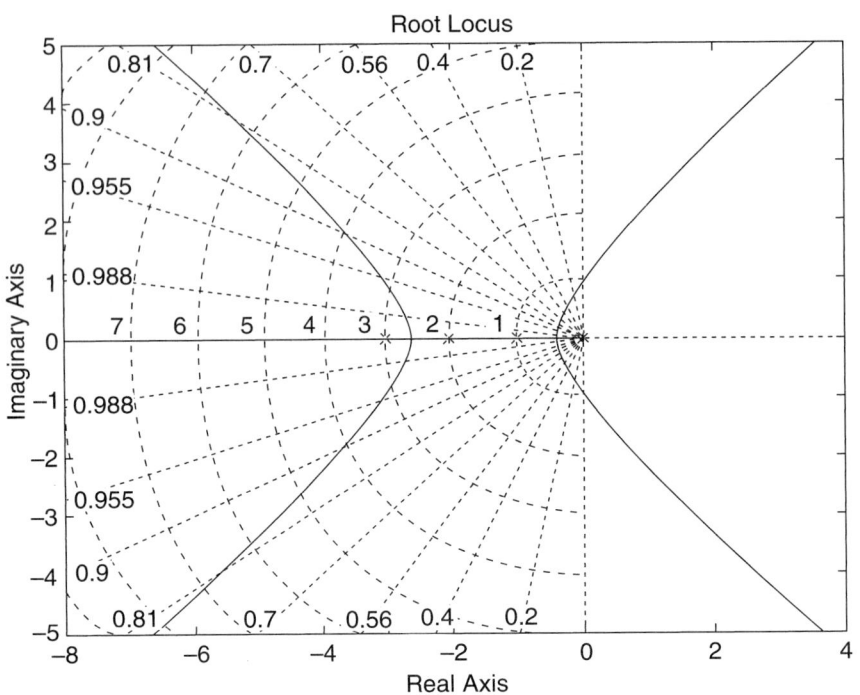

(figure 7-23)

Root locus with grid of constant ζ and ω

Adding the following command will allow you to select any point on the root locus and display the value of the poles and the gain at the poles in the workspace.

```
[k, poles] = rlocfind(sys)        % Requests specific pole information and gain
```

For that example, select the root locus at the $j\omega$ crossing point. You probably will not get the exact crossing point, but the value will be very close. The poles selected are shown as four red crosses on the root-locus plot. The Routh table for this system predicted a gain of 10 for marginal stability; and the selected point bears this out, as shown in Figure 7-24 and the workspace.

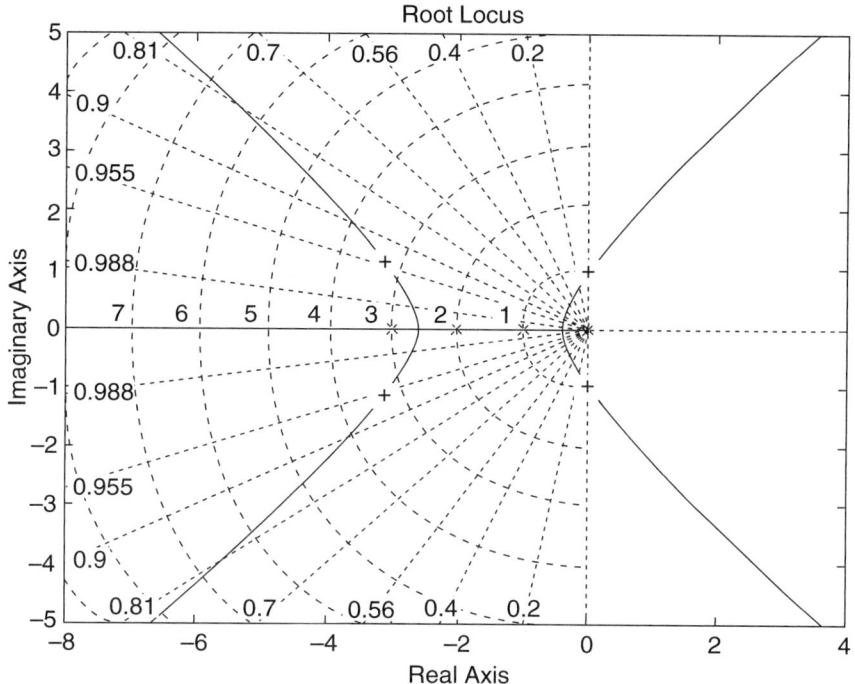

(figure 7-24)

Pole values at $j\omega$ crossing point, using [z, poles] command

You can plot several different values of poles by rerunning the [k, poles] line as many times as needed.

The general root locus is not always proportioned the way you would like it to be to get the information you need. Also, you may be interested in the values of ζ in a specific range. **MATLAB** allows you to do that by specifying the plot parameters. That requires a larger set of instructions, but yields the desired information.

Example 7-9

Consider a control system whose $G(s)$ is

$$G(s) = \frac{K(s + 3)}{s(s + 20)(s^2 + 4s + 8)}$$

The root locus drawn by the simple method above is shown in Figure 7-25.

Example 7-9 (continued)

It is clear that a large amount of the graph is not really necessary and that it reduces the detail in the root locus. There is no important information between −4 and −8 as well as between 1 and 2 on the σ axis. Further, it would be nice to have lines of constant ζ and ω so you can estimate the values of K, ω, and ζ directly from the graph.

To set up the root locus, you must define several things to set the limits of the graph. First, you need to set two limits of gain, $K1$ and $K2$. The reason is that one value may not provide enough calculation points near the breakaway and break-in points. The first gain specifies a small increment to provide enough points for calculation of the breakaway and break-in points. The second gain provides a larger increment of gain after the breakaway or break-in points where fewer points are needed. The general form for the various parameters that will be specified is

> **variable = start value: increment: stop value**

To begin, declare the scale factor, then the gains. Then specify the numerator and denominator of $G(s)$.

```
sf = 1;                          % Scale factor multiplier
K1 = 0: .0025: 10;               % Gain close to breakaway and break-in points
K2 = 11: .05: 100               % Gain far from breakaway and break-in points
K = [K1 K2];                     % Conjoin the gains
Num = [0 0 0 1 3];               % Numerator of G(s)
Den = [1 6 16 16 0];             % Denominator of G(s)
[ r, K] = rlocus(Num, Den, K);   % Set root locus in terms of K
plot(real(r) imag(r), K)         % Plot the root locus
X = [−4*sf 1*sf]; Y = [−3*sf 3*sf];   % Set axis scales
Z = [0 0]; line[X, Z], line[Z, Y]     % Set Z axis to zero, draw axes
Axis([X Y])                      % Plot the axes
Z = .1:.1:.5; W = 1:4;           % Set values of ζ and ω
grid, sgrid(z, w)                % Plot rectangular grid and constant ζ and ω
hold on                          % Hold the plot in place
po = roots(Den);                 % Factor denominator for poles
plot(real(po), imag(po), 'x')    % Mark the poles with an x
ze = roots(Num);                 % Factor denominator for zeros
plot(real(ze), imag(ze), 'o')    % Mark the zeros with an o
hold off                         % Remove the hold
[k, poles] = rlocfind(Num, Den)  % Use cursor to find values of K, ζ and ω
```

The completed root locus with all grid lines is shown in Figure 7-26.

(figure 7-25)

Simple root locus for Example 7-9

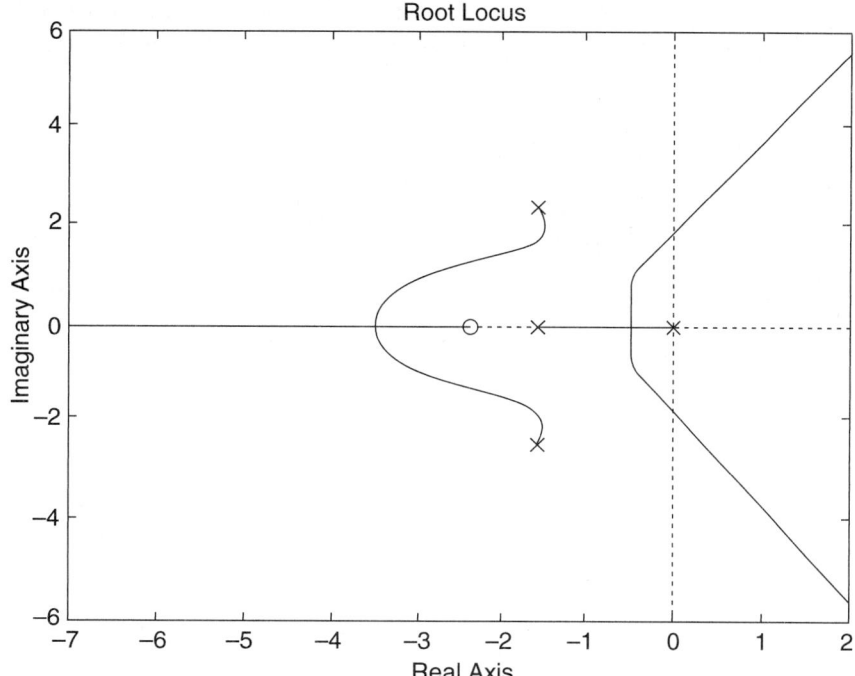

(figure 7-26)

Reconfigured root locus for Example 7-9

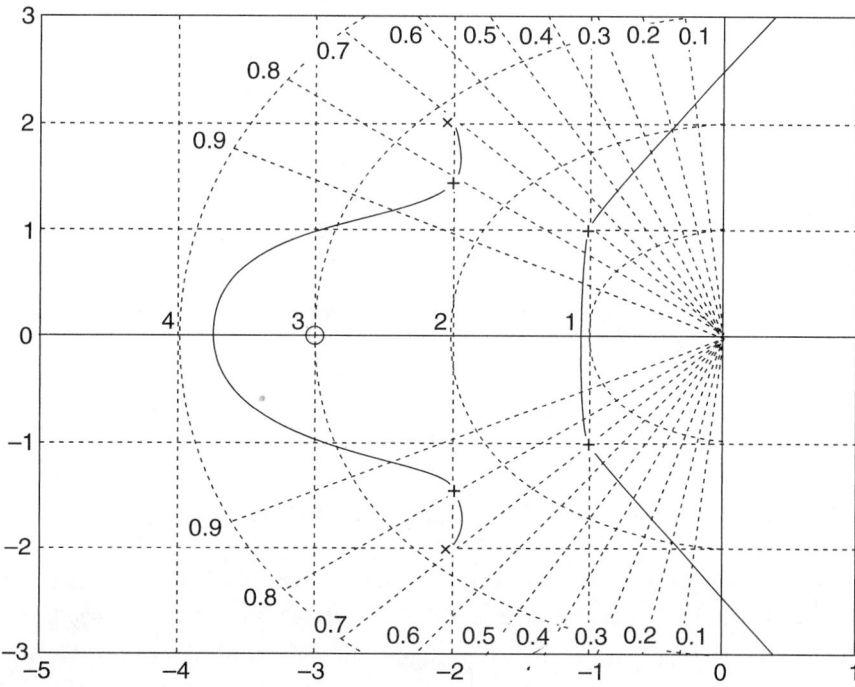

Since this chapter has discussed the CRL, you can use **MATLAB** to produce one. Using Example 7-8, the method is to do a simple root-locus plot in the following manner:

n = −[1 −4]; d = [1 5 0];   % Define the numerator with a negative sign and the denominator in the usual way.

rlocus(n,d)   % Plot the root locus.

sgrid   % Add lines of constant ζ and ω

The root locus is shown in Figure 7-27.

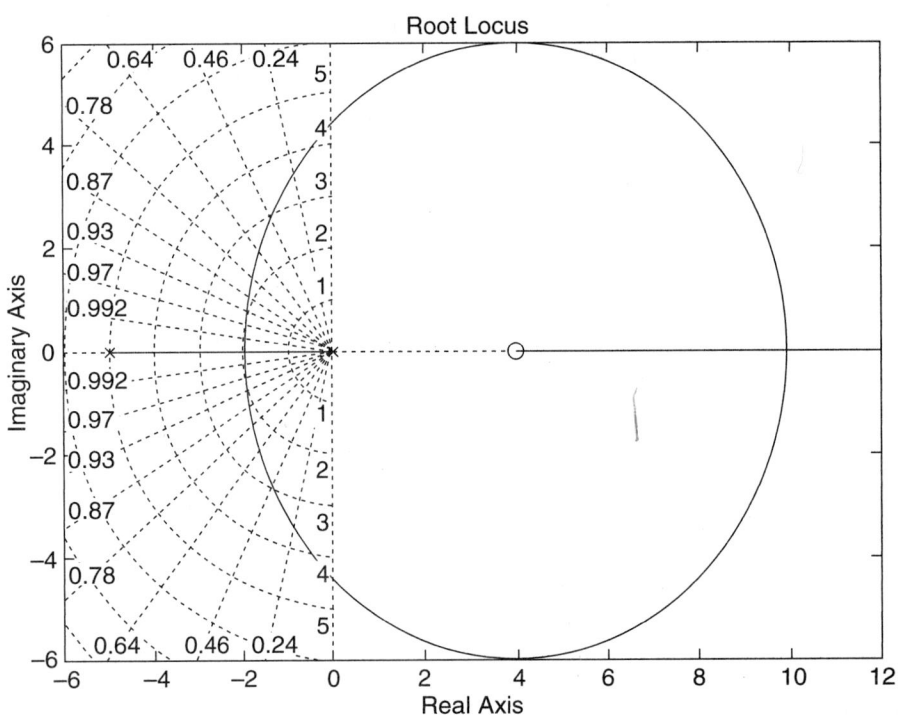

(figure 7-27)
Complementary root locus

Summary

➤ We have shown the concept of how the roots of an open-loop characteristic equation move about in the s-plane as some variable, usually a system gain, is varied.

➤ The variation, usually called the root migration, provides insight into the stability of the system.

➤ The damping ratio of any point on the root-locus plot is determined by the angle of a vector from the origin to the point on the closed-loop graph.

➤ The system gain at a point is determined by the vectors from the poles and zeros to the selected point on the graph.

➤ The departure or arrival of the root locus at an open-loop pole or zero provides insight into whether the system is tending toward stability or instability as the root locus leaves the open-loop pole or zero.

➤ The root locus tends asymptotically toward lines at angles to the real axis in the s-plane.

➤ The concept of breakaway and break-in points is defined as the point at which the closed-loop root-locus departs from the real axis on a segment of the root locus.

➤ Breakaway points are not constrained to the real axis.

➤ The Routh table has been integrated with the root locus to determine $j\omega$ crossing points, and gain at the $j\omega$ crossing point on the closed-loop root locus.

➤ Negative gain is usually not allowed; however, the complementary root locus may predict stable operation for some values of negative gain.

➤ It is necessary to modify the rules of construction for the angle criterion for the complementary root-locus

➤ **MATLAB** can be used to determine the root locus in simple form.

➤ Methods are available in **MATLAB** to produce a root locus to one's own specifications.

➤ Lines of constant ζ and ω can be inserted as part of the **MATLAB** script.

➤ The use of a cursor command is used to determine the approximate gain, damping ratio, and system radian velocity.

➤ While it is a powerful tool in itself, the root-locus method combined with the Routh table does not give information about the frequency response of the system.

1.   W. R. Evans, "Graphical Analysis of Control Systems," Trans. AIEE, Vol. 67, pp, 547–551, 1948.

Problems

P7.1   For the following open-loop transfer functions, find the angle of departure or arrival from or to the designated pole or zero.

a. $G(s) = \dfrac{K(s + 3)}{(s + 2)(s^2 + 9)}$

   for $K > 0$, pole at $s = j3$

b. $G(s) = \dfrac{K(s^2 + 4s + 8)}{s(s^3 + 7s^2 + 10s)}$

   for $K > 0$, zero at $s = -2 + j2$

c. $G(s) = \dfrac{K}{s^2(s^2 + 6s + 13)}$

   for $K > 0$, pole at $s = -3 + j2$

d. $G(s) = \dfrac{K(s^2 + 8s + 20)}{s(s+2)(s+4)}$

   for $K > 0$, zero at $-4 + j2$

P7.2 For the transfer functions in Problem P7.1, provide the breakaway and break-in points, the asymptote angles, and the centroid of asymptotes if they exist.

P7.3 Use each of the following characteristic equations of linear systems to construct the root locus.

a. $s^3 + 6s^2 + 11s + 2K = 0$

b. $s^4 + 5s^3 + (5 + K)s^2 + (3 + 2K)s + 3K = 0$

c. $s^3 + (10 + K)s^2 + (K - 5)s + 5K = 0$

d. $s^4 + (2 + K)s^3 + (15 + 2K)s^2 + 10Ks + 20 = 0$

P7.4 For the following open-loop transfer functions, determine the breakaway and break-in points, asymptotes, and $j\omega$ crossing points. Show the gain at the crossing points and the value of K that produces a damping ratio of 0.707 (if applicable), using the dominant second-order expression. Assume unity gain systems.

a. $G(s) = \dfrac{K(s + 4)}{s(s + 2)}$

b. $G(s) = \dfrac{K(s + 8)}{s(s + 4)(s + 2)}$

P7.5 For the following unity feedback systems, use the angle criteria to determine if the points assigned are on the root locus.

a. $G(s) = \dfrac{K}{s(s + 4)(s + 6)}$

   $(K > 0)$, $s = -1 + j2.63$

b. $G(s) = \dfrac{K(s + 8)}{s(s - 2)(s + 5)}$

   $(K < 0)$, $s = -3.75 + j1$

c. $G(s) = \dfrac{K(s - 2)}{s^2(s^2 + 8s + 12)}$

   $(K > 0)$, $s = 2 + j9$

d. $G(s) = \dfrac{K(s^2 + 4s + 13)}{s^3}$

   $(K > 0)$, $s = -1 - j2$

P7.6 For the system shown, the following controller functions have been suggested. Using the root locus, determine which system produces the best transient response. Explain your reasoning.

a. $G_c(s) = K$

b. $G_c(s) = \dfrac{K(s + 2)}{(s - 1)}$

c. $G_c(s) = \dfrac{K(s + 1)}{(s + 6)}$

(figure P7-6)

Control system for Problem P7.6

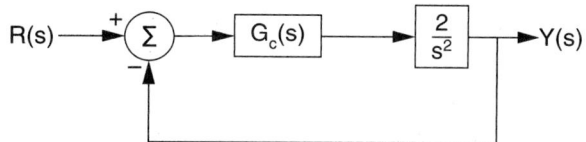

P7.7 The block diagram of a control system with tachometer feedback is shown in Figure P7-7.

a. Construct the root locus for $K = 10$, for $0 < K_t < \infty$.

b. Construct the root locus for $K_t = 5$ and $0 < K < \infty$.

(figure P7-7)

Control system for Problem P7.7

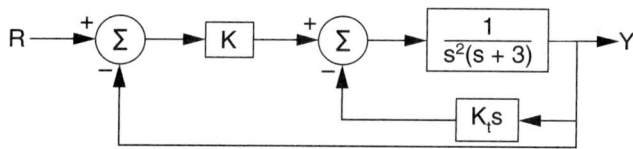

P7.8 Consider the following loop transfer function of a control system.

$$G(s)H(s) = \frac{K(s^2 + 6s + 13)}{s^2 + 9s + 20}$$

a. Plot the root locus for $0 \le K \le \infty$.

b. Plot the complementary root locus for $-\infty \le K \le 0$.

P7.9 A PM servomotor and its load has the following specifications:
$J = (J_M + J_L) = 0.0002$, $B = (B_M + B_I) = 0.025$, $L_a = 0.005$,
$R_a = 5$, $K_T = 7.5$, and $K_b = 0.01$. The characteristic equation is

$$s^3 + \left(\frac{R_a}{L_a} + \frac{B}{J}\right)s^2 + \left(\frac{KK_T + K_bK_T + R_aB}{L_aJ}\right)s + \frac{400KK_T}{L_aJ} = 0$$

Construct the root locus for $0 < K < \infty$.

P7.10   For the following loop transfer function, determine the asymptote angles, the centroid of asymptotes, the breakaway points, and the angle of departure from the poles. Then plot the root locus for $0 < K < \infty$.

$$G(s)H(s) = \frac{K(s + 1)}{s(s + 4)(s^2 + 4s + 13)}$$

P7.11   A unity gain feedback control system has the following forward path transfer function:

$$G(s) = \frac{K(s + \alpha)}{s(s + 1)(s^2 + 4s + 13)} \qquad \alpha > 1$$

Use the root locus to determine which value of α will allow a ζ of 0.7 to be attained for a = 1.5, 2, 2.5, 3.

P7.12   For the pole-zero map shown below, use the angle criteria to determine the angle of departure at the poles $-4 \pm j2$.

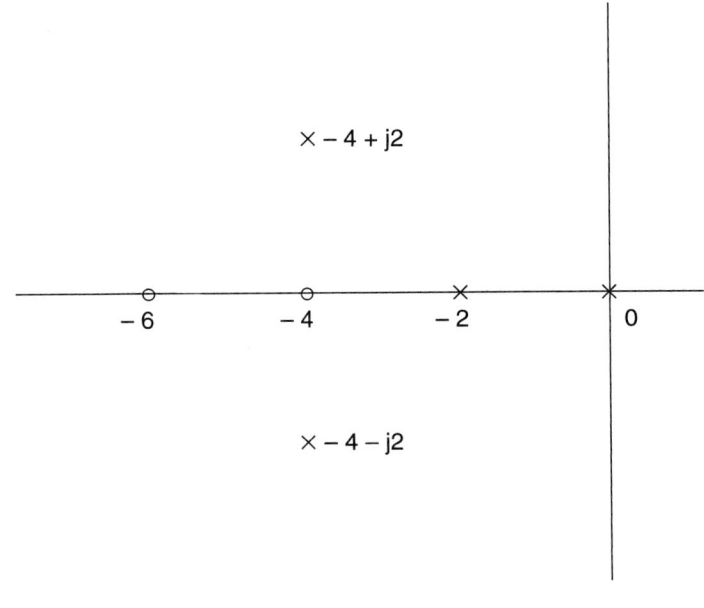

(figure P7-12)
Pole-zero map for Problem P7.12

P7.13   For the following forward path gain of a system, determine values for the second-order expression that places poles at $-2 \pm j3$. Also determine the asymptote angles and the centroid of asymptotes. Construct the root locus and show the asymptotes.

$$G(s) = \frac{10K}{s(s + 2)(s^2 + 2\zeta\omega_n s + \omega_n^2)}$$

P7.14   Repeat Problem P7.13, adding a zero at −1, −2, −3, and −4. Observe the effect on the root locus of adding a zero to the system.

P7.15   For the following forward path transfer functions, determine the value of K at all breakaway and break-in points.

a. $G(s) = \dfrac{K(s + 5)}{s(s + 2)(s + 3)}$

b. $G(s) = \dfrac{K(s^2 + 5s + 6)}{s(s + 1)(s + 4)(s + 5)}$

P7.16   Given the following forward transfer function, plot the root locus and determine the minimum value of ζ attainable with this system.

$$G(s) = \frac{K(s + 2)}{s(s + 1)}$$

P7.17   Given the following forward transfer function, construct the root locus. Determine any breakaway points, the gain at the breakaway points, and the angle of departure from the complex roots.

$$G(s) = \frac{5K}{s(s^2 + 8s + 20)}$$

P7.18   Consider the system shown below. Construct the root locus of the system and determine a value for k such that the damping ratio of the system is 0.7, using the fact that $\cos\theta = \zeta$ for the criteria. (Hint: k is a nonmultiplying factor. Let $8k = K$.)

(figure P7-18)
Control system for Problem P7.18

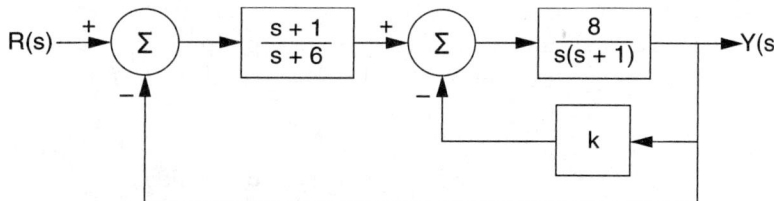

P7.19   For the following forward path transfer function, construct the root locus. Calculate the asymptotes and any breakaway or break-in points.

$$G(s) = \frac{K}{s(s + 2)(s^2 + 8s + 32)}$$

P7.20   Consider the following all-pole open-loop transfer function.

$$G(s) = \frac{K}{s(s + 1)(s^2 + 4s + 13)}$$

Construct the root locus. Determine the asymptote angles, any breakaway/break-in points, and the angle of departure from the complex poles. Show all calculated points on the root-locus plot.

MATLAB Problems

M7.1   For Problem P7.2, plot the root locus using **MATLAB**. Using the cursor, show that the breakaway/break-in points calculated are true. Plot the asymptotes and show that the root locus approaches the asymptotes.

M7.2   For Problem P7.18, use simulink to model the system and provide the step response for the value of k you calculated. Then provide a root locus with the value of k you found in Problem P7.18.

M7.3   For the system shown below, use **MATLAB** to determine a value for maximum value of K_i that allows a damping ratio of 0.7 to be attained. Use a standard grid obtained by right-clicking in the root-locus window.

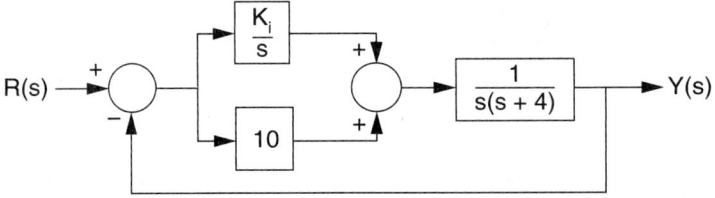

(figure M7-3)
Control system for Problem M7.3

M7.4   For Problem P7.9, use **MATLAB** to generate the root locus. Determine the gain required to produce a damping ratio of 0.707. Then generate a step response for that gain and measure the overshoot produced. Comment on the difference in overshoot produced by the step response at the gain determined on the root locus.

M7.5   For a system whose poles and zeros are located as follows:

Poles at $s_1 = 0$, $s_2 = -1$ $s_3 = -2$ $s_{4.5} = -3 \pm j3$

Zeros at $K(s + -3, \ -4, \ -5, \ -10)$

Generate the step response of the system. Also delineate the role of zero in determining the response of the system to peak response and the system rise-time and settling time.

Chapter 8

Frequency Response Analysis

Chapter Objectives

After completing this chapter, you should be able to:

➤ State the concepts of frequency response and its importance in determining control system stability.

➤ Convert system poles and zeros of an open-loop transfer function to the frequency domain.

➤ Calculate the gain and phase contributions of poles and zeros to the system frequency response.

➤ State the differences between Bode's analysis and the Nyquist, Polar, and Nichols methods for frequency response of a system.

➤ Calculate the gain and phase margins of a system.

➤ State the effect of gain and phase margin on the stability of a system.

➤ Calculate the effect of dead time on system response.

➤ State the effect of lead, lag, and lead-lag compensation of a system.

➤ Show how the different compensation schemes promote stability and speed in a system's response to an input.

➤ Apply Nyquist's stability criterion to control systems.

➤ Apply the Nichols method for relative stability to systems.

➤ Use **MATLAB** and the SISO Design Tool to design and modify a system to meet stability criteria.

8.1 Introduction

All systems respond to periodic signals. A good deal of information about a system is available when step, ramp, or other signals are applied. But when a periodic signal whose frequency can be changed is applied, the effect of the signal is not immediately known, nor is how the system will react as frequency is changed from lower to higher values.

All of the systems you have studied thus far have, as a given, negative feedback. The assumption is that there is no change in the phase of the applied signals. If a periodic signal is applied to a system and the frequency is increased, the phase of the feedback signal will change because of system parameters such as capacitance in the components. If the phase change reaches 180°, the signal at the inverting input of the summing junction will be in phase with the applied signal, resulting in oscillation. For that reason, it is important to analyze the system for its frequency and phase response to periodic signals.

8.2 Bode's Analysis

Hendrik W. Bode (1905–1982) contributed greatly to the concepts of closed- and open-loop stability of networks by developing the methods discussed and used here. While the mathematics of systems were well known, he developed the use of frequency response methods by using the gain and phase margins of the system to predict stability.

In the control of systems, you will use negative feedback. The word *negative* implies that there is a 180° shift in the phase of the returned comparison signal. This inversion of the signal occurs at the summing junction of the control system. As long as the signal remains inverted, the system is stable and will operate properly. It also implies that the open-loop transfer function is stable.

If a second inversion of 180° were to occur to the comparison signal, the comparison would become in phase with the input control signal and the system would oscillate at a predetermined frequency, increasing the output without limit. The second inversion can be caused by active or passive components of the system.

All of that implies that there is a specific frequency response of any system. Simply put, frequency response is the response of the system to signals that vary their frequency with time at a specific rate. As the rate of the signal is increased, additional phase shifts occur and eventually may equal an additional 180° phase shift.

For oscillation to occur in any system, two conditions must exist simultaneously:

1. The open-loop gain must be equal to or greater than unity at the oscillation frequency.
2. The 180° of additional phase shift must occur at or before the unity gain frequency.

s-Plane Relationships to Bode's Analysis

The value of s for any point in the s-plane is $s = \sigma + j\omega$

For Bode's analysis, the value of σ is set equal to zero. Then $s = j\omega$ and $j\omega$ replaces s in all parts of the transfer function being investigated, and the s-plane collapses to a single line.

Parameters Used in Bode's Analysis

There are five parameters that must be considered when doing Bode's analysis:

1. Crossover frequency (ω_c) is the frequency at which the open-loop gain of the system equals unity (1).
2. Bandwidth (BW) is the frequency at which the system response is three dB less than the midband gain.
3. Phase crossover frequency (ω_π) is the frequency at which an additional $180°$ of phase shift occurs in the system.
4. Gain margin (GM) is the gain in dB between ω_c and ω_π.
5. Phase margin (PM) is the angle in degrees between ω_c and ω_π.

Parameters 4 and 5 are system design specifications.

The dB magnitude of gain is calculated from

$$dB = 20log(magnitude) \tag{8.1}$$

The concepts of the Bode plot are as follows:

1. Bode plots are straight-line approximations of frequency response curves.
2. Bode plots are considered to have useful output up to a frequency equal to ω_c.
3. Bode plots decrease or increase by ±20 dB/decade for each order of the system. (First order = ±20 dB/decade, second order = ±40 dB/decade, and so on.)
4. The phase shift associated with a Bode plot is $90°$ per order of the system. (First order = $90°$, second order = $180°$, and so on.)
5. $j\omega$ represents poles or zeros of the system and may occur at any frequency.
6. The origin of the phase/frequency plot is not zero.
7. Any gain associated with a pole or zero moves the curve of frequency response up or down according to the value of the gain.
8. Frequency response plots always have dB as the y-axis coordinates, and the x-axis is always logarithmic.
9. Frequencies where the response curve changes direction are called *corner frequencies*.
10. There is always a small error in dB at the corner frequencies.

Poles and Zeros of Frequency Response Plots

Poles and zeros of frequency response plots are derived from the poles and zeros in the s-plane. When poles and zeros are converted to the frequency plane, they have a magnitude and phase angle in the frequency plane, where in the s-plane they have real and imaginary components. A table of magnitudes and phase of poles and zeros is shown in Table 8-1.

Function	Comment
$\dfrac{1}{s^k} = \dfrac{1}{(j\omega)^k}$, $k = 1, 2, \ldots$	Single or multiple poles at the origin. (The response is similar to a single capacitor and a frequency generator when $k = 1$.)
$s^k = (j\omega)^k$, $k = 1, 2, \ldots$	Single or multiple poles at the origin. (The response is similar to a single inductor and frequency generator when $k = 1$.)
$\dfrac{1}{\left(1 + \dfrac{s}{a}\right)} = \dfrac{1}{\left(1 + \dfrac{j\omega}{a}\right)}$	A single pole at $-a$. (Similar to a first-order RC low-pass filter.)
$\left(1 + \dfrac{s}{a}\right) = \left(1 + \dfrac{j\omega}{a}\right)$	A single zero at $-a$. (Similar to a first-order RC high-pass filter.)

(table 8-1)

Pole-zero representations in the frequency domain

Those functions are useful when the transfer function of the system can be factored into real poles and zeros. The case for complex roots will be discussed shortly.

Bode plots of the magnitude and phase of the four conditions above are shown in Figures 8-1a and b.

Notice that the single pole and zero of 8-1a and b start at origin a, at 0 dB, and decrease or increase at a rate of $(\pm 20 \text{ dB/dec}) \cdot k$ while contributing a constant $\pm k \cdot 90°$ of phase shift.

The single pole and zero, which start at a rad/sec, continue at 0 dB up to a frequency of $10a$ rad/sec and then decrease or increase at a rate of $(\pm 20 \text{ dB/decade}) \cdot k$. In both cases, the phase response is $90°$ for each order of the transfer function. Figures 8-1a and b start at a rad/sec; but if the frequency is reduced from that value ($0.1a$, $0.01a$, and so on), the outputs increase or decrease by $\pm 20 \cdot k$ dB/decade accordingly.

In the case of the single pole or zero at a, the magnitude response is considered to be flat (0 dB) up to the frequency $10a$, which is the bandwidth (-3 dB). During this interval, the phase response will change from $0°$ at a to $\pm 45°$ at $10a$. At a value of $100a$, the phase increases to $90°$. For a first-order system, the phase will continue at $\pm 90°$ to $\omega = \infty$. The amplitude response continues at the rate of ± 20 dB/decade as frequency increases beyond $10a$.

(figure 8-1a)

Frequency response of poles and zeros

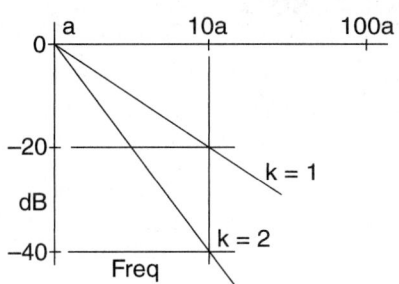

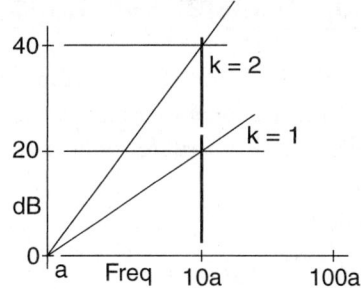

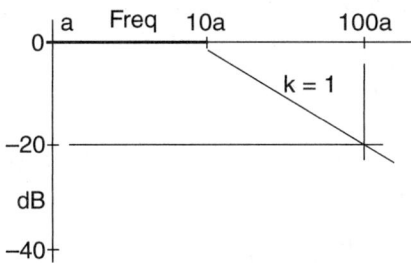

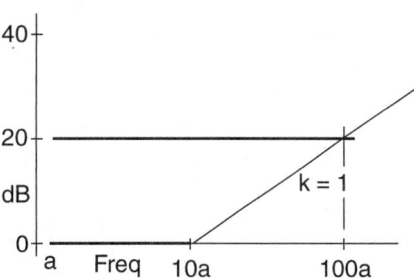

(figure 8-1b)

Phase response of poles and zeros

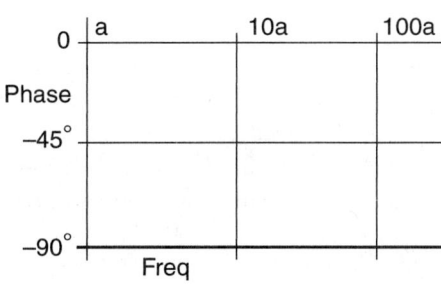

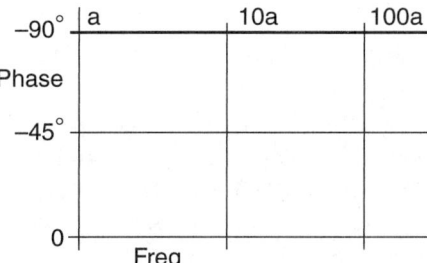

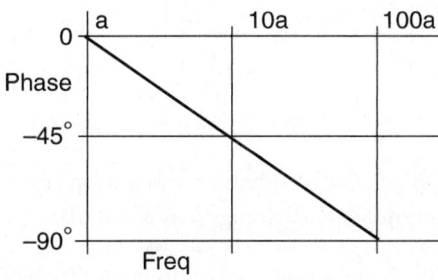

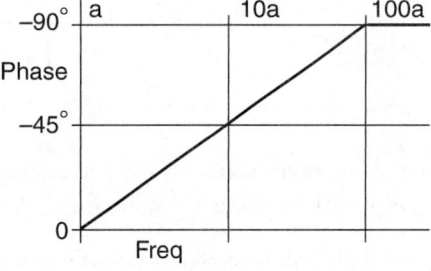

8.3 Loop Transfer Functions for Bode plots

Generally, the loop transfer function will be the open-loop transfer function transformed to the frequency domain, but it can be the closed-loop transfer function also. There is a large difference between the two, and those differences will become apparent.

First-Order Responses

Before Bode's analysis can be used, the open-loop transfer function must be in the right form. Recall that the open-loop transfer function in the s-plane has the form

$$KG(s)H(s) = 0$$

That transfer function can be written in the form of a factored polynomial whose form is

$$\frac{K(s + z_1)(s + z_2)\cdots(s + z_m)}{(s + p_1)(s + p_2)\cdots(s + p_n)} \tag{8.2}$$

Allowing s to equal $j\omega$, the system reduces to

$$\frac{K(j\omega + z_1)(j\omega + z_2)\cdots(j\omega + z_m)}{(j\omega + p_1)(j\omega + p_2)\cdots(j\omega + p_n)} \tag{8.3}$$

At the very least, this form is inconvenient in that the relationship of the cutoff frequency to the actual frequency is not clear. Further, what about K? The factors $j\omega$ are removed from each factor and are used to modify K. In all cases, the value of K is modified by multiplying by z_m and dividing by p_n. When that is done, K becomes a new value, which is called K_B, or the Bode gain of the system. The new form of the system is

$$KG(j\omega)H(j\omega) = \frac{K_B\left(1 + j\dfrac{\omega}{z_1}\right)\left(1 + j\dfrac{\omega}{z_2}\right)\cdots\left(1 + j\dfrac{\omega}{z_m}\right)}{\left(1 + j\dfrac{\omega}{p_1}\right)\left(1 + j\dfrac{\omega}{p_2}\right)\cdots\left(1 + j\dfrac{\omega}{p_n}\right)} \tag{8.4}$$

For the examples given above, the reference line is 0 dB. When an open-loop transfer function is changed to the Bode form, the reference line is changed to K_B in dB.

An example will help clarify how the transformation and the Bode plot are constructed.

Example 8-1

The loop transfer function of a control system is

$$KG(s)H(s) = \frac{60(s + 1)}{(s + 2)(s + 3)}$$

Construct the Bode plot for the function and indicate all corner frequencies and rates of attenuation.

Solution

For the transfer function, the constants multiplying and dividing K are the corner (−3 dB) frequencies for each pole and zero. First, convert the transfer function into the proper form for Bode's analysis. The procedure is as follows:

$$(s + 1) = (j\omega + 1) = \left(1 + j\omega\right)$$

$$(s + 2) = (j\omega + 2) = 2\left(1 + \frac{j\omega}{2}\right)$$

$$(s + 3) = (j\omega + 3) = 3\left(1 + \frac{j\omega}{3}\right)$$

Rewrite the transfer function, using the constants to multiply and divide K.

$$KG(j\omega)H(j\omega) = \frac{60\left(1 + j\omega\right)}{2\left(1 + \frac{j\omega}{2}\right) \cdot 3\left(1 + \frac{j\omega}{3}\right)} = \frac{60 \cdot 1}{2 \cdot 3} \cdot \frac{\left(1 + j\omega\right)}{\left(1 + \frac{j\omega}{2}\right)\left(1 + \frac{j\omega}{3}\right)}$$

$$= \frac{10 \cdot \left(1 + j\omega\right)}{\left(1 + \frac{j\omega}{2}\right)\left(1 + \frac{j\omega}{3}\right)}$$

Since $K_B = 10$, the value of K_B in dB is

$$K_{B(dB)} = 20 \cdot log\left(10\right) = 20 \ dB \tag{8.5}$$

The new reference for the Bode plot is +20 dB.

The corner frequencies for the Bode plot are the denominator values of the factors of the transfer function. The zero starts at 0.1 rad/sec and is flat to 1.0 rad/sec. It then increases at a rate of +20 dB/decade from this frequency. The two poles are flat up to 2 and 3 rad/sec and then decrease at a rate of −20 dB for each corner frequency.

To determine the limits of the Bode plot, use the lowest corner frequency divided by 10. Then use the highest corner frequency times a minimum of 10, where the phase shift is ±90°. For this plot, the lowest corner frequency is 1 rad/sec. Thus, the plot should start at 0.1 rad/sec. This is fortuitous since 1 rad/sec is the start of a cycle of the logarithmic frequency axis of the plot. Usually, that is not the case. Use the start of the cycle for the lowest corner frequency.

The highest corner frequency is 3 rad/sec, where the phase angle is −45° and the pole has a −90° phase shift at 30 rad/sec. Since the plot extends to 30 rad/sec, you should plot the response to at least 100 rad/sec.

Example 8-1 (continued)

Examine the maximum phase shift to see whether the system is capable of oscillation. If it is, then gain and phase margins must be determined for the system. For each zero and pole, the maximum phase shift is ±90° Thus, you have +90° for the zero and −180° for the two poles. The total phase shift is the sum of all pole and zero phase shifts in the system. The total phase shift for this system is −90°.

$$\phi = 90° - 180° = -90°$$

This indicates that the system will not oscillate, so there is no need to concern yourself with a gain or phase margin for the system. Plot the responses as described in the following section.

Amplitude Response

From 0.1 rad/sec to 1.0 rad/sec, the amplitude of the zero is constant at +20 dB. At 1.0 rad/sec, the response of the system starts to rise at a rate of +20 dB/decade (+6dB/octave).

At 2.0 rad/sec, the pole at 2.0 rad/sec begins to decrease at a rate of −20 dB/decade. The two rates cancel each other since they are equal and opposite in phase. The plot continues at +26 dB up to 3.0 rad/sec.

At 3.0 rad/sec, the second pole starts to decrease at a rate of −20 dB/decade and continues at a rate of −20 dB/decade. The first zero and pole cancel each other's response. Figure 8-2 shows the Bode response of this system.

Phase Response

The phase response is not a straight line at 0 dB from *a* to 10*a*. The phase response immediately starts decreasing or increasing at the rate of 45° per decade. At the corner frequency for the pole, the phase is always ±45°. At ten times the corner frequency, the phase shift is ±90°. The phase curve is constructed using those values for each pole and zero.

The phase of the zero at $\omega = 1.0$ rad/sec starts to increase from 0° at 0.1 rad/sec to ±45° at 1 rad/sec, then to +90° at 10 rad/sec. The phase of the pole at $j\omega = 2.0$ rad/sec starts to decrease from 0° at 0.2 rad/sec to −45° at 2.0 rad/sec, then to −90° at 20 rad/sec, canceling the phase contribution of the zero. The phase of the pole at $\omega = 3.0$ rad/sec starts to decrease from 0° at 0.3 rad/sec to −45° at 3.0 rad/sec, then to −90° at 30 rad/sec. The phase response is shown in Figure 8-3.

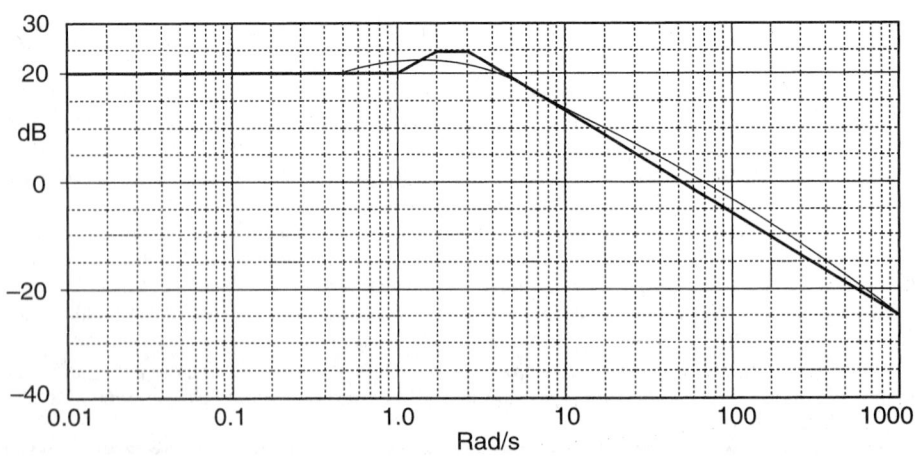

(figure 8-2)

Frequency response of Example 8-1 (logarithmic x-axis)

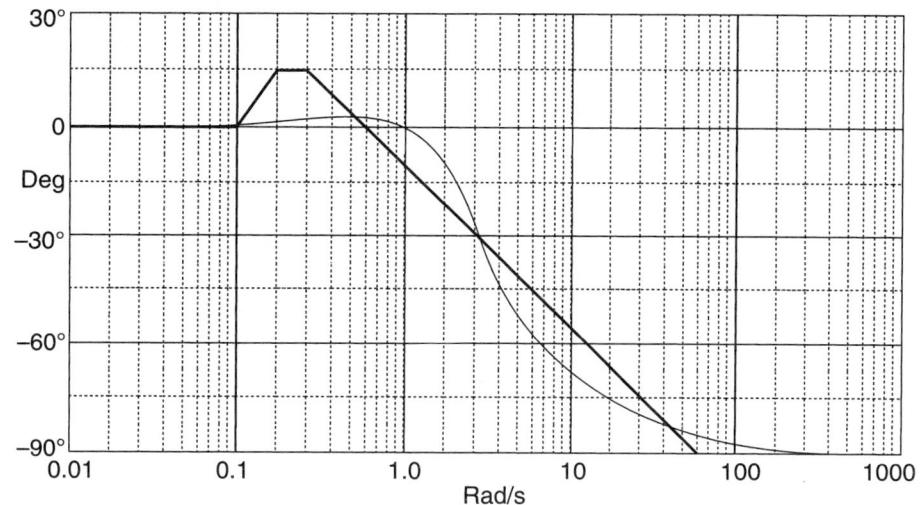

(figure 8-3)

Phase response of Example 8-1 (logarithmic x-axis)

8.4 Second-Order Frequency Response

The second-order system with complex roots presents a different requirement for analysis. The same fundamentals apply, but the system is not factored into its roots. The basic system is

$$D(s)G(s)H(s) = \frac{K\omega_n^2}{s^2 + 2\zeta\omega_n s + \omega_n^2} \tag{8.6}$$

First, replace all s factors with $j\omega$.

$$G(j\omega)H(j\omega) = \frac{K\omega_n^2}{(j\omega)^2 + j2\zeta\omega_n + \omega_n^2}$$

Working only with the denominator, carry out the indicated operations.

$$-\omega^2 + j2\zeta\omega \cdot \omega_n + \omega_n^2 = \omega_n^2 + j2\zeta\omega \cdot \omega_n - \omega^2$$

Finally, dividing all terms in the numerator and denominator by ω_n^2, the function becomes

$$G(j\omega)H(j\omega) = \frac{K}{1 + j2\zeta\left(\dfrac{\omega}{\omega_n}\right) - \left(\dfrac{\omega}{\omega_n}\right)^2}$$

$$G(j\omega)H(j\omega) = \frac{K}{\left[1 - \left(\dfrac{\omega}{\omega_n}\right)^2\right] + j2\zeta\left(\dfrac{\omega}{\omega_n}\right)} \tag{8.7}$$

Notice that ζ is a key player in the equation. Its value has a profound effect on the frequency response of the system. The smaller ζ is, the higher the peak is in the frequency response near the cutoff frequency. For damping ratios less than about 0.2, a marked tendency toward oscillation is noted in the step response and the frequency of oscillation is approximately the same as the peak frequency.

Several parameters of interest can be determined from the magnitude of the transfer function. If you denote the magnitude as M divided by K and then take the inverse of the magnitude, you can easily determine the quantities needed, such as the frequency of the peak and the minimum damping ratio to produce a noticeable peak. The inverse of the magnitude is

$$\left(\frac{K}{M}\right)^2 = \left[1 - \left(\frac{\omega}{\omega_n}\right)^2\right]^2 + \left[2\zeta\left(\frac{\omega}{\omega_n}\right)\right]^2$$

If you take the derivative of the inverse of the magnitude and look for a minimum, the minimum occurs when

$$\left(\frac{\omega}{\omega_n}\right)^2 = 1 - 2\zeta^2$$

From that, the peak is determined by setting $\omega = \omega_{pk}$ and

$$\omega_{pk} = \omega_n \sqrt{1 - 2\zeta^2} \qquad (\zeta < 0.707)$$

If you substitute ω_{pk} into the expression for gain, the maximum amplitude of the peak is

$$\frac{M_{pk}}{K} = \frac{1}{2\zeta\sqrt{1 - \zeta^2}}$$

And the dB rise of the peak over the DC gain is calculated from

$$20 \cdot \log\left(\frac{M_{pk}}{K}\right) = 20 \cdot \log\left(\frac{1}{2\zeta\sqrt{1 - \zeta^2}}\right)$$

In addition, regardless of the fact that the roots of the function are complex, each root contributes 90° of phase shift. Thus, a second-order system has 180° of phase shift. Any additional zeros or poles will add or subtract from that value.

To draw this function, you use the same concepts as for single pole systems, except that in this case, you will have a −40dB/decade (−20dB/decade per order) roll off since the system is second-order.

In most cases, the asymptotic plot of the second-order function will be satisfactory for the approximation of the frequency response of the system. If more accuracy is required, software such as **MATLAB** or another product may be used to plot the response. The frequency response of a second-order system for $\omega_n = 1.0$ and various damping ratios is shown in Figure 8-4.

(figure 8-4)

Second-order frequency response vs. ζ

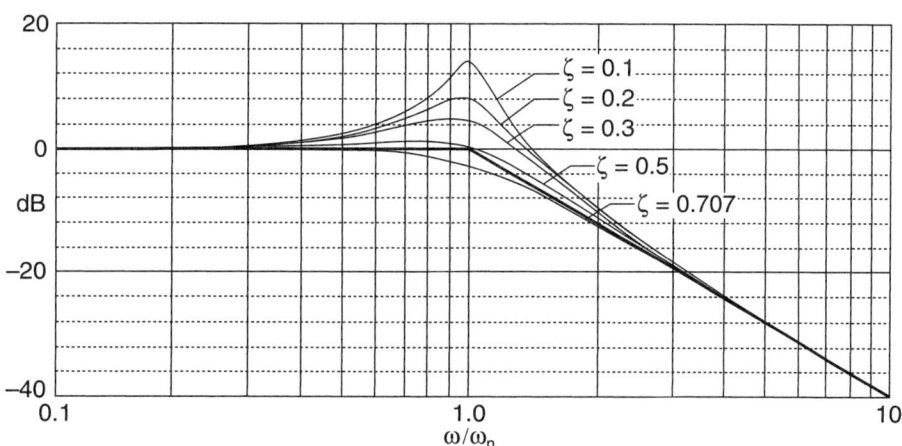

The curves from the largest peak have damping ratios of 0.1, 0.2, 0.3, 0.5, and 0.707. The heavy black line is the asymptotic approximation of the system. Notice that the peak in the response is neglected, which works for most analyses.

The phase response of the second-order system is given from the equation

$$\theta = tan^{-1}\left[\frac{2\zeta\dfrac{\omega}{\omega_n}}{1 + \left(\dfrac{\omega}{\omega_n}\right)^2}\right] \tag{8.8}$$

In all cases, the maximum phase response is −180°, as shown in Figure 8-5.

Notice that as the damping ratio gets smaller, the transition from 0 to 180° gets steeper.

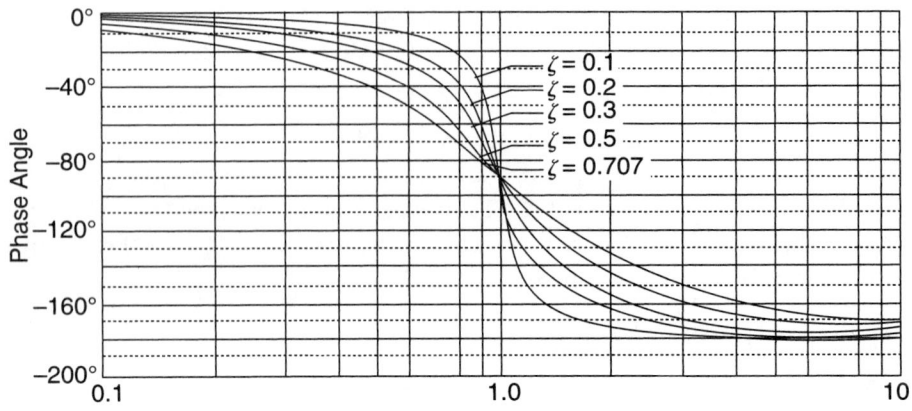

(figure 8-5)

Second-order phase response vs. ζ

8.5 Gain and Phase Margins

The gain and phase margins of a system are critical measurements for the stability of a system. If the gain in the system is greater than unity at the frequency where the phase is $-180°$, the system is unstable.

From the fact that the total system phase response is the sum of all phase shifts of the poles and zeros in the system, the system's ability to oscillate is immediately determined. If the system can oscillate, the gain and phase margins must be determined. There are no set minimum values for the margins, but small margins $(< 20°)$ produce excessive overshoot in the response. (The closer ω_c and ω_π are together, the smaller the value of ζ.)

The gain and phase margins are calculated by observing the gain and phase at two points on the graphs. The gain margin is calculated from the difference in gain between 0 dB, ω_c, and the value of gain at the $-180°$ and ω_π crossing point. Negative decibel values are designated as positive gain margins. The phase margin is calculated by observing the difference between $180°$ and the phase at the 0 dB magnitude crossing point. An example of a system capable of oscillation is shown in Example 8-2.

Example 8-2

The system shown in Figure 8-6 is to be investigated in the frequency domain for stability.

Example 8-2 (continued)

Solution

The closed-loop transfer function of the system is

$$\frac{Y(s)}{R(s)} = \frac{200K}{s^4 + 18s^3 + 180s^2 + 800s + 200K}$$

The Routh table for this system indicates that it is stable for gains less than 30.12.

s^4	1	180	$200K$
s^3	18	800	
s^2	135.6	$200K$	
s^1	$800 - 26.6K$		$0 < K \le 30.12$
s^0	$200K$		

If the required gain, K, is 20, plot all responses and determine the gain and phase margins.

Solution

Since this is a fourth-order system, the total phase shift is –360°. The open-loop transfer function is

$$G(s) = \frac{200K}{s(s + 8)(s^2 + 10s + 100)} = \frac{200K}{s^4 + 18s^3 + 180s^2 + 800s}$$

Each part of the open-loop transfer function can be analyzed separately, using the superposition theorem.

The gain block has a constant gain of +26 dB and contributes no phase to the system; the integrator ($1/j\omega$) has a –20 dB/decade decrease and a constant –90° phase shift with increasing frequency. The integrator output is 0 dB at 1.0 rad/sec. The 2/(s +8) block, which is $0.25[1/(1 + j(\omega/8))]$ in the frequency domain, has a constant response at –12 dB up to 8 rad/sec and then rolls off at a rate of –20 dB/decade. The second-order function has complex roots with a damping factor, ζ, of 0.5. The Bode plot has a peak of about 10 dB, but is considered flat up to its cutoff frequency of 10 rad/sec where it rolls off at –40dB/decade.

The gain block and integrator gain are shown in Figure 8-7. (The phase is not shown since it is a constant –90°.) The gain and phase for each of the other blocks are shown in Figures 8-8 through 8-11.

(figure 8-6)

Fourth-order system for Example 8-2

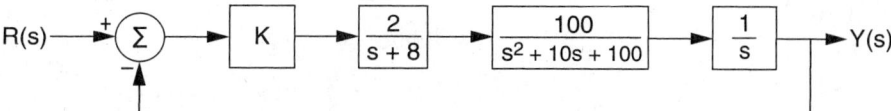

Example 8-2 (continued)

Analysis of the composite plot, Figure 8-12, shows that the gain is –3.56 dB at the 180° phase crossing point of 6.67 rad/s, ω_π, and the phase is –152°, at the 0 dB crossing point, ω_c, for a phase margin of 28°, indicating that the system is stable.

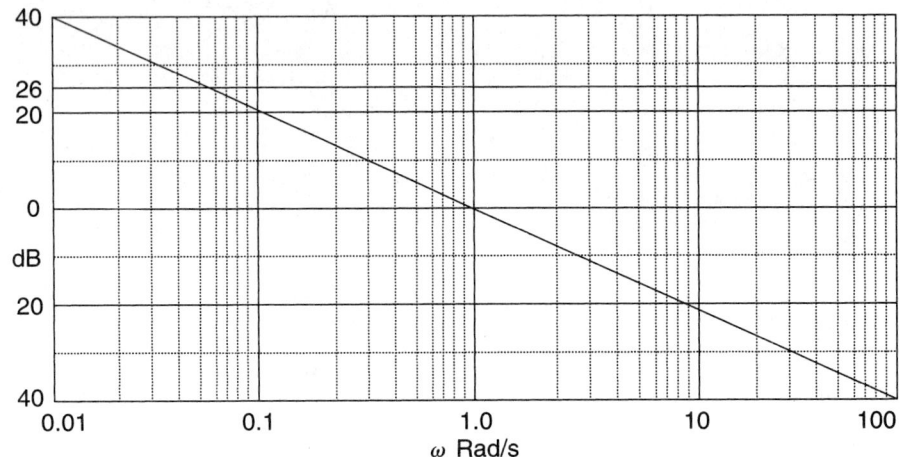

(figure 8-7)

Magnitude plot of 1/s and
$M(\omega) = 26\ dB$

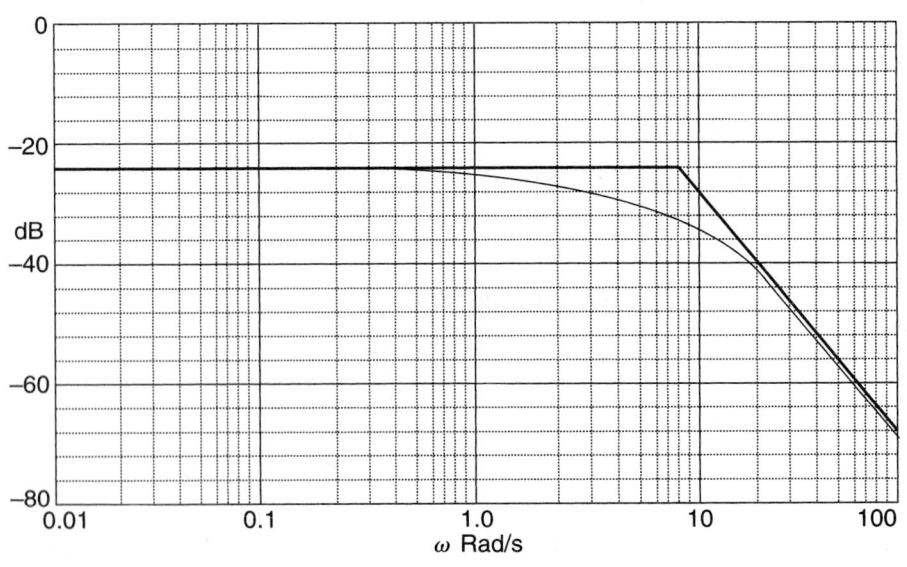

(figure 8-8)

Magnitude plot of
$\dfrac{2}{s + 8}$

Example 8-2 (continued)

Figure 8-13 shows the closed-loop frequency response of the system with a large peak of about 9.0 dB in the response at about 5.76 rad/s.

(figure 8-9)

Phase plot of

$$\frac{2}{s+8}$$

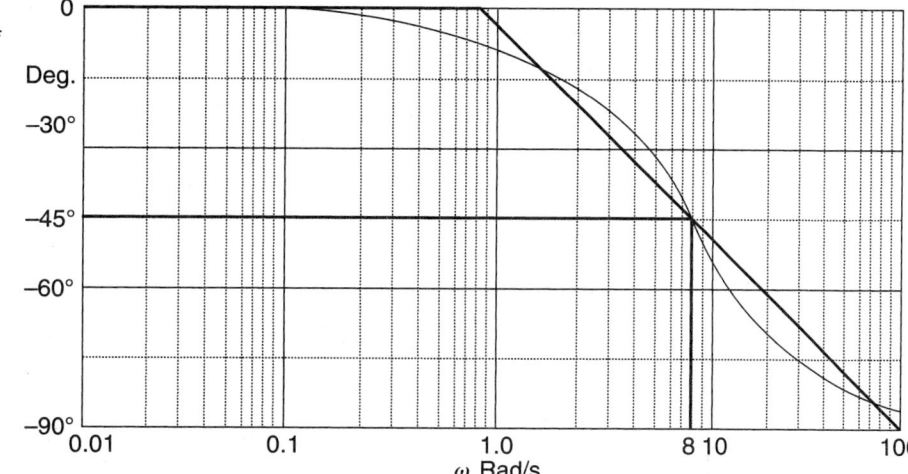

(figure 8-10)

Magnitude plot of

$$\frac{100}{s^2 + 10s + 100}$$

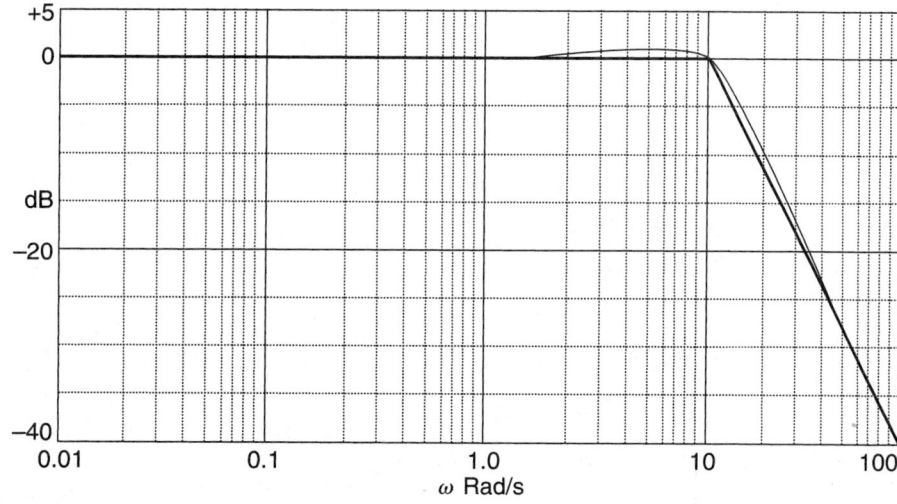

Example 8-2 (continued)

The closer the gain and phase margin frequencies are, the more overshoot there is in the system. When the frequencies become the same, sustained oscillation occurs in a bounded form and the system is called marginally stable.

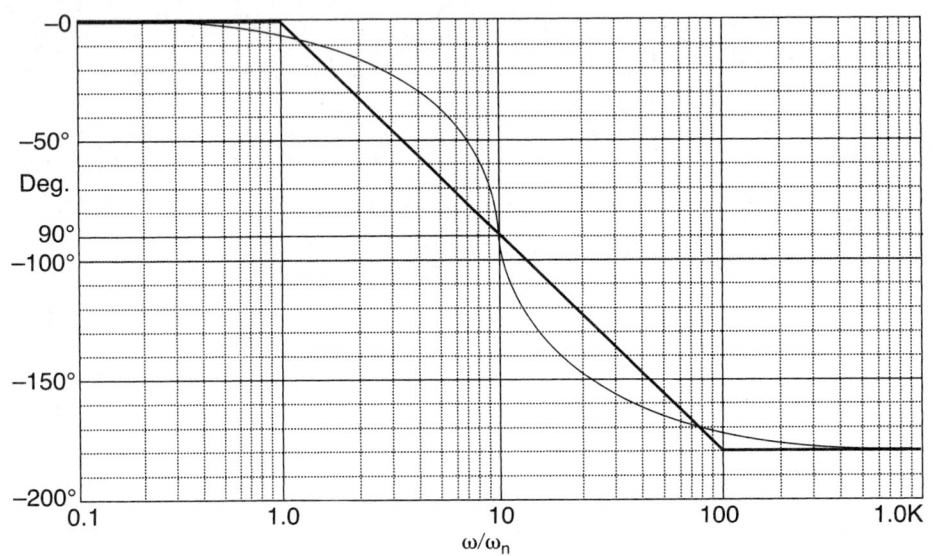

(figure 8-11)

Phase plot of

$$\frac{100}{s^2 + 10s + 100}$$

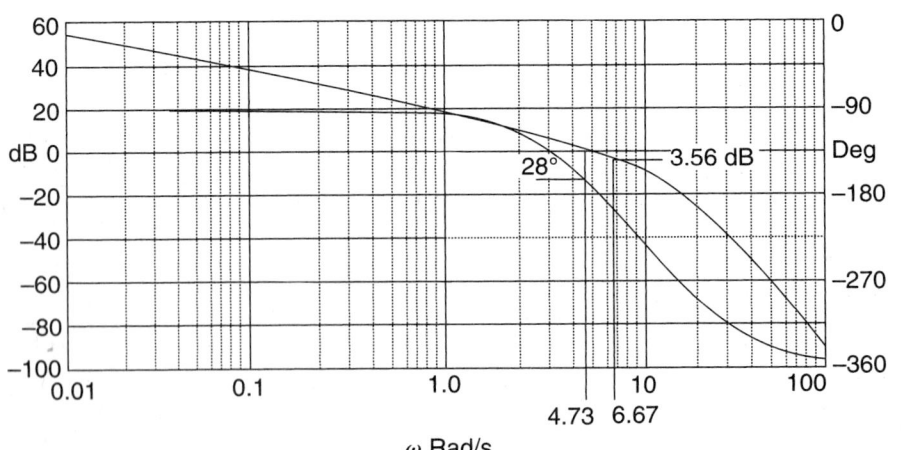

(figure 8-12)

Magnitude plot of

$$\frac{Y(s)}{R(s)} =$$

$$\frac{4000}{s^4 + 18s^3 + 180s^2 + 800s}$$

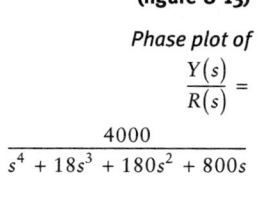

(figure 8-13)

Phase plot of

$$\frac{Y(s)}{R(s)} =$$

$$\frac{4000}{s^4 + 18s^3 + 180s^2 + 800s}$$

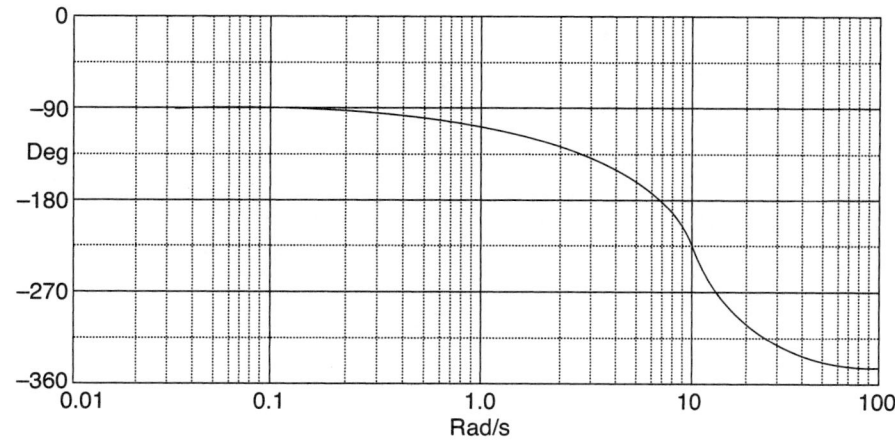

Dead Time

Another phenomenon that can affect a system is a delay between the time a signal is applied and the time the system responds to the change in input. The delay manifests itself as a phase change in the system. If the delay time is substantial, a large amount of phase shift is added to the system response. The added phase shift affects the phase margin in a negative manner. The delay in time is referred to as *dead time*, or T_D. The dead time can cause an otherwise very stable system to become unstable. It is important then to investigate the effect of delay time on a control system.

The Laplace transform of a time-shifted function $f(t - T)$, where $T > 0$ is

$$\mathcal{L}\big[f(t - T)u_s(t - T)\big] = e^{-sT_D}F(s) \tag{8.9}$$

Replacing s with $j\omega$, and T with T_D yields the Euler equation

$$e^{-j\omega T} = \cos(\omega \cdot T_D) - \sin(\omega \cdot T_D) \tag{8.10}$$

The magnitude of the sine and cosine functions is 1.0 from

$$Magn = \sqrt{\cos^2(\omega \cdot T_D) + \sin^2(\omega \cdot T_D)} = 1 \tag{8.11}$$

And the phase angle is given by

$$\theta = \tan^{-1}\left(\frac{-\sin(\omega \cdot T_D)}{\cos(\omega \cdot T_D)}\right) = \tan^{-1}\big[-\tan(\omega \cdot T_D)\big] = -\omega \cdot T_D \tag{8.12}$$

In polar form, the equation is

$$1 \; \angle \; - \omega \cdot T_D \tag{8.13}$$

If ω is set to ω_c and $\omega_c \cdot T_D$ is set equal to the maximum phase angle, an approximation of the maximum amount of time delay the system can tolerate is calculated as follows:

$$T_D \; = \; \frac{\theta_{max}}{\omega_c \cdot (180 \; / \; \pi)^\circ} \tag{8.14}$$

Where θ_{max} is the phase margin of the system.

The magnitude imposed by the delay time is always unity; it is the phase angle that changes. The phase produced by the dead time adds to the phase of the system without a delay time. An example will show the effect of delay time on the phase of a control system.

Example 8-3

Consider the system of Example 8-1, a component that introduces 0.02 seconds of delay time to the system. The original system had gain and phase margins of infinity and was stable. The addition of the delay time introduces additional phase, but no change in the Bode magnitude other than a delay time for the start-up of the system response. The open-loop transfer function of the system with the delay is

$$G(s)H(s) \; = \; \frac{60 \cdot (s + 1)}{(s + 2)(s + 3)} \cdot e^{-0.02s}$$

Solution

The system phase response with the delay time and the change in gain and phase margins are shown in Figure 8-14. (Refer to Figures 8-2 and 8-3 for the original gain and phase of the system.)

Clearly, the delay time had a negative effect on the phase and gain margins. The phase contributed to the system at 10, 30, and 100 rad/sec. The total phase is shown in Table 8-2. The gain and phase crossover points are at 59.9 rad/sec and 81 rad/sec, respectively.

(figure 8-14)

Gain and phase margins for

$e^{-sT_D} = 0.02\ s$

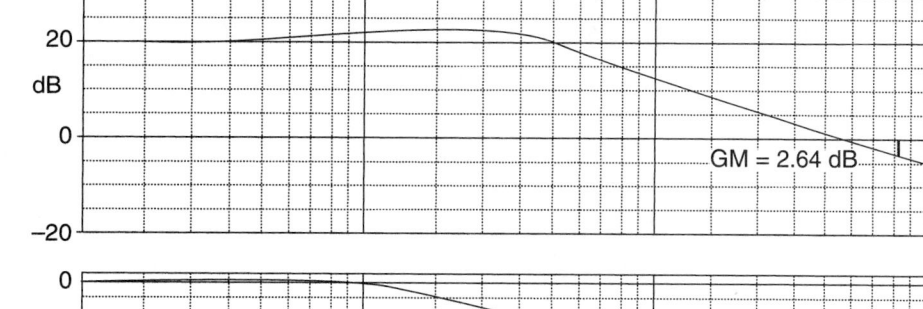

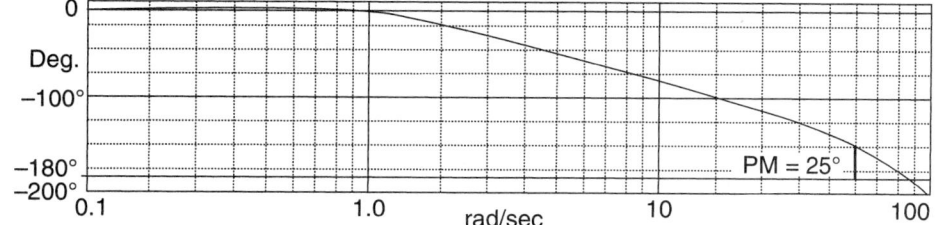

(table 8-2)

Phase contribution of

$e^{-sT_D} = 0.02\ s$

ω	10 rad/sec	30 rad/sec	100 rad/sec
θ_{sys}	−67.7°	−82.3°	−87.7°
ϕ_{delay}	−11.5°	−34.4°	−114.6°
total	−79.2°	−116.7°	−202.3°

8.6 System Compensation to Provide Greater Stability and Faster Response

Essentially, there are three types of compensation schemes: phase-lead, phase-lag, and phase lead-lag compensation. Each type of scheme involves the placement of an additional pole and zero, or both, to a system.

Phase-Lead Compensation

A general equation for lead compensation using opamps is shown in Equation 8.15 and its circuit in Figure 8-15. The second amplifier is needed for phase inversion and for supplying the required gain.

$$G_C(s) = K\frac{(s + a)}{(s + b)} \qquad b > a$$

$$G_C(j\omega) = K\frac{j\omega + \omega_0}{j\omega + k\omega_0} = K \cdot \frac{1 + j\dfrac{\omega}{\omega_0}}{1 + j\dfrac{\omega}{k\omega_0}} \qquad k > 1 \tag{8.15}$$

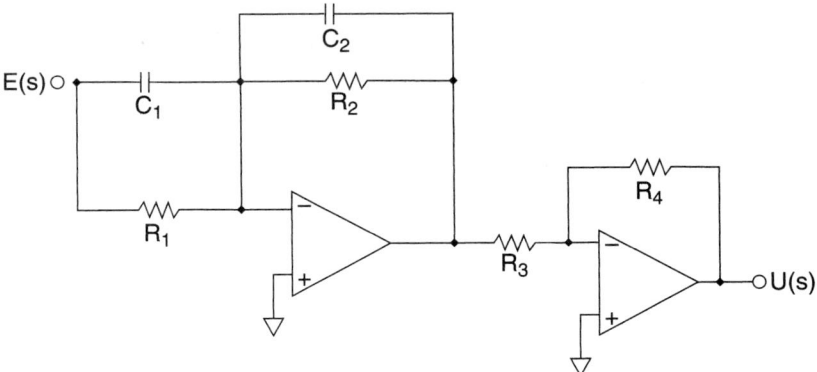

(figure 8-15)

Schematic diagram of lead or lag controller

The pole to be added is determined by the components in the feedback loop, and the zero is formed in the input loop. In the frequency domain, the system has the following form:

$$P_{\text{Lead}}(j\omega) = \left(\frac{a}{b}\right) \cdot \frac{\left(1 + \dfrac{j\omega}{\omega_0}\right)}{\left(1 + \dfrac{j\omega}{k\omega_0}\right)} = \left(\frac{1}{k}\right) \cdot \frac{\left(1 + \dfrac{j\omega}{\omega_0}\right)}{\left(1 + \dfrac{j\omega}{k\omega_0}\right)} \tag{8.16}$$

The factor (a/b) is the lead factor and determines the amount of phase lead the circuit can apply to the system. In general, the maximum amount of phase shift that can be achieved in this configuration is about $55°$ to $60°$. If more phase than the maximum for a single section is required, two or more lead compensators can be used in cascade. The maximum phase occurs at a geometric mean frequency of $\omega_{\text{max}} = \sqrt{a \cdot b}$. The maximum phase shift is found from

$$\phi_{\text{max}} = 90° - 2 \cdot \tan^{-1}\left(\sqrt{\frac{a}{b}}\right) \text{ degrees} \tag{8.17}$$

Another method of determining the maximum phase is by taking the tangent of the difference of two angles.

$$\tan(\phi_m) = \phi_1 - \phi_2 = \frac{\tan\phi_1 - \tan\phi_2}{1 + (\tan\phi_1)(\tan\phi_2)} = \frac{k - 1}{2\sqrt{k}} \tag{8.18}$$

The magnitude and phase curves for several ratios of a/b for Equations 8.16 and 8.17 are shown in Figures 8-16 and 8-17.

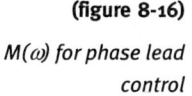

(figure 8-16)

M(ω) for phase lead control

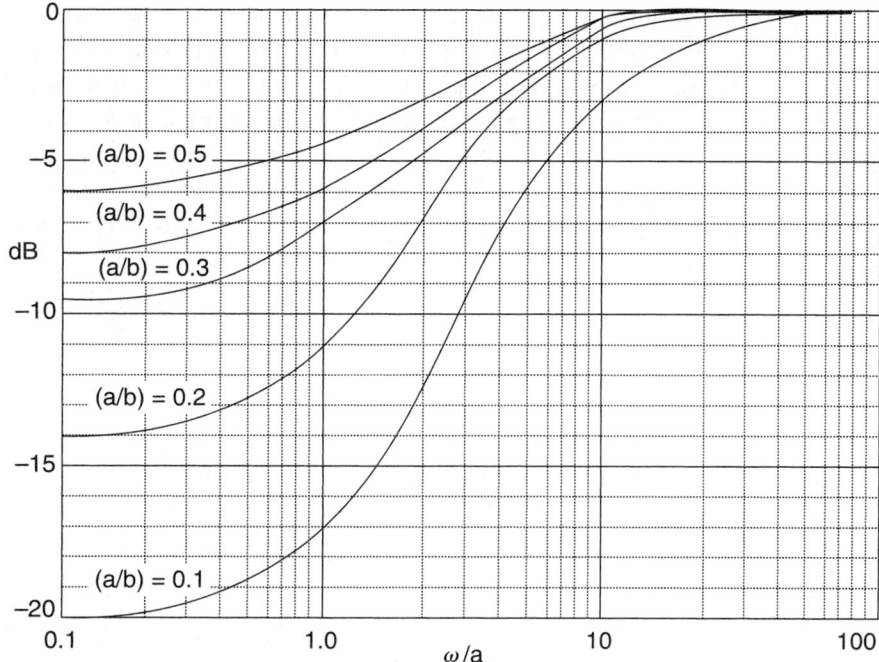

(figure 8-17)

$\phi(\omega)$ for phase lead control

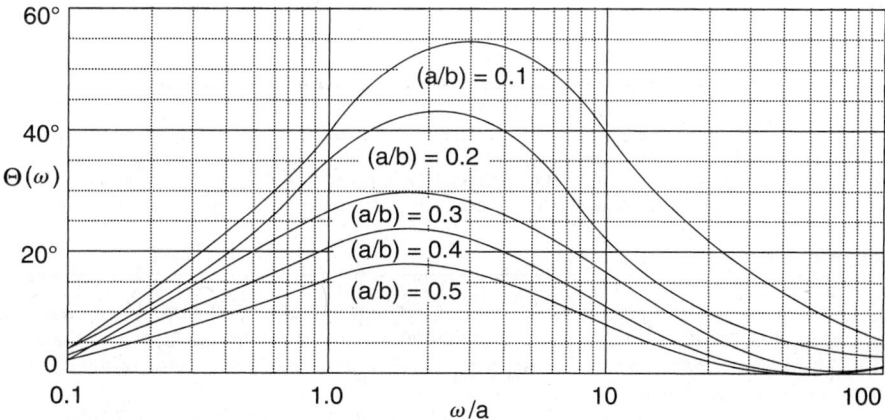

From studying the circuit, it is apparent that there is a loss of gain in the low frequency range; therefore, some gain, K, must be added to the circuit to return the output, $U(s)$, to 1.0. At high frequencies, the gain applied to the compensator shifts the magnitude curve up by the amount of the gain, which increases the frequency response of the system. Since as s approaches zero, the system output tends toward a/b, the gain required by the system is $K = b/a$. The total variation in dB of the gain for the controller is $20 \cdot \log k$.

For this type of compensator, the zero is placed to the right of the pole (closer to the origin) in the s-plane. The separation of the pole and zero is important—when the pole is placed very far from zero, the compensator becomes essentially a differentiator and is subject to noise problems. The separation of the pole and zero should be 10 or less $(k \leq 10)$.

In general, it is desirable to place the zero to the left of two dominant poles. The effect of adding a pole and zero to a second-order system is shown in Figure 8-18.

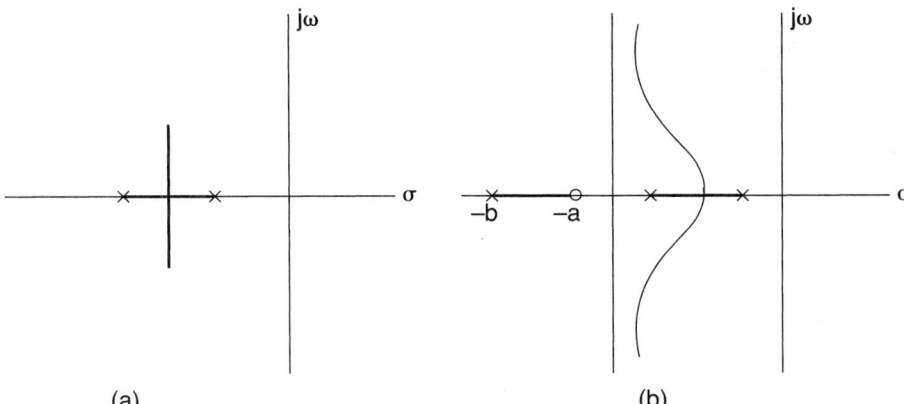

(figure 8-18)

Change in root locus produced by phase lead control

(a)   (b)

An example will help clarify the use of the lead controller.

Example 8-4

The input to the control system shown in Figure 8-19 is a ramp of 0.5 m/s. The steady-state error must be equal to or less than 0.02 meters. In addition, the phase margin must be at least 45° ± 5°.

Solution

The error requirement translates to $K \geq 25$ for proportional control only. The open-loop magnitude and phase curves for $K = 25$ are shown in Figure 8-20. The crossover frequency for the magnitude plot is 8.4 rad/sec, yielding a phase margin of 20°. Since the maximum phase of the system is 180°, the gain margin is infinite.

The system does not meet the phase margin requirement. It has an excessive overshoot of 57.5 percent, and the risetime of the closed-loop system is 0.137 seconds. To meet the specifications, 25° of phase needs to be added to the system. A lead controller is decided on to provide the additional phase.

Lead compensation increases the bandwidth of the system, changing the crossover frequency, which also increases the phase-lead requirement. One method of setting the lead compensator frequency is to assume that the new crossover frequency will occur one octave higher than the uncompensated system crossover frequency, or 16.8 rad/sec. Based on the phase curve for the system, that contributes 9° of additional phase, increasing the total phase to be added to the system to 34°.

Example 8-4 (continued)

If you arbitrarily assign zero to be at $s = -4.0$, the value for the pole can be calculated from Equation 8.14. The modified equation is

$$b = \frac{a}{\tan^2\left(\dfrac{90° - \phi_m}{2}\right)} = \frac{4}{\tan^2\left(\dfrac{90° - 34°}{2}\right)} = 14.15$$

The pole at $s = -14.15$ requires that a gain of 3.5 be added to the overall gain of the system. The complete system is shown in Figure 8-21; the phase margin, in Figure 8-22.

Satisfying the requirements, the error is 0.02 meters and the phase margin is 41.8°. An analysis of the step response of the closed-loop system shows an overshoot of 28.5 percent, and the risetime of the system is reduced to 0.09 seconds—an improvement in both.

(figure 8-19)

System for Example 8-4

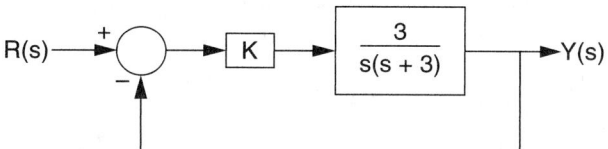

(figure 8-20)

Frequency and phase of uncompensated system of Example 8-4

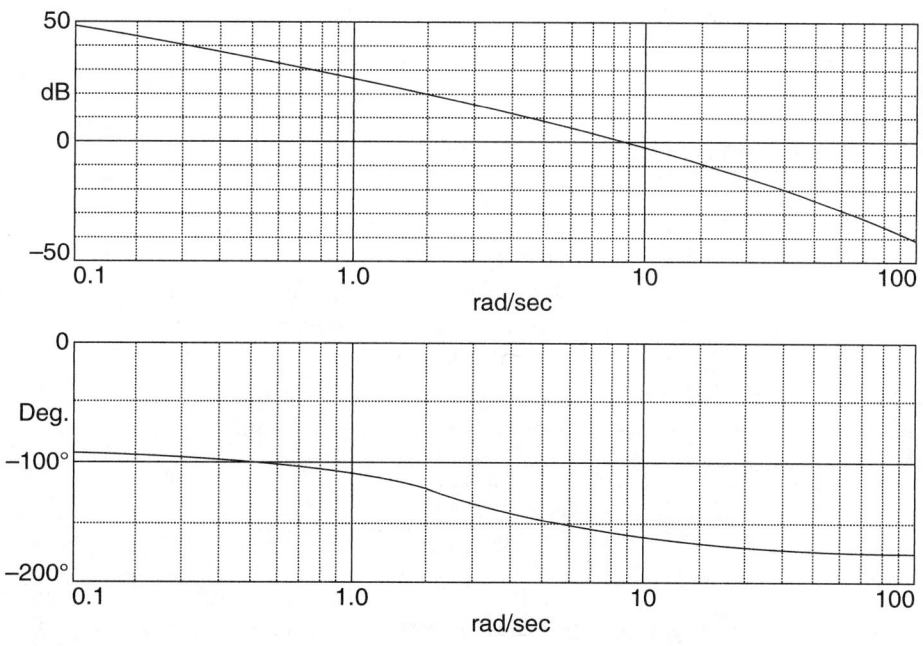

(figure 8-21)

Compensated system for Example 8-4

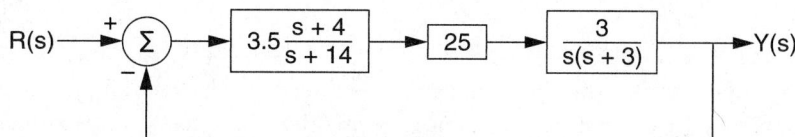

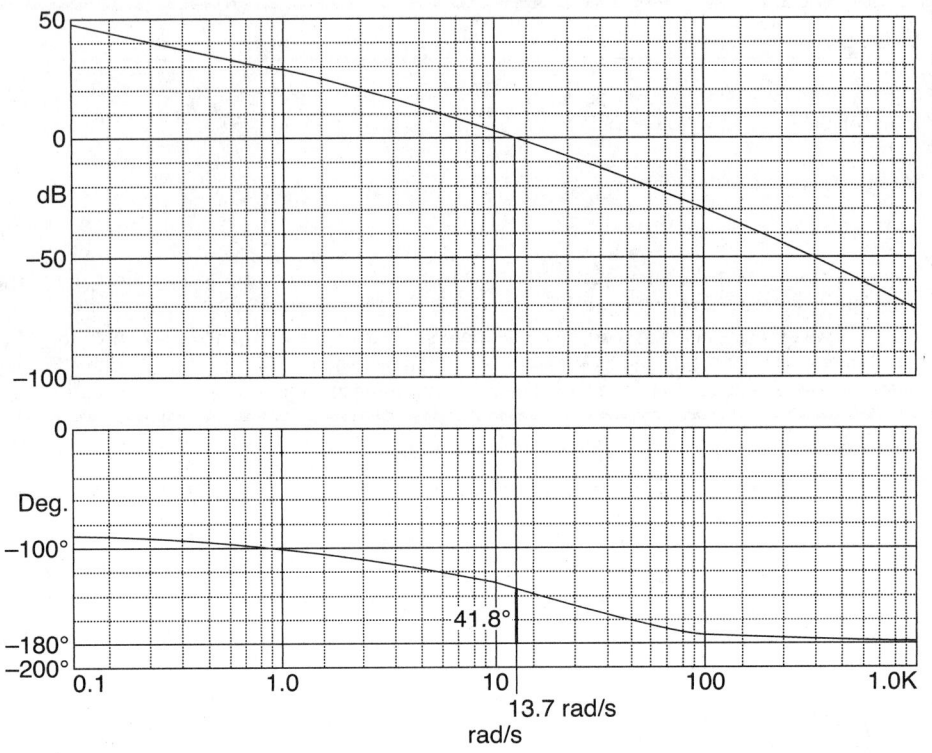

Phase-Lag Compensation

The phase-lag controller places the pole closer to the origin. That has the effect of decreasing the bandwidth of a system and reducing the phase of the system, thereby slowing down the system. The benefit of using the lag compensator is that it is essentially an integrator and, as such, acts to reduce the system error. The same general circuit configuration shown in Figure 8-15 is used for the lag compensator. For this type of controller, the value of gain, b/a, is less than 1. This type of compensation should be used only for type zero and type one systems. The equations of this controller are shown in Equation 8.19

$$G_C(s) = K\frac{(s + a)}{(s + b)} \qquad a > b$$

$$G_C(j\omega) = K\frac{1 + j\dfrac{\omega}{k\omega_0}}{1 + j\dfrac{\omega}{\omega_0}} \qquad k > 1 \tag{8.19}$$

In this type of controller, the phase change is less important than the amplitude change. The purpose of the controller is to reduce the crossover frequency, thereby increasing the phase margin to the required value. However, a small amount of excess phase change contributed by the lag controller must be accounted for. The gain and phase curves for several values of a/b are shown in Figures 8-23 and 8-24.

(figure 8-23)

M(ω) for phase-lag compensator (Magnitude Curve)

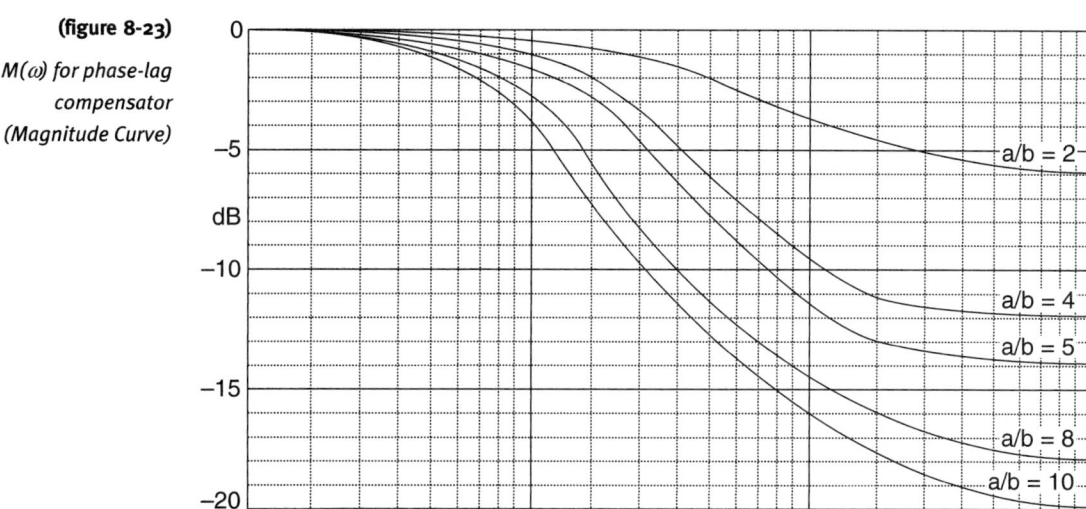

(figure 8-24)

ϕ(ω) for phase-lag compensator (Phase Curve)

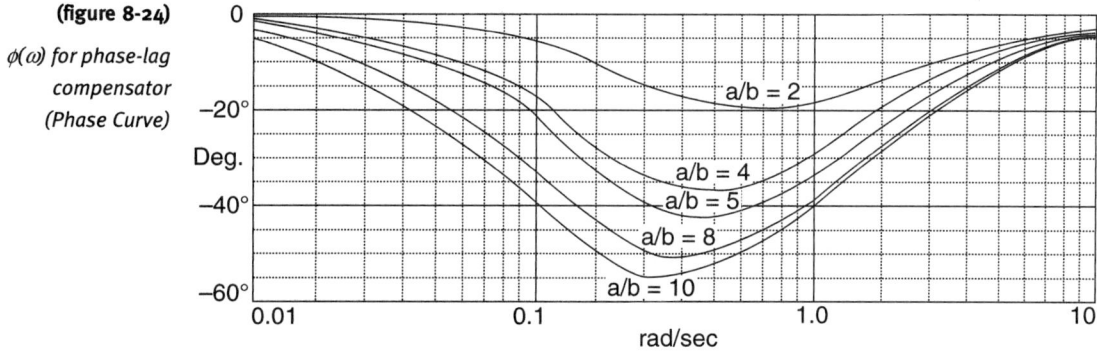

An example will demonstrate the use of this type of controller.

Example 8-5

Consider the control system shown in Figure 8-25. This system must exhibit an error constant, K_p, of 100 s^{-1} for a step input, and the phase margin must be at least 50°. To satisfy the error requirement, $K = 20$.

Example 8-5 (continued)

Solution

The uncompensated gain and phase plots are shown in Figures 8-26 and 8-27. The plots clearly show that the system is unstable. The phase margin is −53°, and the gain crossover frequency is about 450 rad/sec. Because the rate of phase change is fast, lead compensation is not an option as the required phase lead would be greater than that available from a single lead circuit.

To meet the phase margin requirement, it is necessary to change the crossover frequency to the frequency where the gain margin is −55°. The additional 5 degrees is for the small residual phase shift the lag circuit contributes to the compensation scheme. The frequency at which the phase is −(180° − 55°) = −125°) is 90 rad/sec, and the gain at the new crossover frequency is about 33 dB. The maximum attenuation of the lag compensation scheme must then be −33 dB. From the equation for gain variation ($k = 10^{dB/20}$), the value of k is 44.7. As a rule of thumb, you can place the zero at one-tenth of the new crossover, or 9 rad/sec. That places the pole at 9/44.7 = 0.2 rad/sec, close to the origin of the s-plane. The required value of K is b/a = 0.022. The lag controller then has the configuration

$$G_C(s) = 0.022\frac{(s + 9)}{(s + 0.2)}$$

The new block diagram is shown in Figure 8-28; and the compensated gain and phase curves are shown in Figures 8-29 and 8-30, respectively.

The curves show that the new crossover frequency is 90 rad/sec and that the phase margin is 50°.

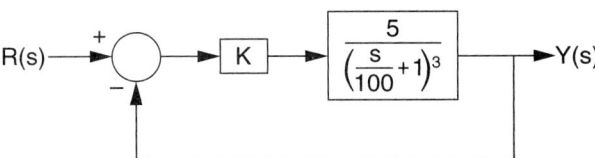

(figure 8-25)

Uncompensated system for Example 8-5

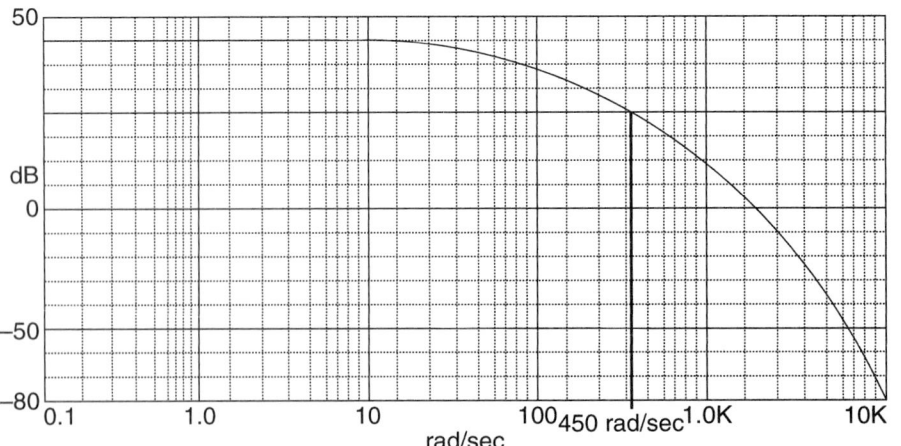

(figure 8-26)

Uncompensated gain of Example 8-5

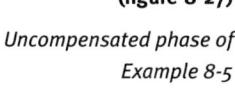

(figure 8-27)

Uncompensated phase of Example 8-5

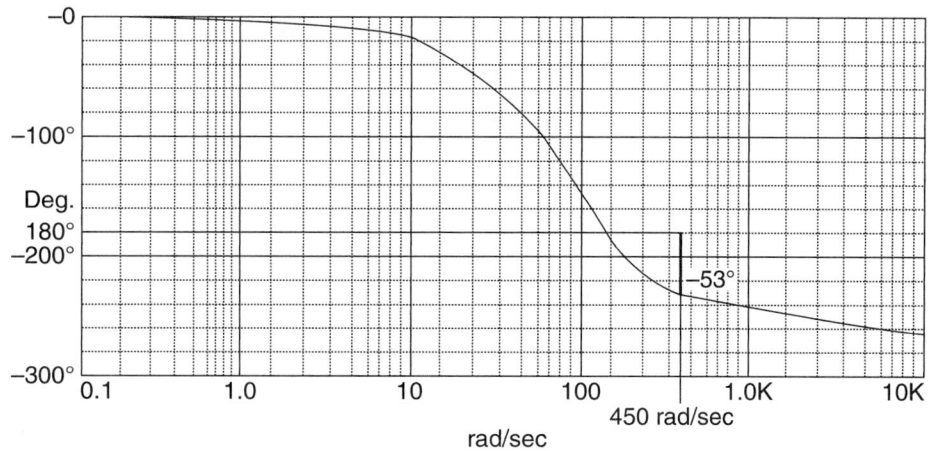

(figure 8-28)

Compensated system for Example 8-5

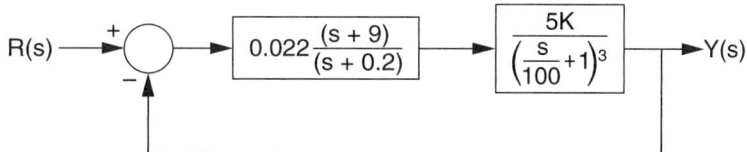

(figure 8-29)

Compensated gain of Example 8-5

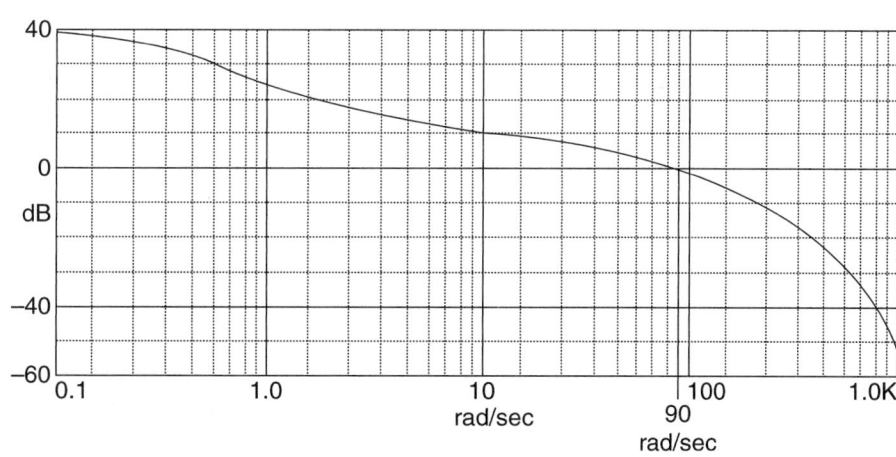

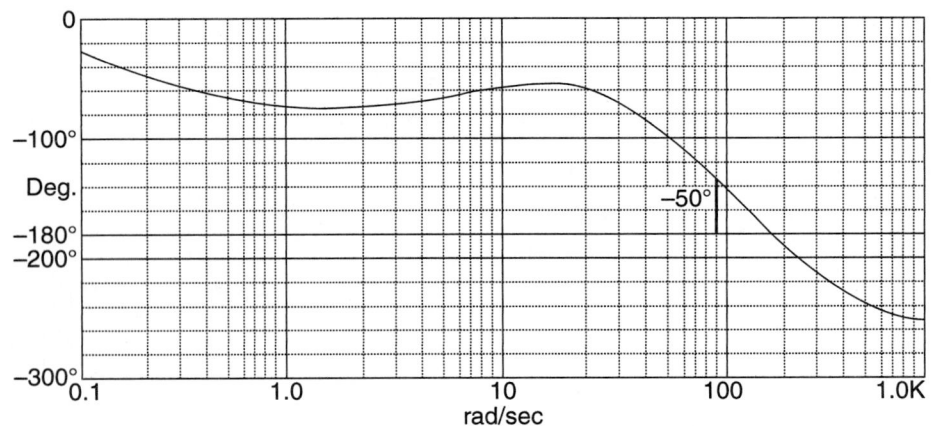

(figure 8-30)

Compensated phase of Example 8-5

Phase Lead-Lag Compensation

In the previous two sections, you learned that phase-lead control increases the natural frequency, increases the damping, and improves risetime of a system. Phase lag reduces the bandwidth of the system and slows the system down while decreasing the error inherent in the system. Both systems have advantages that are useful in the design of control systems. In some cases, it is desirable to use both types of systems in a single controller to obtain the benefits of each. The lead-lag controller has the general form

$$G_C(s) = K\frac{(s + a_{\text{Lead}})(s + a_{\text{Lag}})}{(s + b_{\text{Lead}})(s + b_{\text{Lag}})} \qquad a_{\text{Lead}} < b_{\text{Lead}}, a_{\text{Lag}} > b_{\text{Lag}} \qquad (8.20)$$

An opamp schematic for one type of lead-lag controller is shown in Figure 8-31 and the transfer function of the controller in Equation 8.21.

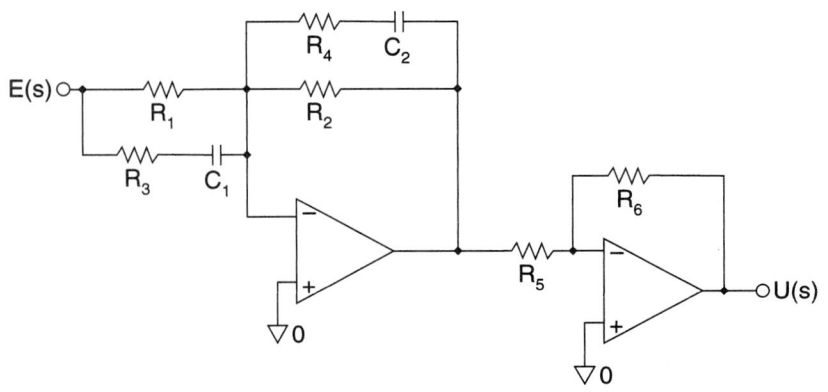

(figure 8-31)

Schematic diagram for lead-lag controller

$$\frac{U(s)}{E(s)} = -\frac{R_2}{R_1} \cdot \underbrace{\frac{(R_1 + R_3)C_1s + 1}{R_3C_1s + 1}}_{\text{Lead}} \cdot \underbrace{\frac{R_4C_2s + 1}{(R_2 + R_4)C_2s + 1}}_{\text{Lag}} \quad (8.21)$$

You can define time constants for the numerator and denominator for the lead and lag circuits as

Numerator:

$$\tau_1 = (R_1 + R_3)C_1 \qquad \frac{\tau_1}{k_1} = R_3C_1$$

Denominator:

$$\tau_2 = R_4C_2 \qquad k_2\tau_2 = (R_2 + R_4)C_2$$

Where

$$k_1 = \frac{R_1 + R_3}{R_3} > 1, \qquad k_2 = \frac{R_2 + R_4}{R_4} > 1, \qquad K_C = \frac{R_2R_4R_6}{R_1R_3R_5} \cdot \frac{R_1 + R_3}{R_2 + R_4}$$

From which the transfer function becomes

$$\frac{U(s)}{E(s)} = K_C \cdot \frac{k_2}{k_1} \left(\frac{\tau_1 s + 1}{\frac{\tau_1}{k_1}s + 1} \right) \cdot \left(\frac{\tau_2 s + 1}{k_2\tau_2 s + 1} \right) = K_c \cdot \left(\frac{s + \frac{1}{\tau_1}}{s + \frac{k_1}{\tau_1}} \right) \cdot \left(\frac{s + \frac{1}{\tau_2}}{s + \frac{1}{k_2\tau_2}} \right) \quad (8.22)$$

Note that k_1 is often made equal to k_2. Figure 8-32 shows the gain curves and Figure 8-33 shows the phase curve of this type of control.

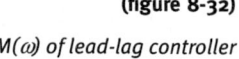

(figure 8-32)

$M(\omega)$ of lead-lag controller

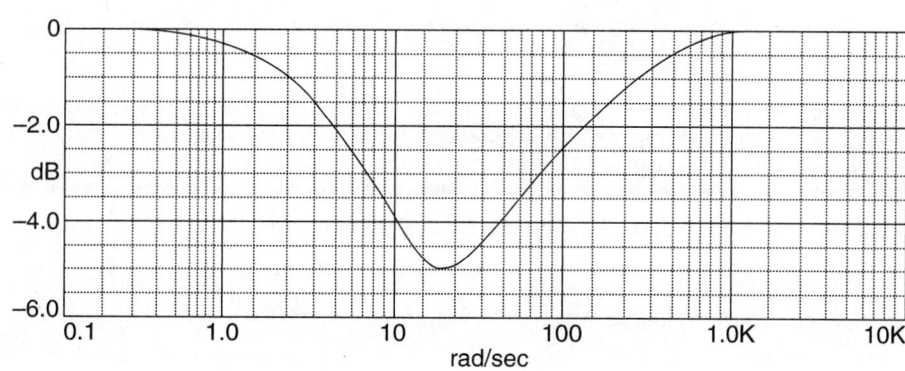

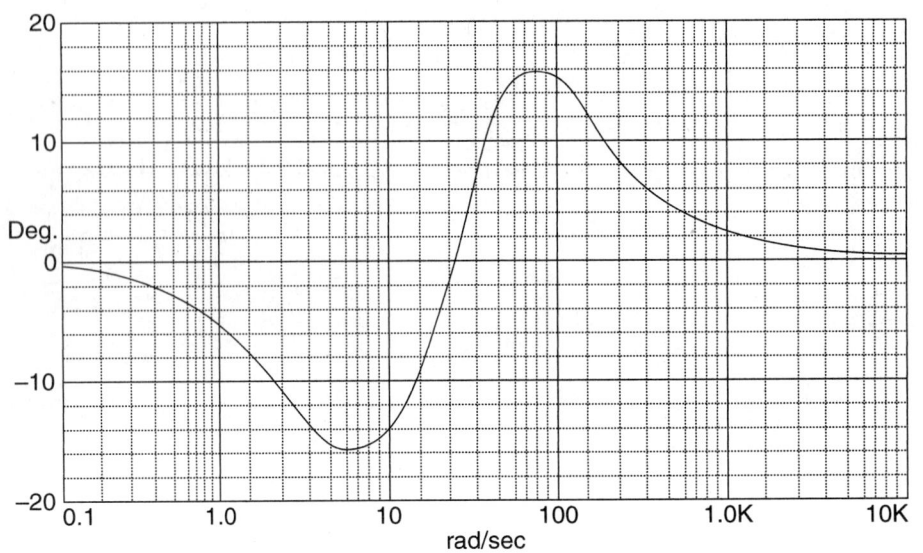

(figure 8-33)

$\phi(\omega)$ for lead-lag
controller

Example 8-6 shows the use of this type of circuit when k_1 is not equal to k_2.

Example 8-6

Consider the system shown in Figure 8-34.

Solution

The open-loop transfer function is

$$G(s) = \frac{5}{s(s + 1)}$$

The closed-loop poles of the system are $0.5 \pm j2.18$, and the damping ratio is 0.224 with a velocity constant of $K_V = 5$ sec^{-1}.

The goal is to make the damping ratio equal to 0.5 and increase the undamped radian velocity to 5.0 rad/sec and the error constant to 50 sec^{-1}. The values of k^1 and k^2 do not need to be equal for this design.

From the specifications, the new poles must be located at $-2.5 \pm j4.333$. The poles are calculated from the fact that the damping ratio, ζ, is the cosine of the angle between the origin of the s-plane and the poles. The hypotenuse of the triangle formed must be 5.0.

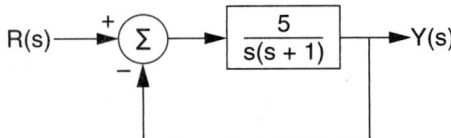

(figure 8-34)

*Uncompensated system for
Example 8-6*

Example 8-6 (continued)

The excess phase contributed by the new pole values is

$$\angle \frac{5}{s(s+1)}\bigg|_{s=-2.5+j4.333} = -229°$$

This value assures that the root locus passes through the desired closed-loop poles, which are the dominant poles of the system. The phase-lead portion of the compensation must supply approximately 50°.

Either the lead or lag portion can be designed first, in this case you will design the phase-lead portion of the system first. While there are a number of choices of zero that you could make, select a zero value of $s = -1.0$, which cancels the pole at $s = -1.0$. The cancellation is not complete in that approximations will have to be made for the time constants of the system. Because you selected the zero position, you can determine the value of the pole from Equation 8.17 yielding the pole at $s = -7.72$.

$$\tan\left(\frac{90-50}{2}\right) = \sqrt{\frac{1}{b_{Lead}}}$$

$$b_{Lead} = \frac{1}{(0.36)^2} = 7.72$$

Thus, the phase-lead portion of the compensator has the form

$$K_C \frac{s + \dfrac{1}{\tau_1}}{s + \dfrac{k_1}{\tau_1}} = K_C \frac{s+1}{s+7.72}$$

Equation 8.22 shows that $\tau_1 = 1.0$ and that $k_1 = 7.72$. K_C is found by setting the value of the lead portion equal to 1.

$$\left| K_C \frac{s+1}{s+7.72} \cdot \frac{5}{s(s+1)} \right|_{s=-2.5+j4.333} = 1$$

$$K_C = \left| \frac{s(s+7.72)}{5} \right| = \left| \frac{(-2.5+j4.333)(-2.5+j4.333+7.72)}{5} \right| = 6.36$$

That completes the lead portion design. You will now turn your attention to the lag portion of the compensator. The lag portion is designed to meet the velocity error requirement of the controller. The velocity error is produced by the total system and is given by the error calculations shown in Chapter 6.

$$K_V = \lim_{s \to 0} s \cdot G_C(s)G(s) = \lim_{s \to 0} s \cdot K_C \frac{k_2}{k_1} G(s) \tag{8.23}$$

The calculation of the value of k_2 for this system is

$$K_V = \lim_{s \to 0} s \cdot (6.66) \cdot \frac{k_2}{7.55} \cdot \frac{5}{s(s+1)} = 4.41k_2 = 50$$

$$k_2 = \frac{50}{4.41} = 11.34$$

Example 8-6 (continued)

Finally, the value of τ_2 chosen is large enough that the following conditions are met:

$$\left| \frac{s + \dfrac{1}{\tau_2}}{s + \dfrac{1}{11.34\tau_2}} \right|_{s = -2.5 + j4.333} \simeq 1$$

$$-5° < \angle \left. \frac{s + \dfrac{1}{\tau_2}}{s + \dfrac{1}{11.34\tau_2}} \right|_{s = -2.5 + j4.333} < 0°$$

For the two criteria, values of τ_2 equal to or greater than 5.0 will meet the criteria. You then select $\tau_2 = 5.0$. The compensator transfer function is

$$G_C(s) = 6.66 \cdot \left(\frac{s + 1}{s + 7.55} \right) \cdot \left(\frac{s + \dfrac{1}{5}}{s + \dfrac{1}{5 \cdot 11.34}} \right) = 6.66 \cdot \left(\frac{s + 1}{s + 7.55} \right) \cdot \left(\frac{s + .2}{s + 0.0176} \right)$$

$$= \frac{10 \cdot (s + 1)(5s + 1)}{(0.1325s + 1)(56.7s + 1)}$$

The total loop transfer function is

$$G_C(s)G(s) = \frac{50 \cdot (s + .2)}{s(s + 7.55)(s + 0.0176)}$$

The magnitude and phase curves for the system, showing the phase margin to be 52°, which is slightly larger than required, are shown in Figures 8-35 and 8-36, respectively.

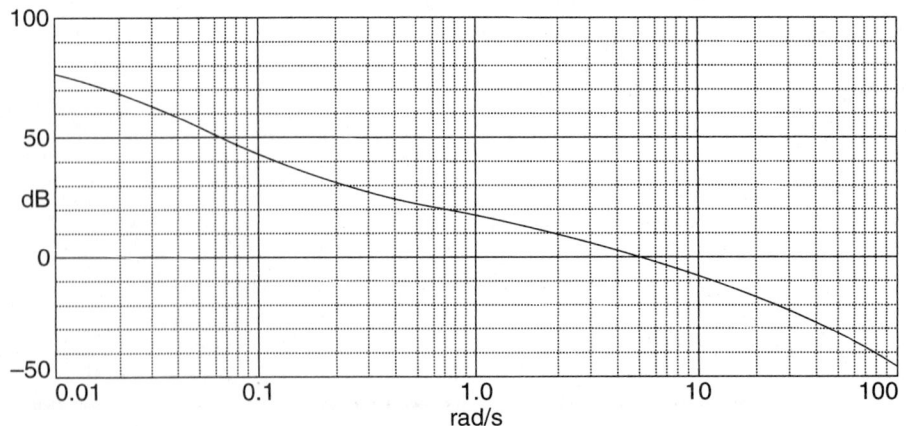

(figure 8-35)

Compensated M(ω) of Example 8-6

Example 8-6 (continued)

The step output of the system is shown in Figure 8-37 for the original system and for the compensated system. The figure shows that the damping ratio is correct since the overshoot is about 16 percent. The system specifications are met or exceeded, completing the design.

(figure 8-36)

Compensated φ(ω) of Example 8-6

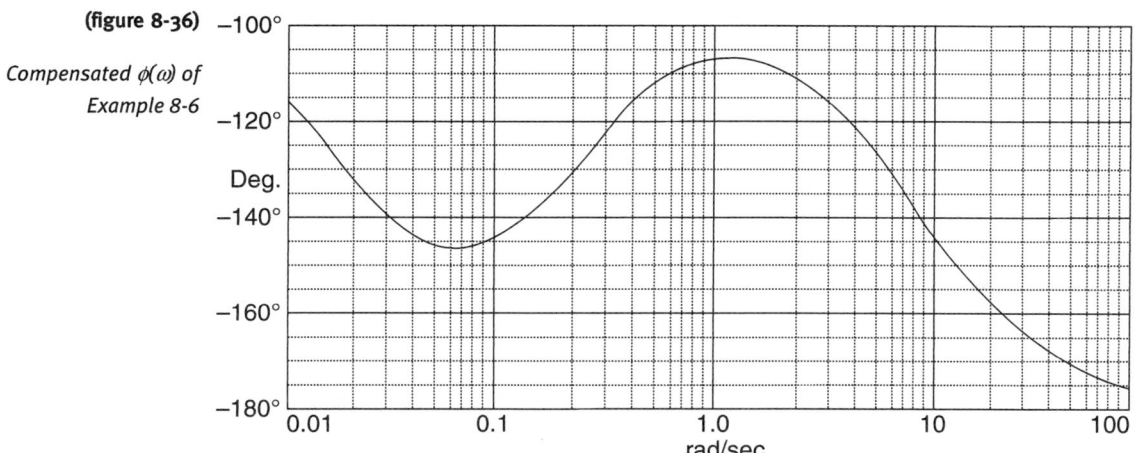

(figure 8-37)

Step response of uncompensated and compensated systems of Example 8-6

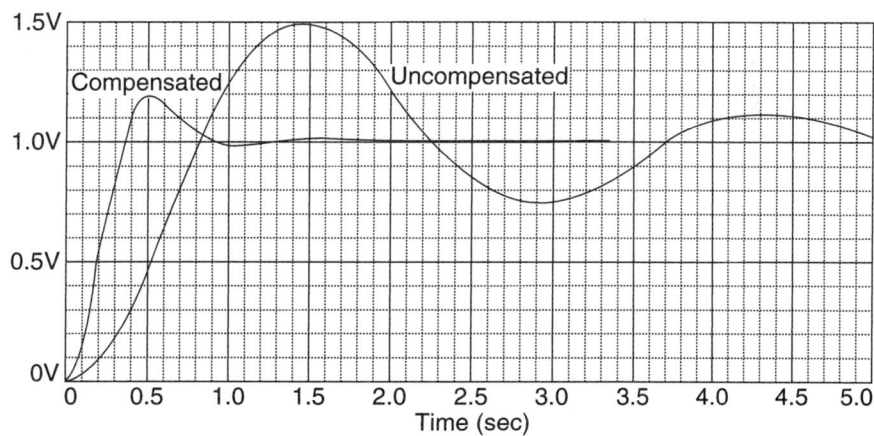

8.7 Nyquist Criteria for Stability of Control Systems

The Routh table and root-locus methods relate the absolute stability of a system when the characteristic equation is a finite polynomial in s. If delay time is included in the system, an approximate solution can be achieved by using a few terms of the expansion of e^{-sT}; but the solution is only approximate.

Nyquist analysis is a graphical procedure that relates absolute and relative stability of systems, using frequency response methods. The Nyquist method handles time delays directly without the necessity of approximation.

Conformal Mapping of Complex Variables

Real values can be graphed in simple Cartesian coordinates with x (real) as the abscissa and $y = f(x)$ as the ordinate. Functions of complex variables such as $F(s)$ where $s = \sigma + j\omega$ cannot be mapped on a single set of coordinates because there are two independent quantities, the real and the imaginary components of s. Thus, a single line cannot represent s. To plot $F(s)$, two graphs are required—one relating $j\omega$ versus σ, which is called the s-plane, and one representing the imaginary part of $F(s)$ with respect to the real part of $F(s)$, which is called the $F(s)$ plane. Generally, only very specific points are plotted from one plane to another.

Corresponding points from one plane to the other is called **mapping** the points. For Nyquist stability plots, the path taken by the mapping is called the *Nyquist path*.

For the special case where $\sigma = 0$, the s-plane degenerates into a single line and $F(j\omega)$ is represented in the $F(j\omega)$ plane as a locus of points with $j\omega$ as a parameter. Polar plots of systems are constructed in the $P(j\omega)$ plane taken from the line $s = j\omega$ in the s-plane. As an example, given the s-plane function $F(s) = s^2 + 4$, where a point $s_0 = -4 + j6$, the position in the $F(s)$ plane is $F(s_0) = -16 - j48$, the mapping is shown in Figure 8-38.

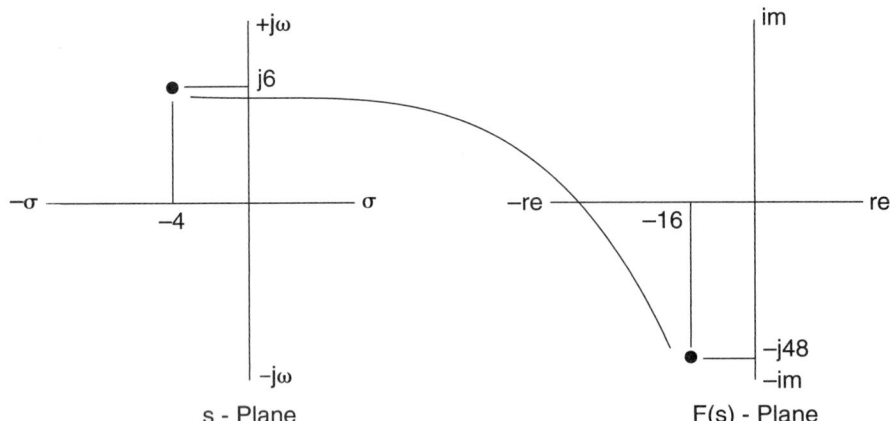

s - Plane   F(s) - Plane

(figure 8-38)

Mapping of point $s_0 = -4 + j6$ from the s-plane to the F(s) plane

Polar Plots of Systems

If σ is allowed to be zero, then plots of a control system can be made using only the magnitude and phase angle of the system. The plot of the system is called a *polar plot*.

Polar plots of systems are constructed by allowing the frequency to vary from zero to infinity and by plotting the magnitude and phase angle at various frequencies on a polar chart. As an example, consider the system given by

$$G(j\omega)H(j\omega) = \frac{5}{\left(\dfrac{j\omega}{4} + 1\right)^3}$$

The magnitude and phase of this system is shown in Table 8-3, and the polar plot is shown in Figure 8-39.

(table 8-3)

Values of $M(\omega)$ and $\phi(\omega)$

ω	$M(\omega)$	$\phi(\omega)$
0	5	0°
1	4.56	−42°
2	3.58	−79.7°
3	2.56	−110.6°
4	1.77	−136°
5	1.22	−154°
6	0.85	−169°
7	0.61	−180.8°
8	0.45	−190.3°
9	0.33	−198.1
10	0.26	−204.6°
20	0.04	−236°
∞	0	−270°

The gain margin is found from the polar plot as the point at which the curve crosses the −180° line, and the phase margin is where the polar plot crosses the magnitude 1.0 line. The example plot crosses the −180° line at $M = 0.624$. The gain margin is given by $GM = 20 \cdot \log\left(\dfrac{1}{0.624}\right) = 4.09$ dB. That point is labeled as a on the plot. The plot crosses the magnitude 1.0 line at −162°, for a phase margin of 18°, indicating the system is stable. An important criterion to remember is that the plot should not cross the 180° line at a value greater than 1, which indicates an unstable system.

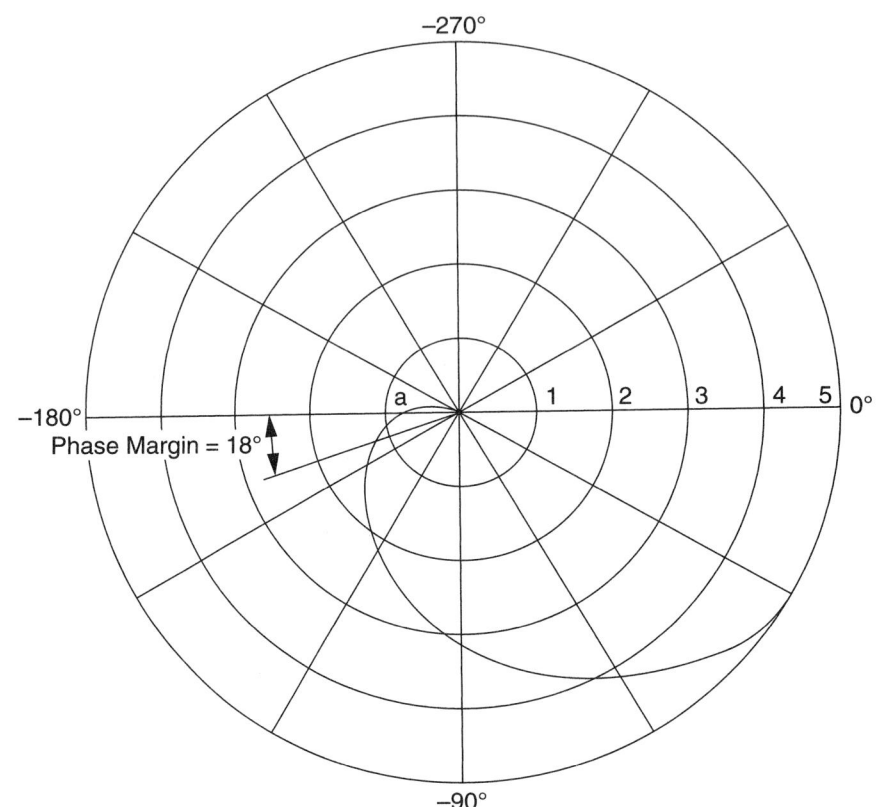

(figure 8-39)

Polar plot of

$$G(j\omega)H(j\omega) = \frac{5}{\left(\dfrac{j\omega}{4} + 1\right)^3}$$

Nyquist Stability Criterion

The Nyquist criterion is an attractive frequency response method that gives information on the absolute and relative stabilities of a system. In addition, it allows you to develop information on the degree of instability of a system and on systems with pure time delays that cannot be treated using the Routh-Hurwitz criterion and that are difficult to analyze using the root-locus method. The Nyquist criterion also gives an indication of how the system stability can be improved.

Before starting the evaluation of the stability criterion, it is important to develop the concept of moving complex variable values from one plane to another. If you consider a ratio of polynomials in factored form where the poles and zeros are defined in the s-plane,

$$F(s) = \frac{K(s + z_1)(s + z_2)\cdots(s + z_n)}{s^n(s + p_1)(s + p_2)\cdots(s + p_m)}e^{-T_D s} \qquad (8.24)$$

If an arbitrary closed contour is drawn in the s-plane, enclosing some of the poles and zeros but not passing through any pole or zero; if points are selected on the closed contour in the s-plane; and if values from the poles and zeros are plotted in the $F(s)$ plane, another closed contour will be generated. The contour generated is called the image of the contour in the s-plane. Further, the closed contour in the s-plane must be single-valued; it must map to one and only one point in the $F(s)$ plane. Such a contour is shown in Figure 8-40. While the contour in the s-plane may be simple, the contour in the $F(s)$ plane may not be and may encircle some specific points of interest, such as the origin, one or more times.

(figure 8-40)

Mapping of contour in the s-plane to another complex plane

From looking at the figure, it is apparent that the poles and zeros located inside the closed contour in the s-plane will travel a total of $2\pi(Z - P)$ radians and that the pole located outside the contour will travel zero radians. From complex variable theory, the *principle of the argument* reveals that the number of encirclements of the origin is given by

$$N = Z - P \qquad (8.25)$$

Where N is the number of encirclements of the origin, Z is the number of zeros encircled by the closed path, and P is the number of poles encircled by the closed path.

In Figure 8-40a, with the travel direction shown, the contour in the $F(s)$ plane is sketched. From the formula, $N = Z - P$; $N = -1$; and the number of encirclements is one, as shown in Figure 8-40b. Also, the direction of travel is opposite that of the contour shown in Figure 8-40a.

To apply that concept to stability concerns, a particular path, called the Nyquist path, in the s-plane is chosen. Of particular interest is the possibility of zeros of

$1 + G(s)H(s)$ in the RHP. For that reason, the Nyquist path usually encircles the entire RHP, usually in a clockwise direction. Since the limits of a plane are essentially infinite, the plot starts at $-\infty$ on the $j\omega$ axis; traverses the plane to $+\infty$; and encircles the plane (with a radius approaching infinity), returning to $-\infty$ as shown in Figure 8-41.

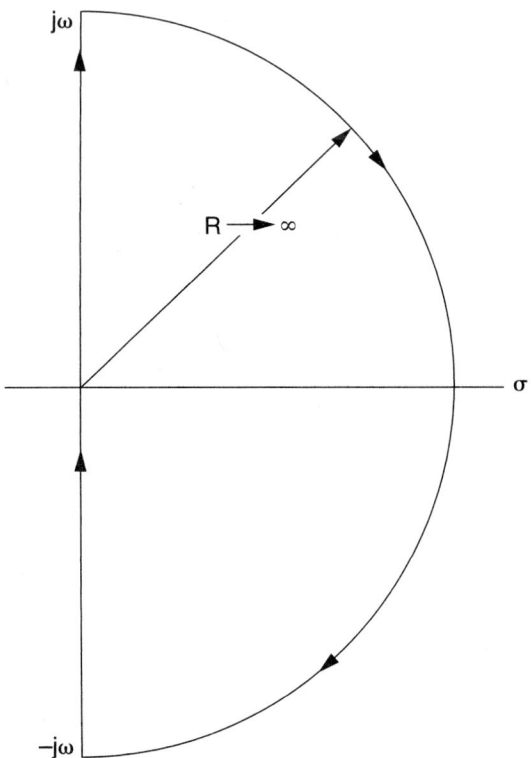

(figure 8-41)

The Nyquist path in the s-plane

Evaluation of $F(s) = 1 + G(s)H(s)$ could be done, but it is somewhat simpler to evaluate only $G(s)H(s)$, which has the effect of shifting the contour one unit to the left. The concern of not encircling the origin now shifts to not encircling the $-1 + j0$ point in the $F(s)$ plane.

That changes the considerations for Equation 8.25 to the following:

N = number of encirclements of the $-1 + j0$ point in the $F(s)$ plane.

Z = number of zeros of the characteristic equation in the RHP that are encircled by the contour in the $F(s)$ plane.

P = number of poles of the characteristic equation in the RHP that are encircled by the contour in the $F(s)$ plane.

Based on those considerations, a system is stable when Z is equal to zero. That can be further broken down into two possible scenarios:

1. $P = 0$: A system is stable when the Nyquist path is mapped into the $F(s)$ plane and there are no encirclements of the $-1 + j0$ point.

2. $P \neq 0$: A system is stable when the Nyquist path is mapped into the $F(s)$ plane and there are $-P$ encirclements of the $-1 + j0$ point.

The concept of $-P$ encirclements is that the direction of the contour in the $F(s)$ plane is opposite that of the Nyquist path in the s-plane.

Application of the Nyquist Stability Criterion

Consider the system shown in Figure 8-42 with the Nyquist path in the s-plane shown in Figure 8-43.

(figure 8-42)

System model for Nyquist path

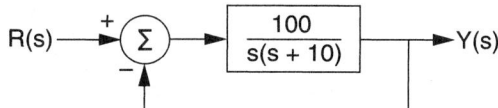

(figure 8-43)

Nyquist path in the s-plane

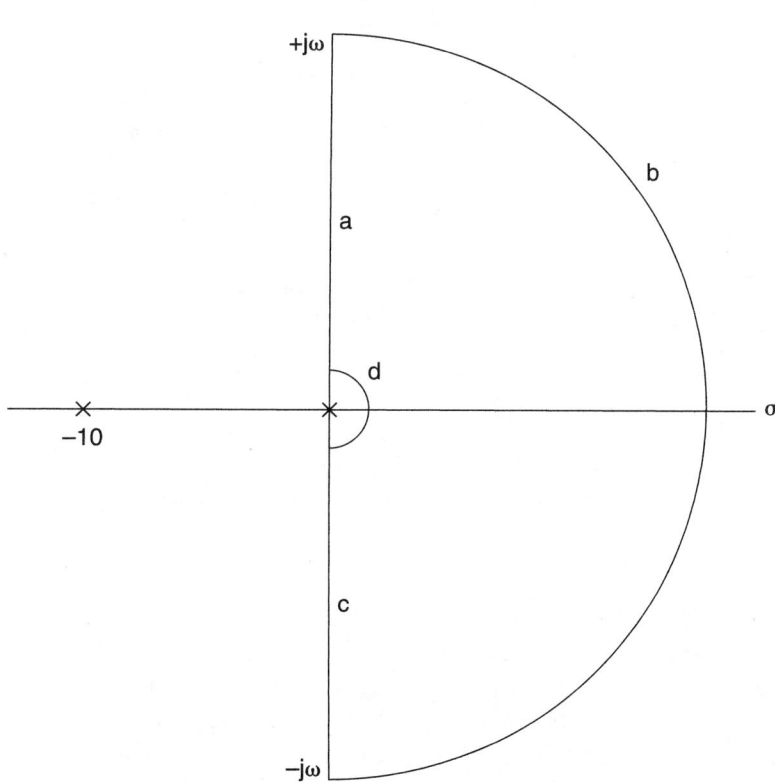

In the evaluation of P, the Nyquist path may either include or exclude the pole at zero. To make this function type one as above and to ensure that there are no poles in the RHP, a small detour around the pole at zero is constructed in the Nyquist path. Evaluation of the path in the $F(s)$ plane is accomplished by separating the transfer function into its real and imaginary components.

$$G(j\omega)H(j\omega) = \frac{100}{j\omega(j\omega + 10)} = \frac{-100(\omega^2 + j10\omega)}{\omega^2(\omega^2 + 100)}$$

$$Re\,G(\omega)H(j\omega) = \frac{-100}{(\omega^2 + 100)} \qquad Im\,G(j\omega)H(j\omega) = -j\frac{1000}{\omega(\omega^2 + 100)}$$

Evaluation of the real and imaginary components at various frequencies gives the plot of Figure 8-44.

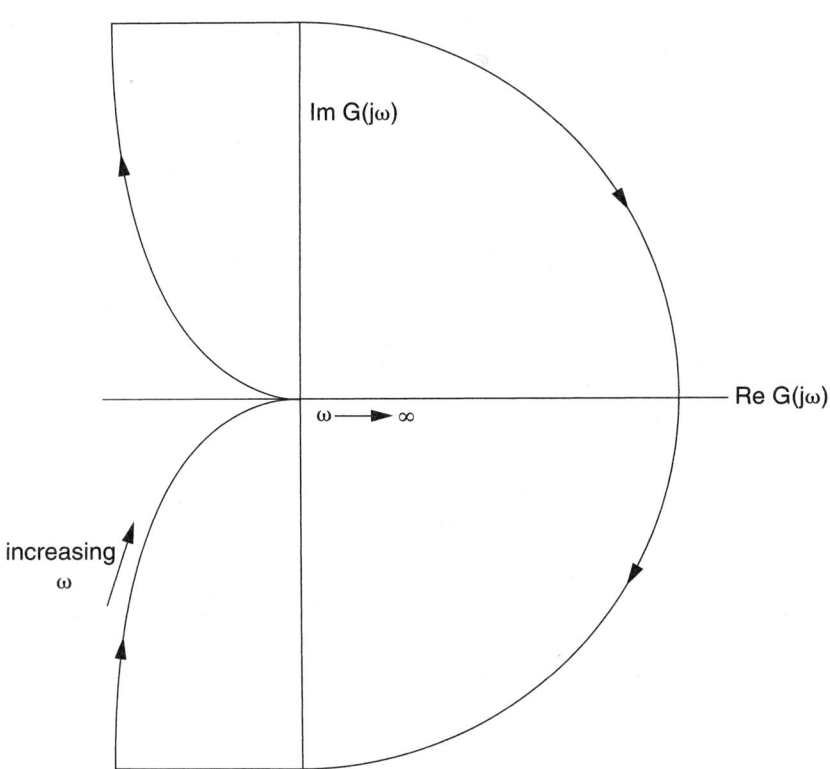

(figure 8-44)

Nyquist plot of Figure 8-43 in the F(s) plane

This plot in the $F(s)$ plane does not cross the negative real axis. Rather, it starts at a frequency of zero radians, at infinity, and terminates at the origin where frequency is equal to infinity. The system is stable with a gain margin of infinity, and no further computation is required. The upper portion of the curve is the mirror image of the lower portion. A short table of the results of calculations for this system is shown in Table 8-4.

	ω	$G(s)H(s)$ Re	$G(s)H(s)$ Im
(table 8-4)	1	−0.99	−j9.9
Nyquist values for	2	−0.96	−j4.91
$G(j\omega)H(j\omega) =$	5	−0.5	−j1.6
$\dfrac{100}{j\omega(j\omega + 10)}$	10	−0.5	−j0.5
	20	−0.2	−j0.1

Gain and Phase Margins from the Nyquist Plot

To obtain the gain and phase margins from the Nyquist plot, it is necessary to expand the plot in the region of the origin to see the crossing point and to add a unit circle drawn from the origin through the point −1 + j0.

To show the use of the plot for gain and phase margins, you will increase the order of the system of Figure 8-44 to have a second-order term.

Example 8-7

The transfer function of Figure 8-42 becomes

$$G(s)H(s) = \frac{100}{s(s + 10)^2}$$

Solution

Setting $s = j\omega$, the transfer function is

$$G(j\omega)H(j\omega) = \frac{100}{(j\omega)^3 + 20(j\omega)^2 + 100j\omega} = \frac{100}{-20\omega^2 - j\omega(\omega^2 - 100)}$$

$$= \frac{100[-20\omega^2 + j\omega(\omega^2 - 100)]}{400\omega^4 + \omega^2(\omega^2 - 100)^2} = \frac{100[-20\omega + j(\omega^2 - 100)]}{\omega\left[400\omega^2 + (\omega^2 - 100)^2\right]}$$

From which the real and imaginary components are

$$Re\ G(j\omega)H(j\omega) = -\frac{2000}{400\omega^2 + (\omega^2 - 100)^2}$$

$$Im\ G(j\omega)H(j\omega) = j\frac{100 \cdot (\omega^2 - 100)}{\omega\left[400\omega^2 + (\omega^2 - 100)^2\right]}$$

Example 8-7 (continued)

The first step in constructing the Nyquist plot is to determine the frequency at which the plot crosses the negative real axis. That is done by setting the imaginary part equal to zero and equating $\omega^2 - 100 = 0$ to obtain $\omega = 10$. Next is determining $Re\ G(j\omega)H(j\omega)$ at that point, which yields a value of -0.05. The gain required to cause the system to cross the $-1 + j0$ point can be calculated from $1/l.05l = 20$. Thus, the gain margin is $20*\log(20) = 26$ dB. Last is drawing a circle from the origin through the $-1 + j0$ point and calculating the angle from the origin to the point where the plot crosses the unit circle. Another method for finding the frequency where the gain of the system is 1.0, which is probably easier, is to use a Bode plot to determine the 0 dB frequency of the plot and the phase angle at that frequency. For this system, the phase margin is 79°.

Relative Stability and the Nichols Chart

Underdamped systems that yield acceptable gain and phase margins determined from open-loop frequency response plots may still have excessive peak values, M_r, in the closed-loop frequency response plot. Further, when working with the polar data of the Nyquist plot, simple changes in loop gain cause the Nyquist plot to have to be recalculated. Often, in system design, changes must be made to the loop gain and other components may be added to the system. In design work involving M_r and bandwidth, it is easier to work directly with the magnitude and phase of a system.

The plot is made using rectangular coordinates. The magnitude is plotted on the ordinate; the phase angle, on the abscissa. This type of plot, suggested by N. B. Nichols, is called the Nichols chart. The data for the plot is obtained from the Bode plot of magnitude and phase. One advantage to this type of plot is that if the system gain is changed, the plot simply slides up or down depending on the change, but does not change its shape. Similarly, if some phase is added, the plot will slide horizontally with no change in shape. The plot of Example 8-7, a third-order system showing the gain and phase margins, is presented in Figure 8-45.

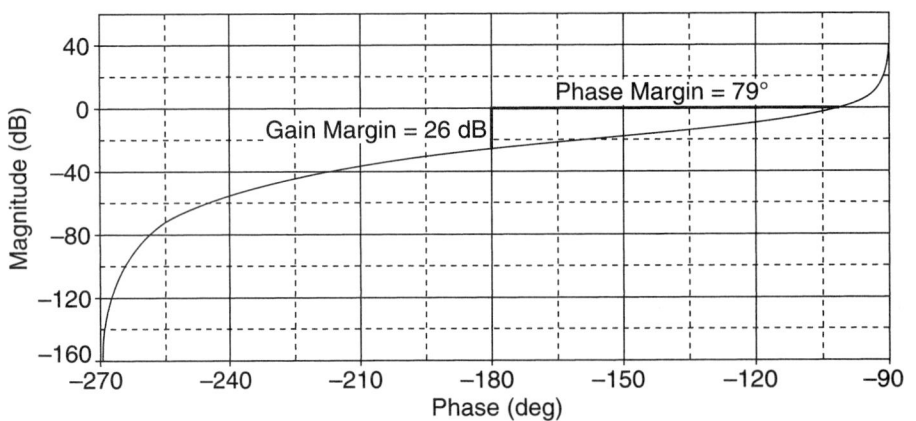

(figure 8-45)

Magnitude-phase chart showing gain and phase margin

While the gain and phase margins are obtained directly from the plot, determining the maximum peak of the response is not an easy task.

Closed-loop gain contours can be mapped into the Nichols chart, which makes finding the resonant peak (M_r) of the response a simple task; you find the closed-loop gain contour to which the magnitude-phase contour is tangent. The Nichols chart is shown in Figure 8-46.

(figure 8-46)

Nichols chart

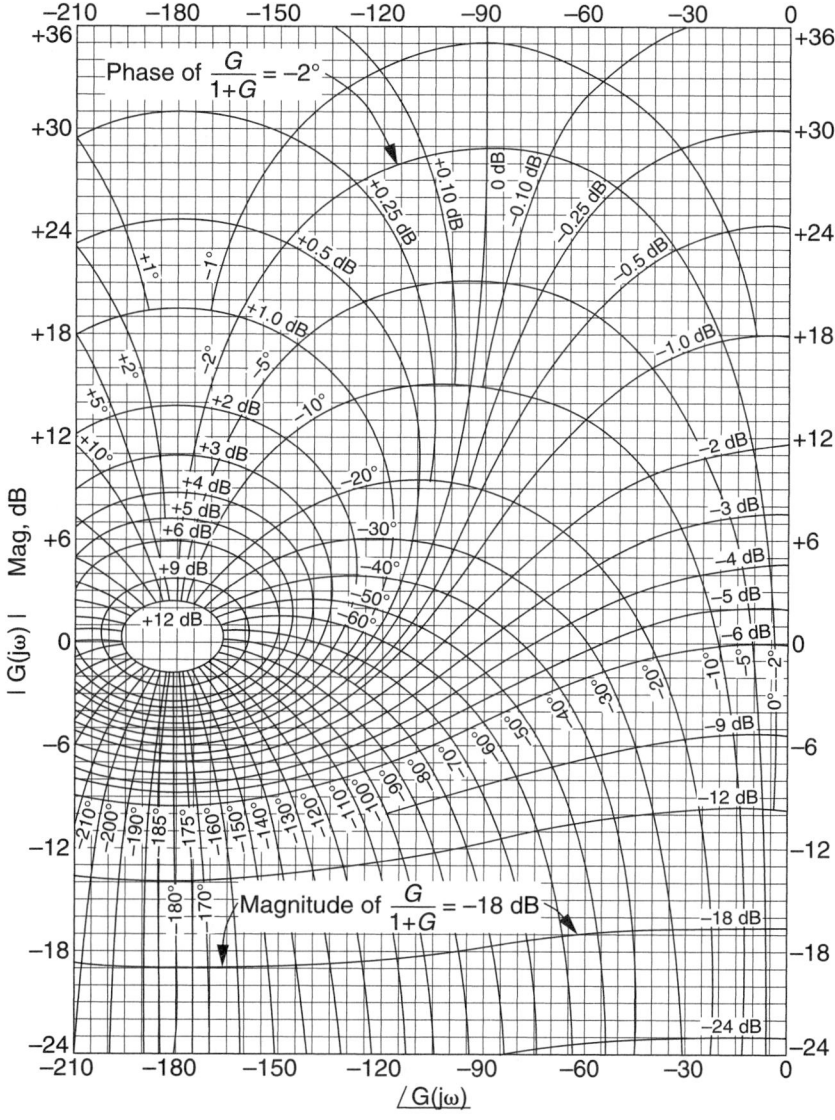

The system of Figure 8-46 has no peak in its closed-loop magnitude plot; but if the numerator of the system is changed from 100 to 750, a +6 dB peak is observed in the closed-loop magnitude plot. The system is plotted on a Nichols chart with this gain. Note the change in gain and phase margins. In addition, the plot is tangent to the +6 dB contour on the chart, which is the value of M_r determined from the closed-loop Bode plot. The Nichols chart for the system is shown in Figure 8-47.

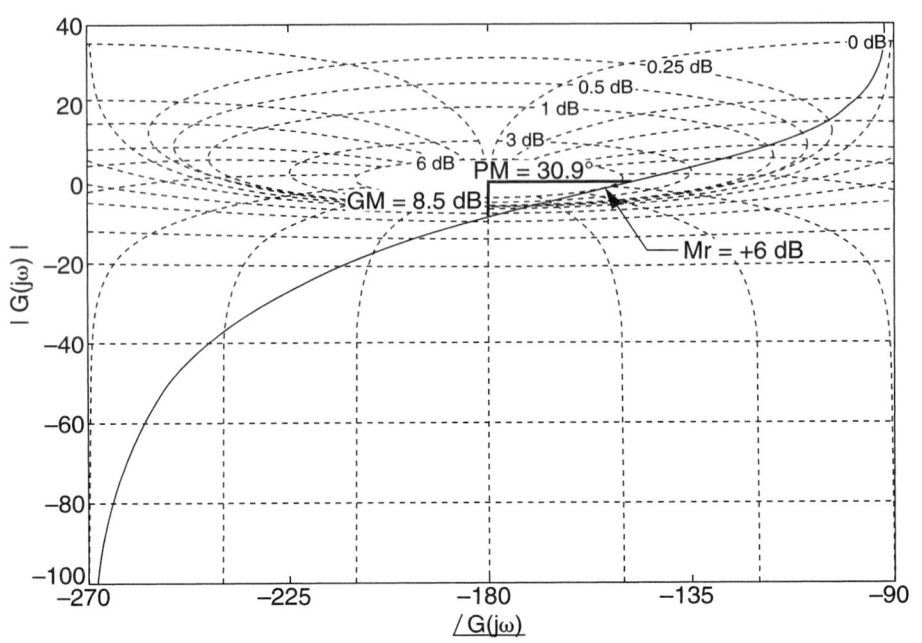

(figure 8-47)

Nichols chart for system of Figure 8-45

Application of the Nichols Chart to Systems with Pure Time Delay

From the investigation of dead time, you learned that the magnitude of a system is unaffected, but the phase changes dramatically. A pure time delay is one that is caused by a physical separation between the point where the process performs an operation and the point where the measurement of the result of the operation is done. The pure time delay is a result only of distance and time.

An example of a pure delay time is a rolling process to make thin sheet metal controlled to tight thickness requirements. The process is shown in Figure 8-48. There is clearly a distance between where the pressure is applied to thin the metal and what the thickness of the measuring point is.

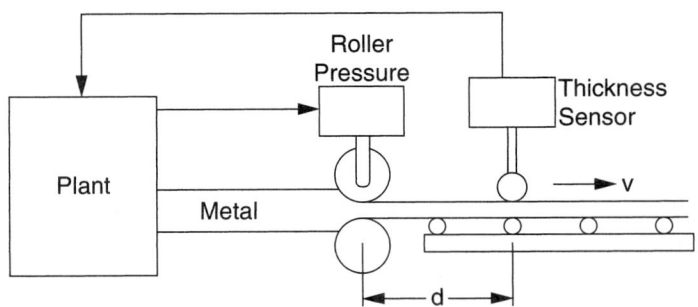

(figure 8-48)

Process with a time delay

The time delay caused by this process measurement is $T_D = v/d$. If the thickness is designated to be w, then the signal at the sensor is described by $y(t) = w(t - T_D)$. The Laplace transform of the function is then $Y(s) = W(s)e^{-sT_D}$.

$$\frac{Y(s)}{W(s)} = e^{-sT_D} \tag{8.26}$$

If a pole-zero representation of the delay is required, an infinite number of poles and zeros is required; but an approximation of e^{-sT_D} can be made using one of the e^{-x} series expansions, such as

$$e^{-sT_D} = 1 - sT_D + \frac{s^2 T_D^2}{2!} - \frac{s^3 T_D^3}{3!} + \ldots$$

$$\frac{1}{e^{sT_D}} = \frac{1}{1 + sT_D + \frac{s^2 T_D^2}{2!} + \frac{s^3 T_D^3}{3!} + \ldots} \tag{8.27}$$

$$\frac{e^{\frac{-sT_D}{2}}}{e^{\frac{sT_D}{2}}} = \frac{1 - \frac{sT_D}{2} + \ldots}{1 + \frac{sT_D}{2} + \ldots}$$

The last approximation is called the Padé expansion. All three forms of e^{-sT_D} can be truncated to two or three terms to provide an approximation for the time delay. It is easier, however, to use frequency response techniques such as the magnitude and phase plot to determine the rate of phase change directly from the plot of the magnitude and phase of the system.

8.8 MATLAB Frequency Response Design Using MATLAB and the SISO Design Tool

MATLAB has a rich store of tools and devices with which to do Bode, Nyquist, and Nichols designs. There is also a design tool for compensation schemes called the SISO Design Tool. In addition, a number of parameters common to all types of plot are available by right-clicking on the graph. The parameters are the minimum stability criteria, all stability criteria, and peak response.

To begin this part of working with **MATLAB**, you will plot the frequency response of the system in Example 8-1. The numerator and denominator of the transfer function are entered in the usual way.

  n = 60*[1 1]; d = [1 5 6];   % Numerator and denominator of Xfer function

  bode (n,d)   % Command to plot the Bode magnitude and phase

The default plot will show both the magnitude and phase on one plot. The graphs will not have the grid lines for frequency. Right-click inside either of the graphs; a pop-up menu listing choices will appear. Select *Grid* and the lines of frequency will appear. The only parameter of use when you right-click and select *Characteristics* is the Peak Response since this system can have only −90° of phase shift. The magnitude-phase response with the peak value labeled is shown in Figure 8-49.

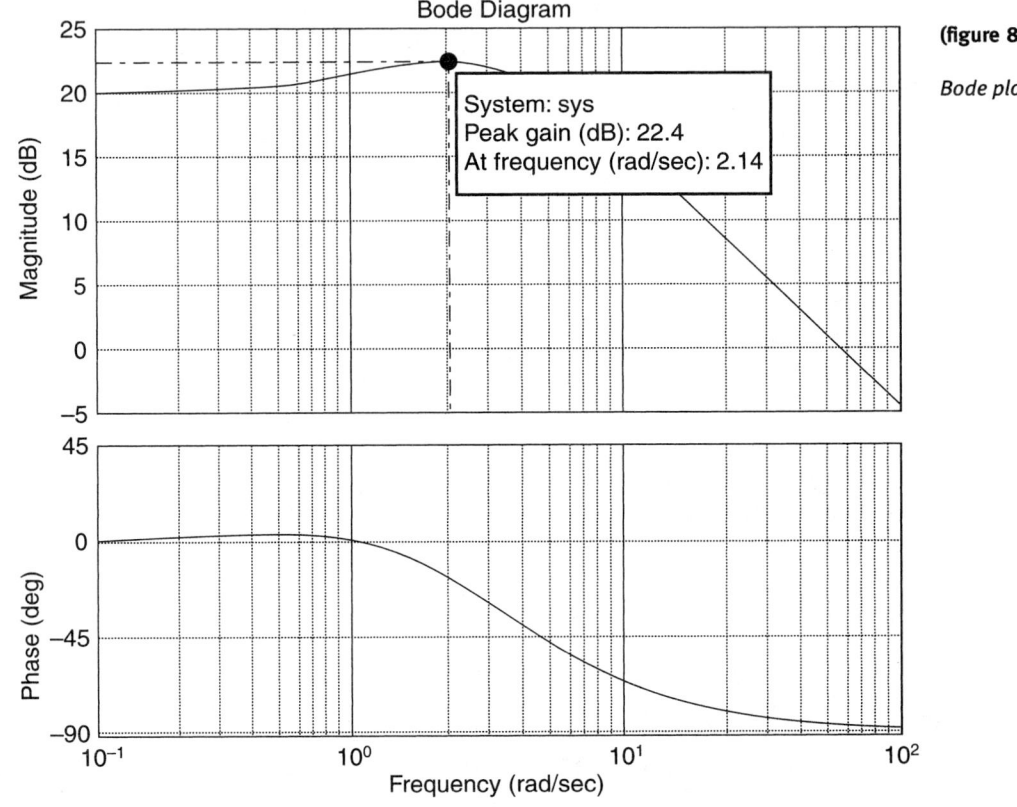

(figure 8-49)

Bode plot of Example 8-1

The Nyquist plot is constructed in a similar way by adding the line

    Nyquist(n,d)   % Generates the Nyquist plot

The plot is shown in Figure 8-50.

The Nichols plot is constructed using

    nichols(n,d)   % Generates the Nichols plot

The plot is shown in Figure 8-51.

Sometimes it is desirable to show both the open- and closed-loop frequency response of a system on the same graph. That is accomplished by setting up the plots manually for both the frequency and phase responses. Since more lines are required than in default

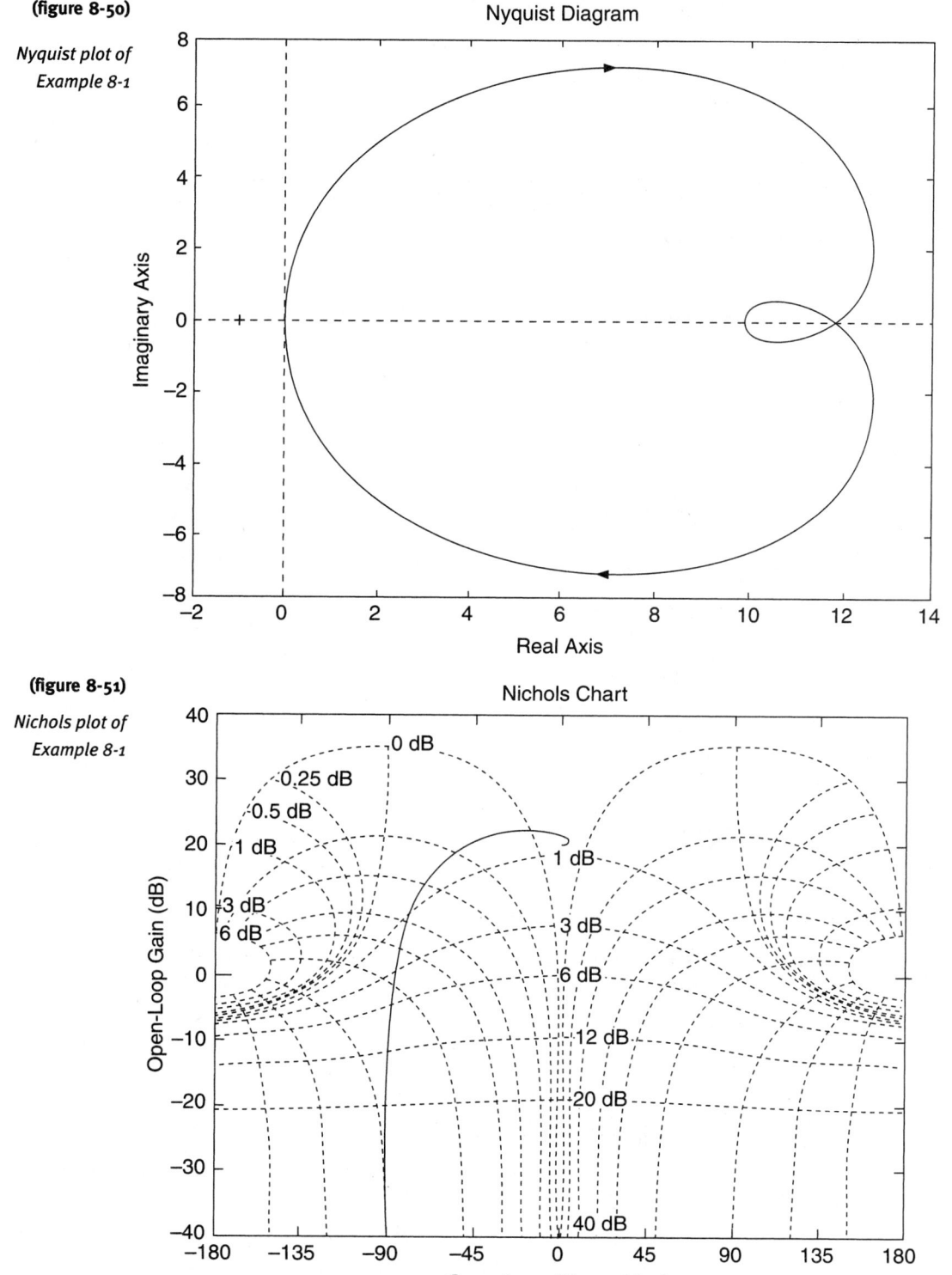

(figure 8-50)

Nyquist plot of Example 8-1

(figure 8-51)

Nichols plot of Example 8-1

plots, it is appropriate to create what is called an M-file or script. An M-file is nothing more than a set of commands that perform a specific set of calculations. The file is saved with the extension .m and, with a minimum of trouble, can be modified as needed for other calculations.

To make the M-file, click the File menu, then click New, then click M-File. The editor will open, and you can start typing the commands. The comments below are not necessary, but some may be useful when making other M-files. When you finish typing the instructions, save the file with a name of your choice. To run the file, type the name of the file in the workspace.

% M-file to show open and closed-loop frequency and phase response of Example % 8-5. The title of the file is "twoplot."

num = 2.2e6*[1 9];	% Numerator for open- and closed-loop gain
d1 = [1 300 3.006e4 1.006e6 2e5];	% Denominator for open-loop gain
d2 = [1 300 3.006e4 3.2006e6 1.98e7];	% Denominator for closed-loop gain
wn = logspace(−2, 3, 400);	% Set up for min and max frequency. −2 = 0.01 Hz, 3 = 1000 Hz and 400 is the number of points calculated.
[mag1, ph1, wn] = bode(num, d1, wn);	% Open-loop gain and phase vectors
[mag2, ph2, wn] = bode(num, d2, wn);	% Closed-loop gain and phase vectors
dB1 = 20*log10(mag1); dB2 = 20*log10(mag2);	% Convert magnitude to dB.
[gm, pm] = margin(n,d1)	% Calculate gain and phase margins for open-loop transfer function.
figure(1)	% First graph
semilogx(wn, dB1, wn, dB2), grid	% Set graph one to be logarithmic on horizontal axis with the frequencies noted above and the gains determined by dB1 and dB2.
axis([.01 1000 −60 20])	% Set axis limits.
xlabel('Freq (Rad/s)'), ylabel('Gain (dB)')	% Print axis labels.
pause	% Wait for keypress to go to figure 2.
figure(2)	% Second graph

semilogx(wn,ph1,wn,ph2), grid   % Logarithmic graph for frequency
                                               and phase

axis([.01 1000 –270 30])   % Set axis limits for phase plot.

set(gca,'ytick', [–270:30:30])   % Specify the y-axis scale (optional).

xlabel('Freq (Rad/s)'), ylabel('Phase (Deg)') % Label axes.

When the M-file is run, the magnitude plot for the open- and closed-loop gain is displayed. To continue with the phase plot, press any active key; the phase plots are displayed. Note that there are no right-click options available on these graphs since you have already set axis and grid limits.

Use of the SISO Design Tool

The SISO Design Tool is part of the control system toolbox. This tool is particularly useful in the design of compensation systems. To use it, you must first create a transfer function and then call the SISO Design Tool. Let's reexamine Example 8-4 using the SISO Design Tool.

Example 8-8

The conditions remain the same; the input to the control system is a ramp of 0.5 m/s. The steady-state error must be equal to or less than 0.02 meters. In addition, the phase margin must be at least $45° \pm 5°$. For the error specification, the gain is $K = 25$.

Solution

First, create a transfer function for the system with the following script:

```
n = [0 0 75]; d = [ 1 3 0];  % Numerator and denominator vectors
sys_tf = tf(n,d)             % Create the OL transfer function.
```

After creating the transfer function, open the design tool window by typing **sisotool(sys_tf).** The **SISO Design Tool** opens with the Controls and Estimation Tools Manager and the root-locus, gain magnitude, and phase in three graphs of the system open-loop response displayed. The window is shown in Figure 8-52a and 8-52b.

The plots of frequency and phase confirm the results of the open-loop plots of Example 8-4 prior to adding the lead network.

After creating the transfer function, open the design tool window by typing *sisotool(sys_tf)*. The SISO Design Tool opens, displaying the root locus, gain magnitude, and phase in three graphs of the system open-loop response. The window is shown in Figure 8-52.

To add a lead network to the system, use the following procedure: The requirements are that you allow the new frequency crossover point to be approximately one octave higher then the uncompensated system, or 16.8 rad/sec, and add $34°$ of additional phase.

Example 8-8 (continued)

By default, the **SISO Design Tool** assumes the compensator to be in the forward path of the system. At the top of the manager, the feedback system is pictured. The compensator position can be changed by clicking the *Control Architecture* button on the manager.

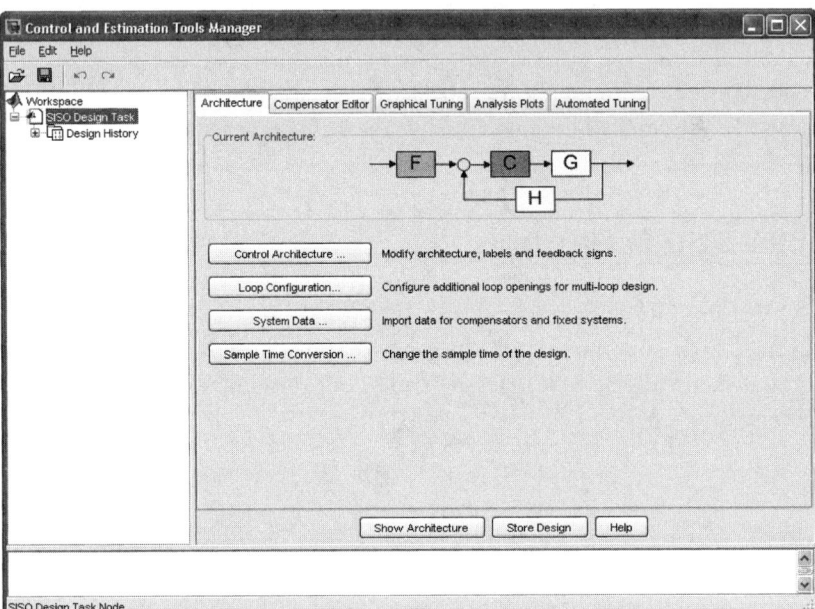

(figure 8-52)

SISO design screen for Example 8-8

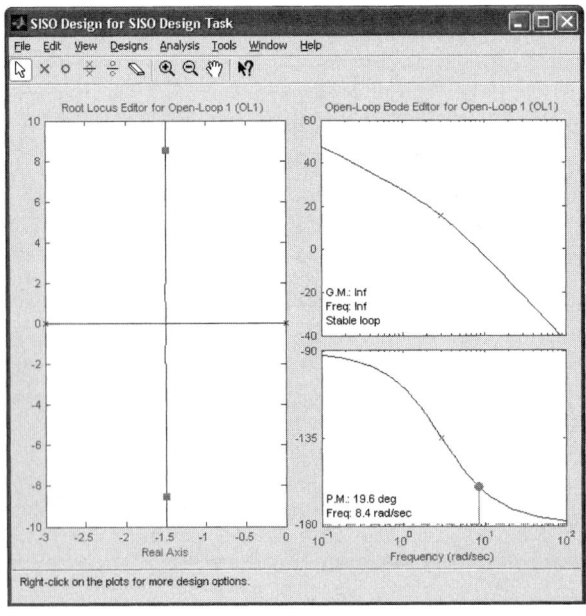

Example 8-8 (continued)

There are four possible configurations. The default values of all functions except **G**, is one. The blocks are: a prefilter, **F**, the compensator, **C**, the feedback block, **H**, and the plant function, **G**. For our system, the lead compensator is in the forward path. The +/– button changes the sign of the feedback, the default is negative. We will use the default configuration.

Before you add the lead network, look at the step response of the system. Hold it on screen for observation as you add and adjust the lead network. Click the Analysis menu and select *Response to Step Command.* The LTI Viewer for SISO Design Tool graph is displayed with the step response shown. Arrange the two graphs as shown in Figure 8-53.

To add a lead compensator, right-click inside the magnitude graph and select *Add Pole/Zero;* then click *Lead.* The cursor has an *x* attached to it, representing the pole. Place the pole close to the rightmost pole (blue *x*) on the chart. Doing so changes the crossover frequency and phase; the root locus changes to reflect the addition of the pole and zero. An example is shown in Figure 8-54. Yours may differ slightly.

(figure 8-53)

Uncompensated response of Example 8-8

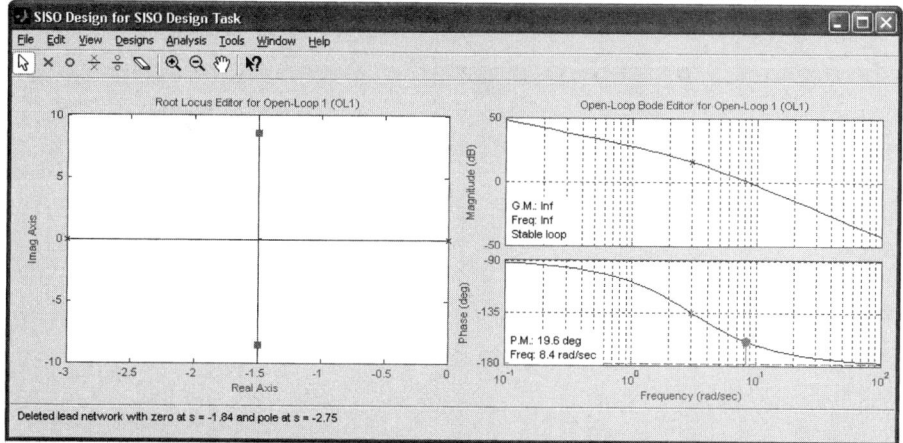

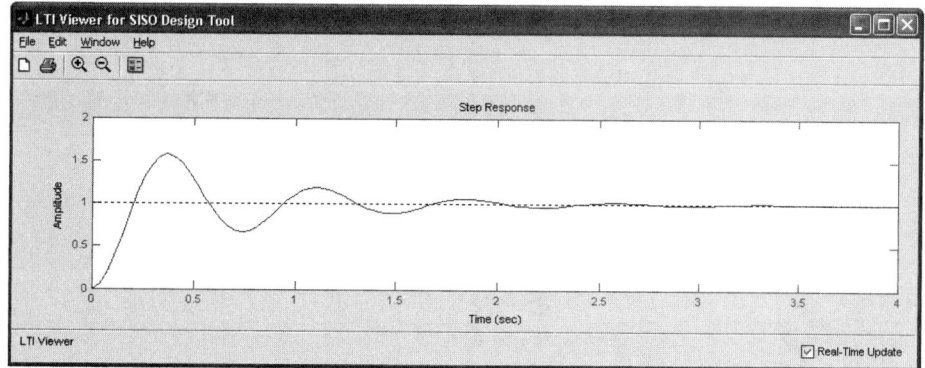

Example 8-8 (continued)

The value of the pole and zero is shown at the bottom of the design tool screen. You can move the pole to the desired position by placing the cursor on the pole. The cursor changes to a hand, and the value of the pole is displayed at the bottom of the screen as you move the pole to the proper value, which is $s = -14$. You may not be able to get -14 exactly, but 13.9 as shown in Figure 8-55 is fine. Then you move the pole to the desired position, which is $s = -4$. As you do so, observe the step response and root-locus change with each movement. When the moves are complete, the panel at the top of the design tool shows the values of the time constants and the overall gain of the compensator. Do not change the gain, as it is correct for the system. The ratio of the time constants provides the gain of the compensator that maintains the output at 1.0. The ratio is approximately 3.5, which is the same as that calculated in Example 8-4. Figure 8-55 shows the design tool representation of the magnitude and phase, the root-locus, and the step response of the system with the compensator added.

The characteristics of the finished design are the same as those presented for Example 8-4, but they were completed in considerably less time.

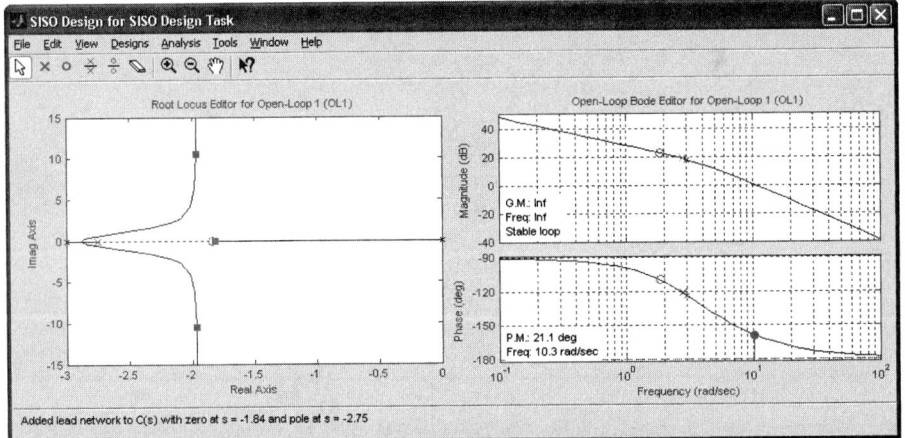

(figure 8-54)

Starting position of pole and zero for Example 8-8

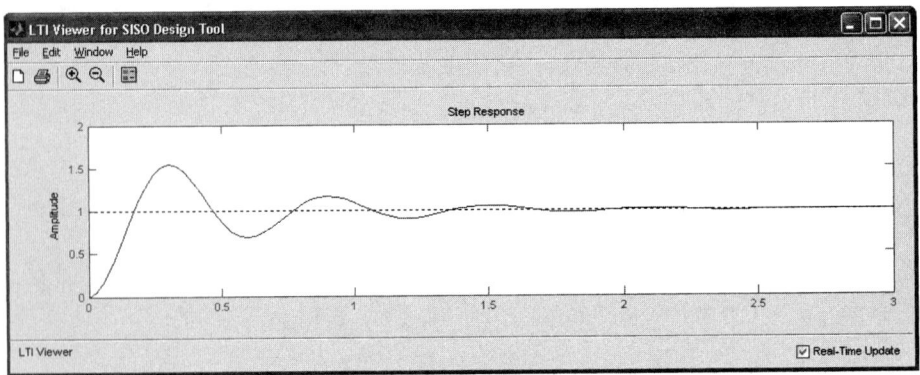

(figure 8-55)

*Compensated system of
Example 8-8*

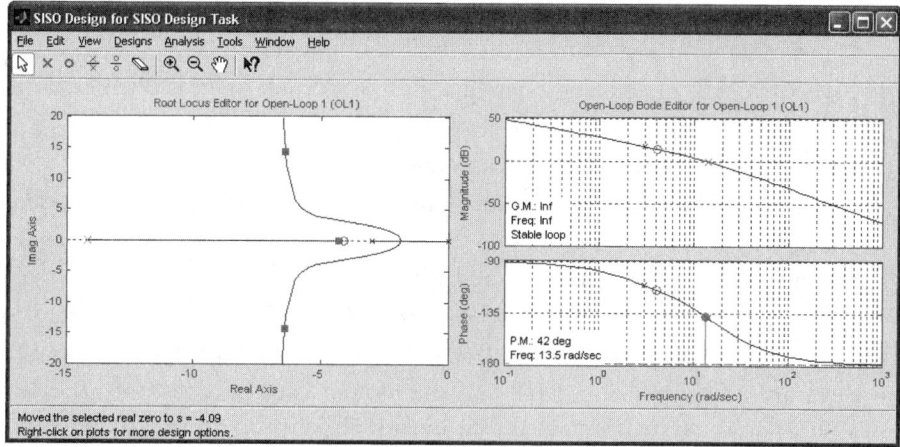

(figure 8-55)

*Compensated system of
Example 8-8*

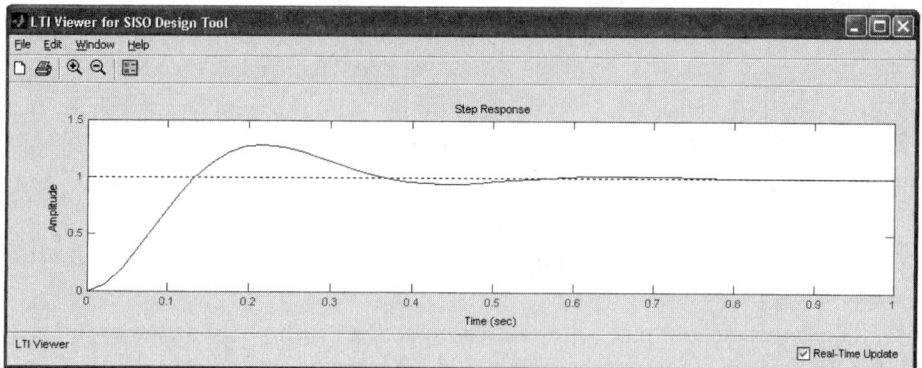

Summary

➤ Methods to determine stability that rely on s-plane parameters are changed to reflect how the system responds to changes in frequency applied to a system.

➤ Bode's analysis shows the frequency response of a system and the gain and phase margins before instability of the system occurs.

➤ Both magnitude and phase of a system in response to changing frequency are generated by this analysis. While it is very useful, no information regarding the overshoot or damping of a system is directly provided by the analysis.

➤ A peak in the response curve can be indicative of the damping ratio. A damping ratio of $\zeta = 0.707$ can be related to a peak of approximately 0.5 dB in the frequency response magnitude curve.

➤ Plotting the Bode magnitude and phase ($M\angle\theta$) on a polar plot allows us to see if a system is unstable at a glance.

➤ The determination of gain and phase margins on the polar plot is often easier than on a Bode plot.

➤ The polar plot also shows that encirclement of the $-1 \angle 180°$ point causes instability.

➤ Using gain and phase margins provides a measure of relative stability of a system.

➤ Reducing a system to a single loop incorporates all cascaded functions around the loop, without the minus sign associated with the feedback signal.

➤ The Nyquist stability criterion is applied by calculating the Nyquist path in the s-plane and plotting the function $G(s)H(s)$ in the $F(s)$ plane on real and imaginary axes.

➤ If there are no poles in the RHP and the plotted contour does not encircle the $-1 + j0$ point, the system is stable. If the open-loop function exhibits P poles in the RHP, the system is stable.

➤ If there are $-P$ encirclements of the $-1 + j0$ point in the $G(s)H(s)$ plane, the system is stable.

➤ In all cases, $N = Z - P$ gives the number of encirclements of the $-1 + j0$ point.

➤ The Nyquist plot does not give an indication of damping ratio or overshoot as the Bode plot does.

➤ Lead, lag and lead-lag compensation using frequency response analysis is used to improve the gain and phase margins of a system.

➤ Lead, lag and lead-lag circuits and the calculations needed to produce the functions have been provided.

➤ If the magnitude and phase are plotted on a chart with phase as the ordinate and magnitude as the abscissa, the gain and phase margins are readily calculated.

➤ Magnitude and phase are useful for systems with a pure time delays.

➤ Frequency response methods are particularly effective in displaying pure time delays.

➤ Time delay does not change the magnitude of the plot, but it causes a large change in phase, which is readily apparent on the phase/magnitude plot.

➤ If contours of constant magnitude in dB are added to the chart, the result is called a Nichols chart.

➤ The percentage of overshoot is readily apparent as the point of tangency of the phase/magnitude curve with a contour of constant magnitude.

➤ **MATLAB** can be used to show all of the types of plots discussed.

➤ The **MATLAB** M-file as a convenient method to store system parameters and output graphical data has been shown.

➤ The SISO Design Tool can be used to develop and design lead, lag, and lead-lag circuits.

Problems

P8.1   For a system whose open-loop transfer function is

$$G(s) = \frac{10K}{(0.5s + 1)^2}$$

a. Make an asymptotic plot of the Bode magnitude and phase if $K = 2.5$.

b. Determine the gain and phase margins from your plot. If it is not possible to determine a gain margin, explain why.

P8.2   Use Bode magnitude and phase plots to determine the gain and phase margins of a system whose open-loop transfer function is

$$G(s) = \frac{K}{s(0.25s + 1)^2}$$

a. Determine the gain and phase margins, using Bode magnitude and phase plots if $K = 1, 5, 8$.

b. What type of number is this system?

P8.3   For the following open-loop transfer function, plot the Bode magnitude and phase for $K = 50, 100$. Determine gain and phase margins. Use the Routh test to determine the gain for which the system becomes unstable. Plot the magnitude and phase at the gain from the Routh test and confirm that the gain and phase margins are approximately zero.

$$G(s) = \frac{K(s + 3)}{s^4 + 12s^3 + 49s^2 + 78s + 40}$$

P8.4   Consider the following characteristic equation:

$$s^4 + 6s^3 + 11s^2 + (6 + K)s + 5K = 0$$

Use the Routh test to determine the maximum gain for which the system becomes unstable. Then provide the open-loop transfer function and plot the Bode gain and phase, using this gain.

P8.5   For the following open-loop transfer functions, construct a Bode plot to determine the phase crossover frequency. Then calculate the maximum dead time the systems can tolerate and show the plot of each function with the dead time on the same plot.

a. $G(s)H(s) = \dfrac{10}{s + 2}$

b. $G(s)H(s) = \dfrac{100}{s(s + 20)}$

c. $G(s)H(s) = \dfrac{60 \cdot (s + 2)}{s(s + 1)(s + 4)}$

P8.6   The system shown in Figure P8-6 has a step response that exhibits excessive overshoot (39.8 percent) and a phase margin of 32.1° when $K = 1$. It is required that the system have no more than 5 percent overshoot. Design a lead compensator to provide the desired response. The new crossover frequency should be $\geq$ 30 rad/sec, and the phase margin should be $\geq$ 60°. Show the opamp circuit of the compensator and the component values required to produce the required specifications.

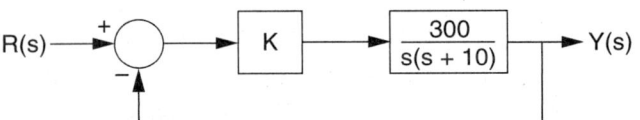

(figure P8-6)

Control system for Problem P8.6

P8.7   The system shown in Figure P8-7 must have a ramp error constant of $K_v = 10$ and no more than 20 percent overshoot. To satisfy the error constant, the gain must be $K = 3.33$, which results in >50 percent overshoot. Design a lag compensator to provide the specifications.

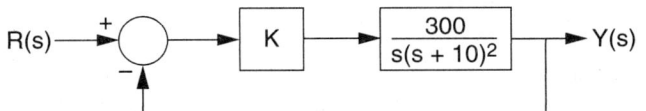

(figure P8-7)

Control system for Problem P8.7

P8.8   Given a unity feedback system with open-loop transfer function

$$G(s) = \frac{K}{s(s + 2)(s + 3)}$$

Design a lead-lag compensator to provide the following specifications:

   Overshoot less than or equal to 10 percent

   Phase margin of at least 40°

   Velocity error constant $K_v = 15$

P8.9   For the following type zero system, plot the magnitude and phase angle on a polar chart. Then determine the phase margin and gain margin of the system.

$$G(j\omega)H(j\omega) = \frac{5}{\left(\dfrac{j\omega}{5} + 1\right)^3}$$

P8.10   For the following system, plot the Nyquist path in the $F(s)$ plane. Does any part of the Nyquist plot cross the imaginary axis?

$$G(s)H(s) = \frac{10 \cdot (s + 2)}{(s + 3)(s + 6)}$$

P8.11   For the following single-loop system, construct Nyquist plots for $K =$, 2.0, 4.0, 8.0. Determine the gain and phase margins for each gain. At $K = 8.0$, the system passes through the $-1 + j0$ point. Is the system stable, conditionally stable, or unstable?

$$G(j\omega)H(j\omega) = \frac{K}{\left(\dfrac{j\omega}{2} + 1\right)^3}$$

P8.12   For the following type one system, determine the real and imaginary components of the system. Determine the gain and phase margins of the system from the plot.

$$G(j\omega)H(j\omega) = \frac{1}{j\omega\left(\dfrac{j\omega}{2} + 1\right)^3}$$

P8.13   For the following type zero system, plot the Nyquist path in the $F(s)$ plane. Determine the gain and phase margins of the system from the plot.

$$G(s)H(s) = \frac{120}{(s + 2)(s + 3)(s + 5)}$$

P8.14   From the Nyquist diagram in Figure P8-14, estimate the gain and phase margins graphically.

(figure P8-14)

Nichols chart for Problem P8.14

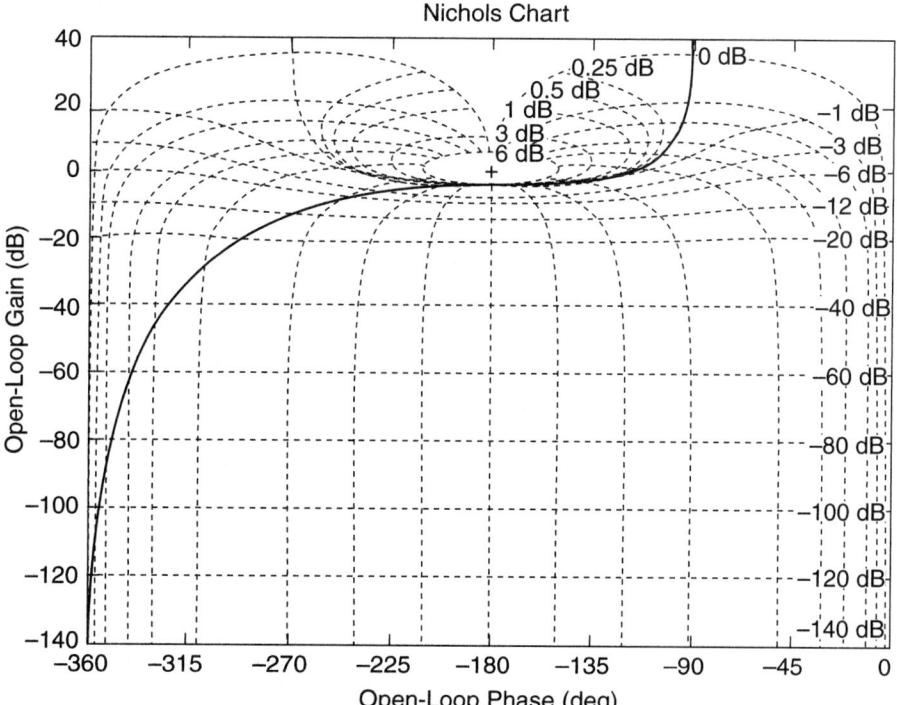

P8.15   Sketch the log magnitude and phase curves for the following open-loop functions. Show the gain margin and phase margin for each of the systems. Then show those points on the plots.

a. $G(s)H(s) = \dfrac{10}{s + 5}$

b. $G(s)H(s) = \dfrac{10}{s(s + 5)}$

c. $G(s)H(s) = \dfrac{100}{s(s + 4)(s + 5)}$

d. $G(s)H(s) = \dfrac{10 \cdot (s + 2)}{s\left(\dfrac{s}{4} + 1\right)^3}$

MATLAB Problems

M8.1   Use **MATLAB** to check the results for Problems P8.1–P8.5.

M8.2   For the following system, write an M-file to produce the Nyquist plot of the system.

$$G(s)H(s) = \dfrac{1}{s\left(\dfrac{s}{10} + 1\right)^3}$$

M8.3   Use the **MATLAB** SISO Design Tool to design a lead controller for the following specifications:

The open-loop transfer function of the system is

$$G(s)H(s) = \dfrac{20}{s\left(\dfrac{s}{5} + 1\right)}$$

Overshoot ≤ 10 percent

Phase margin $\geq 60°$

M8.4   Write an M-file for Problem M8.3 that shows both the uncompensated and compensated system as a Nichols plot.

M8.5   For the following system, produce the closed-loop step response and the Bode plot. Also show the stability margins. Then make a closed-loop Nyquist plot and calculate the gain margin from the plot.

$$\dfrac{Y(s)}{R(s)} = \dfrac{120}{s^3 + 10s^2 + 40s + 120}$$

| # Controller Design and Loop Tuning

Chapter Objectives

After completing this chapter, you should be able to:

➤ Describe various forms of control systems—on-off control, proportional control, proportional-derivative control, proportional-integral control, and proportional-integral-derivative control.

➤ Define the fundamental second and higher order models for plants used in control systems.

➤ Demonstrate how the four types of commonly used controllers are modeled mathematically.

➤ Describe what parameters of the four types of controllers can be manipulated.

➤ Achieve the required error, settling time, and final value required in the design of a controller.

➤ Explain how the time domain and frequency response of the systems being designed can be used to show the stability and the gain and phase margins of the systems.

➤ Explain the concepts of proportional, proportional-derivative, proportional-integral, and proportional-integral-derivative controllers.

➤ Explain how the positioning of poles and zeros of a system can be used to achieve the required response.

➤ Explain the two concepts of Ziegler-Nichols loop tuning as applied to PID controllers.

➤ Explain the conditions under which each of the tuning methods is used.

9.1 Introduction

The purpose of any control system is to provide as near an ideal relationship between a reference input and a plant output as sensor technology and cost will allow. In this book, it is assumed that the system is a single loop system or a multiple loop system that can be reduced to a single loop using the methods introduced thus far.

For the purposes of this chapter, the control system can be considered a controller consisting of various types of control schemes in cascade with a plant function and a feedback scheme determined by the system requirements.

Figure 9-1 shows the basic linear control system configuration. $G_C(s)$ may consist of proportional, integral, or derivative controllers and any compensators that are needed. $G_p(s)$ is a linear model of the driven elements of the system, and $H(s)$ is the feedback that is decided on by factors such as available sensors and cost.

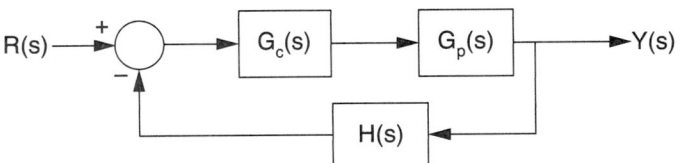

(figure 9-1)

Fundamental system configuration

9.2 On-Off Control

Of all the control schemes, on-off is the simplest. The concept is that the system turns fully on at some predetermined time, temperature, fluid level, and so on, and turns completely off when the desired conditions are met. In general, there is always an overshoot and undershoot when the system reaches the desired conditions. On-off control uses feedback to monitor the conditions of the measured variable. An external sensor measures the condition of the variable and generates an error signal when the measured variable does not equal the setpoint. This type of control is often called bang-bang control due to its all-or-nothing nature.

An example of this type of system is the heating of water in a tank. If the thermostat is set to 150°F and the water is above or below that temperature, the heating element is turned on or off. When the element is turned on and when the water reaches the desired temperature, the heater turns off; but the water temperature continues to rise because the element does not cool instantaneously. When the element is off, the water temperature falls. The heater will come on at the setpoint, but the water temperature will continue to fall because the heater does not heat instantaneously either. While that may be acceptable for heating water or another liquid, it is not acceptable for controlling high-precision systems.

This book will consider the continuous feedback systems needed when precision in the controlled variable is required.

9.3 Proportional Control

The simplest type of system is proportional control. In general, this type of control has only an amplifier in cascade with a plant. The feedback may be unity or some proportion of the output of the system as measured by a sensor at the output.

Second-Order Systems

The general transfer function of the system is given by consideration of Figure 9-1.

$$\frac{Y(s)}{R(s)} = \frac{G_C(s)G_P(s)}{1 + G_C(s)G_P(s)H(s)} \tag{9.1}$$

This type of controller is a simple gain, $G_C(s) = K_P$. If $K_P G_P(s)H(s)$ is much greater than 1, the output of the system is approximately $1/H(s)$.

The plant transfer function can be of any order; but in general, there are two broad types that are encountered. One is similar to the DC motor transfer function and is second-order and type zero.

$$G_P(s) = \frac{A}{s^2 + 2\zeta\omega_n s + \omega_n^2} \tag{9.2}$$

This type of plant usually has an error between the input and the output when a step input is applied. This type of system will not follow a ramp input unless an additional pole is inserted in the overall transfer function.

The second form is one that does not have an error for a step input but does have an error for a ramp input; the system is type one. The form of the transfer function is

$$G_C(s) = \frac{\omega_n^2}{s(s + 2\zeta\omega_n)} \tag{9.3}$$

The only system parameter you have control of in this type of controller is the system damping ratio, which gives some control of the risetime of the system. Unfortunately, the cost of this control is in the overshoot of the system. The greater the value of K_P, the lower the damping ratio and the greater the overshoot. Some examples will illustrate the relationship between K_P, ζ, and the overshoot in the system.

Example 9-1

The system shown in Figure 9-2 is to be designed to have a damping ratio of 0.707. Determine the required value of K_P for that damping ratio.

Solution

First, determine the value of damping ratio for the plant.

$$2\zeta\omega_n = 17 \qquad \omega_n^2 = 60 \qquad \zeta = \frac{17}{2\sqrt{60}} = 1.097$$

The damping ratio shows that the system is moderately slow, with a damping ratio slightly greater than 1. Next, determine the system closed-loop transfer function and calculate K_P for the desired damping ratio.

$$\frac{Y(s)}{R(s)} = \frac{60K_P}{s^2 + 17s + (60 + 60K_P)}$$

The new required value of ω is

$$\omega = \frac{17}{2 \cdot 0.707} = 12.022 \text{ rad/sec}$$
$$\omega^2 = 144.54$$

Note that the new value of ω is not called ω_n. K_P is found from

$$144.54 = 60 + 60K_P \qquad K_P = \frac{144.54 - 60}{60} = 1.409$$

Using the new parameters, the transfer function of the system is

$$\frac{Y(s)}{R(s)} = \frac{84.54}{s^2 + 17s + 144.54}$$

The overshoot of the system is now 4.3 percent, but an error in the output relative to the input is seen. From the parameters, the final value of the system is 0.585 of the input for a large error of 41.5 percent. Use **MATLAB** to write the numerator and denominator of the transfer function; then simulate the step output of the system. The required lines are

n = 84.54; d = [1 17 144.6];

step(n,d)

The response of this system is shown in Figure 9-3.

If the damping ratio is decreased to 0.5, there will be a change in the error and overshoot of the system. The calculations for ω and K_P are left for the reader. The new transfer function is

$$\frac{Y(s)}{R(s)} = \frac{229}{s^2 + 17s + 289}$$

Example 9-1 (continued)

The output of the system is shown in Figure 9-4.

The error is now only 20 percent, but the overshoot is greater than 15 percent. Another interesting anomaly is that the risetime decreases for the increased K_P, but the settling time (5 percent) increased slightly because it is now on the right side of the peak response.

(figure 9-2)

System configuration for Example 9-1

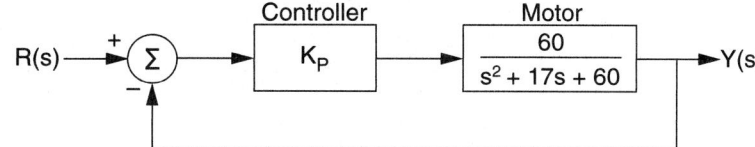

(figure 9-3)

Step output of Example 9-1 with $\zeta = 0.707$

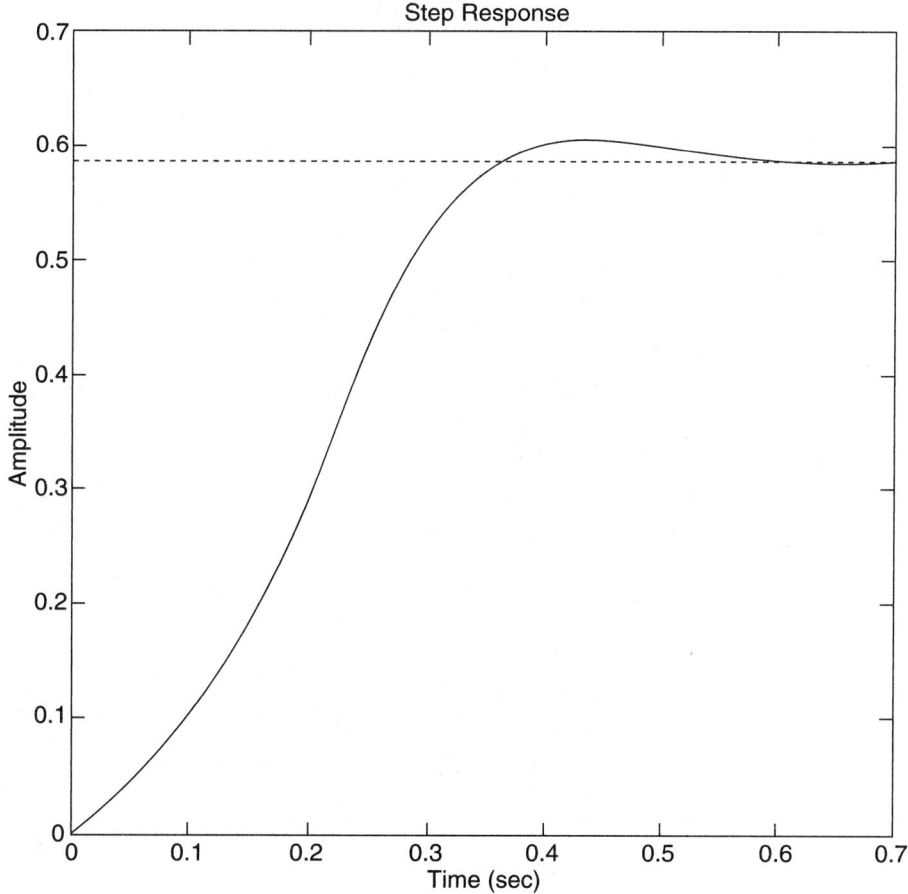

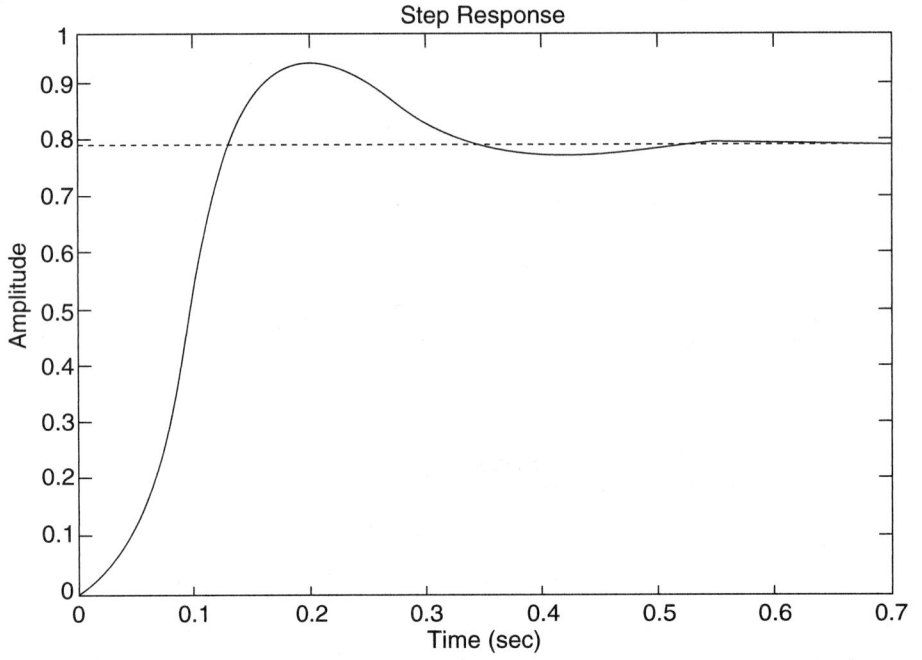

(figure 9-4)

Step output of Example 9-1 with ζ = 0.5

Example 9-2

A DC motor that turns a tracking antenna has the following specifications:

$$K_t = 10.5 \text{ oz-in/A}, \quad B = 0.701 \text{ oz-in/(rad/sec)}, \quad J = 4.4 \cdot 10^{-3} \text{ oz-in-s}^2;$$
$$R_a = 1.11\Omega, \quad L_a = 0, \quad K_b = 0.074 \text{ V/(rad/sec)}. \text{ The gear ratio is}$$
$$N = 1 : 10.$$

The system is to follow a ramp input $tu(t)$ rad/sec with an error less than or equal to 0.075 rad/sec. Determine the transfer function and K_P to satisfy all of the requirements. The block diagram is shown in Figure 9-5.

Solution

First, determine the transfer function of the motor.

$$\frac{V_m(s)}{\Omega_m(s)} = \frac{K_t / R_a J}{s + \dfrac{BR_a + K_t K_b}{R_a J}} = \frac{10.5 / (1.11 \cdot 4.4 \cdot 10^{-3})}{s + \dfrac{0.701 \cdot 1.11 + 10.5 \cdot 0.074}{1.11 \cdot 4.4 \cdot 10^{-3}}} = \frac{2149.88}{s + 318.41}$$

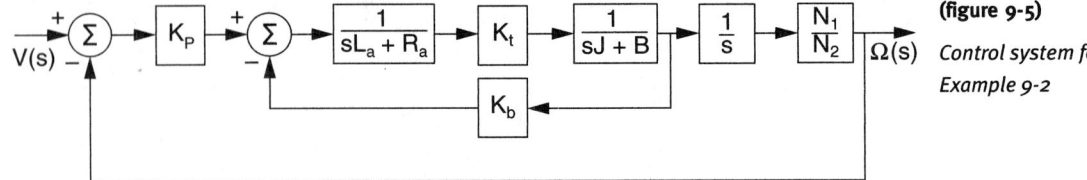

(figure 9-5)

Control system for Example 9-2

Example 9-2 (continued)

To simplify, you can round the transfer function to whole numbers.

$$\frac{V_m(s)}{\Omega_m(s)} = \frac{2150}{s + 318}$$

To determine the value of K_p to satisfy the error requirement, you need the forward path, including the gear ratio, $G(s)$.

$$G(s) = \frac{215}{s(s + 318)}$$

To meet the specifications, K_v must be greater than or equal to 13.333.

$$K_v = \lim_{s \to 0} sG(s) = \frac{215K_p}{318} = 0.6761K_p \geq 13.333$$

$$K_p = \frac{13.333}{0.676} = 19.72$$

You will use $K_p = 20$. The transfer function of the system is

$$\frac{\Omega_m(s)}{V_m(s)} = \frac{4300}{s^2 + 418s + 4300}$$

The output of the system for a ramp input is shown in Figure 9-6.

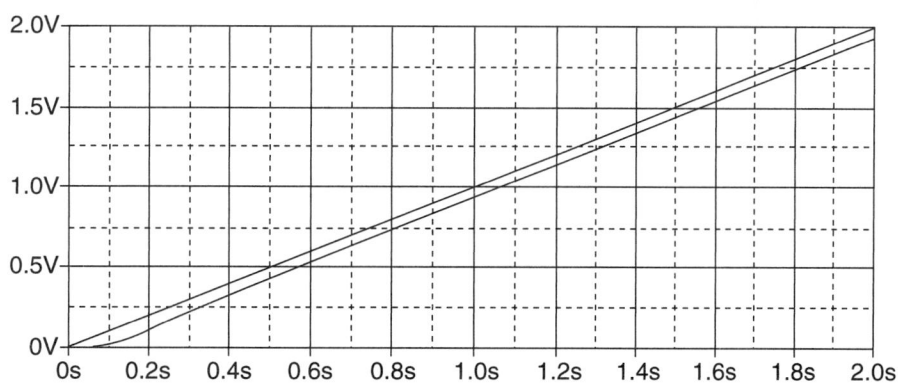

(figure 9-6)

Ramp output of Example 9-2

9.4 Proportional-Derivative (PD) Control

Derivative control is an anticipatory form of control in that it follows the slope of the output of a system. It adds a zero to a system but does not increase the system order.

Control of the system rise and settling times is also obtained by using this type of control. The derivative controller may be implemented using the circuit of Figure 9-7.

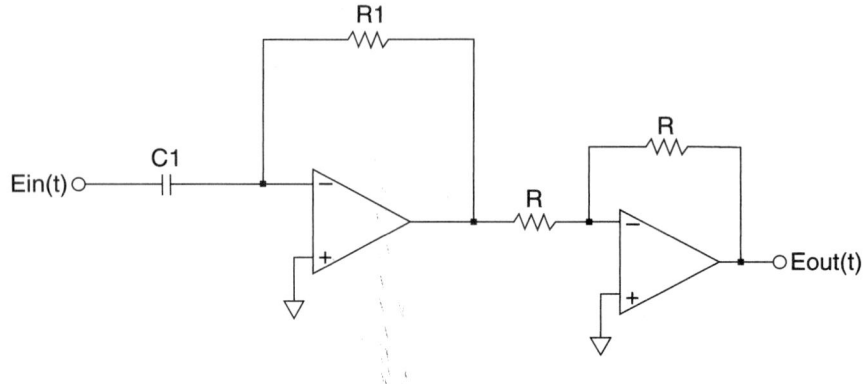

(figure 9-7)

Opamp derivative controller

The transfer function of this system is

$$K_D = sR_1C_1 \qquad (9.4)$$

The second opamp is for inversion to keep the phase of the signal correct. Keep in mind that this is only one method of designing a derivative controller. Other forms are available in the literature.

The PD controller has the mathematical form

$$G(s) = K_D s + K_P \qquad (9.5)$$

and the circuit form shown in Figure 9-8.

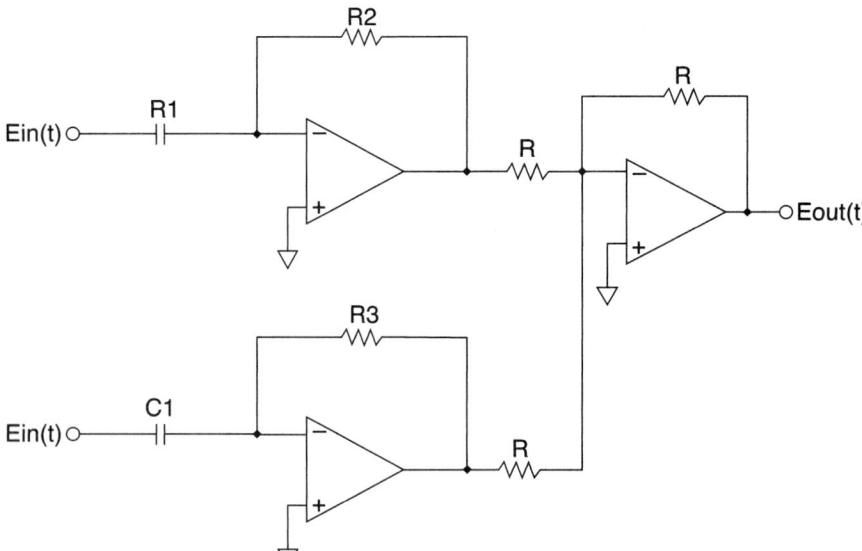

(figure 9-8)

Three-opamp PD controller

The summing amplifier at the output provides the transfer function and the necessary inversion of the input signal.

$$|G_C(s)| = sR_3C_1 + \frac{R_2}{R_1} \tag{9.6}$$

Where $K_D = R_3C_1$ and $K_P = \frac{R_2}{R_1}$.

PD Controllers in the Time Domain

Using the prototype second-order system, you can get insight into how the derivative controller affects the transient response of a system. The prototype system is repeated here.

$$G_P(s) = \frac{\omega_n^2}{s(s + 2\zeta\omega_n)}$$

Consider the system shown in Figure 9-9.

(figure 9-9)

PD controller and second-order plant

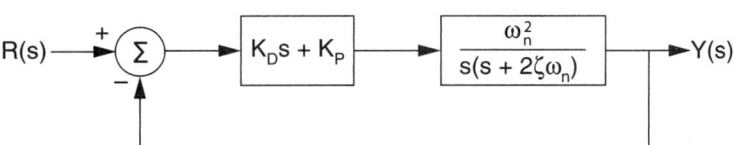

The open-loop transfer function of the system is

$$G_C(s)G_P(s) = \frac{\omega_n^2(K_Ds + K_P)}{s(s + 2\zeta\omega_n)} \tag{9.7}$$

Examination of the transfer function reveals that it is no longer the prototype system and will have no error for a step input. (The system is type one.) The velocity constant, K_V, for a ramp input is determined by $\omega_n^2 K_P/2\zeta\omega_n$. Further, a zero is placed in the system at $s = -K_P/K_D$.

In general, the benefits of the PD controller are an improvement in the damping ratio of a system, an increase in the bandwidth of the system, and a reduction in the risetime and settling time of a system. The detriment of the PD controller is that high-frequency noise can cause difficulties when pure differentiation is used. Therefore, it is prudent to design the controller so that the effect of noise is limited.

Frequency Response Considerations in the PD Controller

In terms of frequency response, the PD controller has a corner frequency set by $\omega_d = K_P/K_D$, with a maximum phase of +90°. The frequency and phase plots of the derivative controller are shown in Figure 9-10.

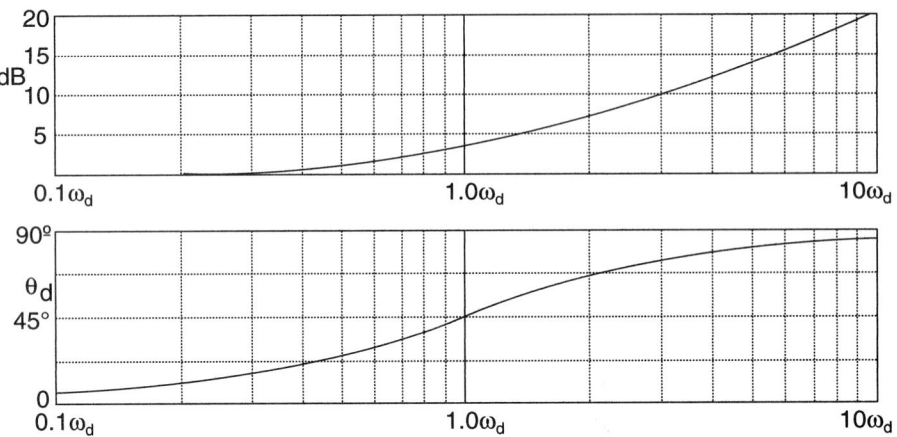

(figure 9-10)

Magnitude and phase response of derivative controller

Figure 9-10 clearly shows that the characteristic of the system is similar to a high-pass filter, which increases the bandwidth and decreases the risetime of the system. The increasing phase indicates that a new crossover frequency can be determined with an accompanying increase in the phase margin at the crossover frequency.

An example will show the benefits of this type of control.

Example 9-3

A motor is to be used in an attitude control system for an aircraft. The motor has the transfer function

$$G_P(s) = \frac{1.952 \cdot 10^5 K}{s(s + 773)}$$

The performance requirements are as follows:

    Steady-state error for a ramp input = 0.0001
    Maximum overshoot $\leq 10\%$
    Risetime ≤ 1 msec
    Settling time ≤ 1.5 msec (5% criteria)

Solution

To satisfy the error requirement for the ramp error, the value of K is

$$K_V = sG_C(s) = \frac{1.952 \cdot 10^5 K}{773} = 252.5K$$

From which K is readily calculated to be 39.6, which provides a numerator value of $7.723 \cdot 10^6$. Calculation of the damping ratio for this system gives a value of $\zeta = 0.154$, with an overshoot of 64.3 percent.

Example 9-3 (continued)

The system of Figure 9-10 is decided on for the controller, from which the open-loop transfer function is

$$G_C(s)G_P(s) = \frac{7.723 \cdot 10^6 (K_D s + K_P)}{s(s + 773)}$$

The closed-loop transfer function of this system is

$$\frac{Y(s)}{R(s)} = \frac{7.723 \cdot 10^6 (K_D s + K_P)}{s^2 + (773 + 7.723 \cdot 10^6 K_D)s + 7.723 \cdot 10^6 K_P}$$

It should be clear from the second term in the denominator that the value of $2\zeta\omega_n$ is changed to a new value. $2\zeta\omega$ and ζ can be controlled by utilizing the following formula:

$$\zeta = \frac{773 + 7.723 \cdot 10^6 K_D}{2 \cdot \sqrt{7.723 \cdot 10^6 K_P}}$$

Since you have already calculated the value of K for the ramp error constant, you can arbitrarily assign K_P the value of 1.0 since K_P directly affects the overshoot in the system.

The effect of K_P will be demonstrated later. If you want the system to have critical damping, $\zeta = 1.0$, the value of K_D is calculated from

$$1 = \frac{773 + 7.723 \cdot 10^6 K_D}{2\sqrt{7.723 \cdot 10^6}} = 0.139 + 1389.5 K_D$$

$$K_D = \frac{1 - 0.139}{1389.5} = 0.00062$$

The closed-loop transfer function of the new system is

$$\frac{Y(s)}{R(s)} = \frac{4788(s + 1613)}{s^2 + 5561s + 7.723 \cdot 10^6}$$

The response of the system without and with derivative control is shown in Figure 9-11.

The overshoot in the system is caused by the placement of the zero at K_P/K_D even though the damping ratio was chosen as 1.0. Calculation of ζ will show the damping ratio to be 1.0.

Finally, assume that the value of K_P is doubled to two; the output of the system is shown in Figure 9-12. Note that the overshoot is now over 15 percent, but the system still meets the other requirements.

Adding a derivative controller to the system shows a gain in the phase margin over the uncompensated system. The defining equations for the uncompensated systems are

$$G_P(s) = \frac{7.723 \cdot 10^6}{s(s + 773)} \qquad G_C(s)G_P(s) = \frac{4784(s + 1614.34)}{s(s + 773)}$$

Example 9-3 (continued)

The gain and phase curves are shown in Figures 9-13 and 9-14, respectively.
The change in gain margin is evident as is the change in crossover frequency.

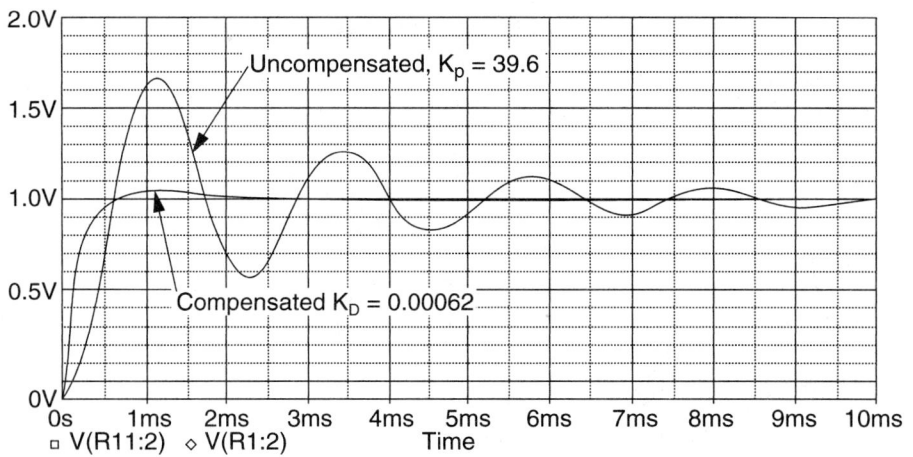

(figure 9-11)

Output of uncompensated and compensated system of Example 9-3

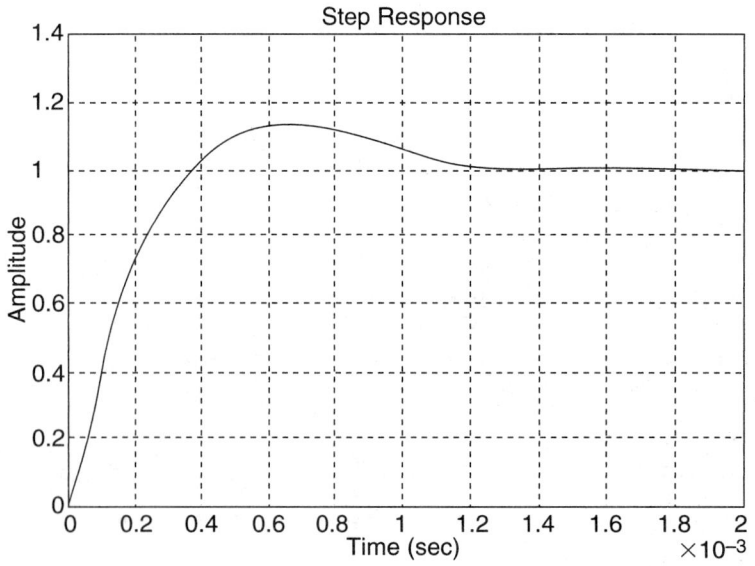

(figure 9-12)

Response of Example 9-3 with K_p doubled

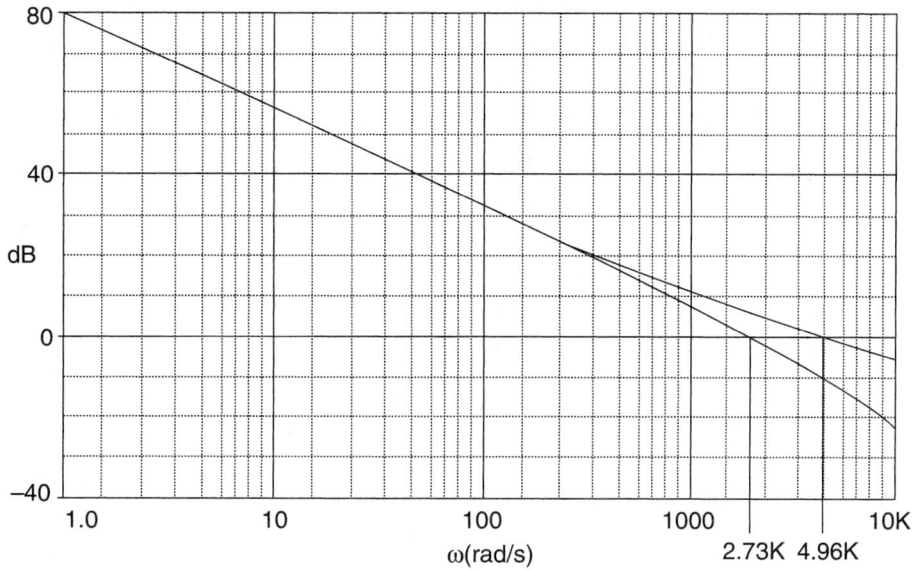

(figure 9-13)

Uncompensated and compensated crossover frequencies for Example 9-3

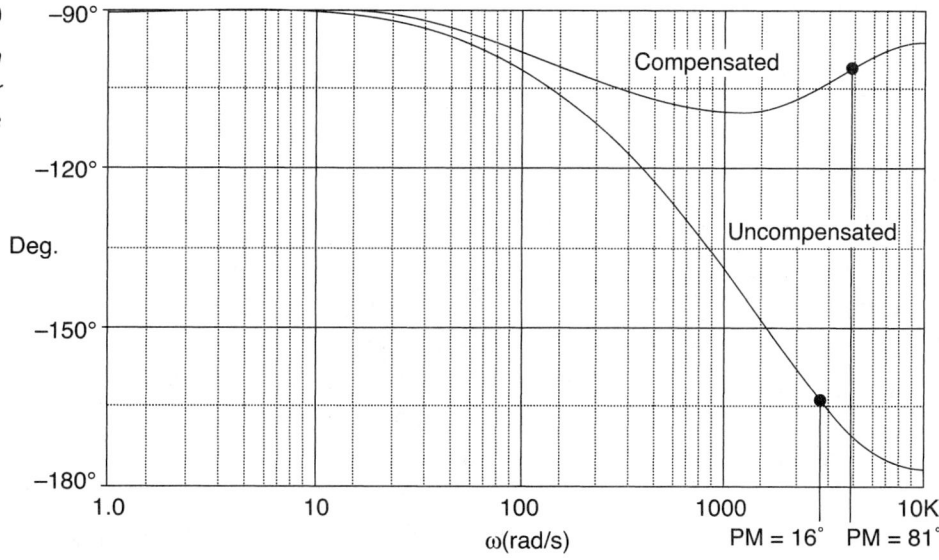

(figure 9-14)

Phase improvement with derivative control for Example 9-3

9.5 Proportional-Integral (PI) Control

You have seen that derivative control improves the damping and transient response of a system. However, it is not useful in eliminating error unless the error varies with time, which is not usually the case with step inputs.

Integral control increases the order of a system by adding a pole at the origin of the s-plane; it adds a zero to the system as well. This type of compensation is often used to eliminate error for a specific form of input or type of controller. This type of compensation is proportional to the time integral of the input to the controller. The transfer function of the compensator is

$$G_C(s) = K_P + \frac{K_I}{s} = \frac{K_P s + K_I}{s} \tag{9.8}$$

The zero is at K_I/K_P and is often placed to effect a cancellation or partial cancellation of a pole to improve the transient response of the system.

A possible disadvantage of this type of control is that if the system is already second-order or higher, the tendency toward oscillation may be increased. The added zero is used to maintain acceptable transient performance in this case.

PI Controllers in the Time Domain

The integral controller has the form shown in Figure 9-15.

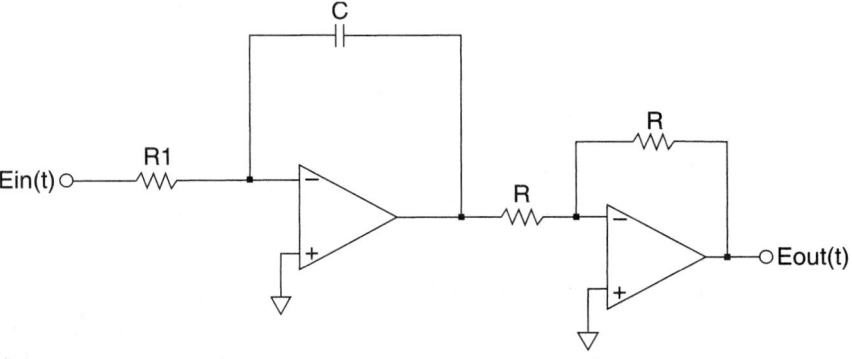

(figure 9-15)

Opamp integral controller

The transfer function for the integral controller is

$$K_I = \frac{1}{R_1 C} \tag{9.9}$$

To design a PI controller, the three-opamp form is needed, as shown in Figure 9-16.

(figure 9-16)

Three-opamp PI controller

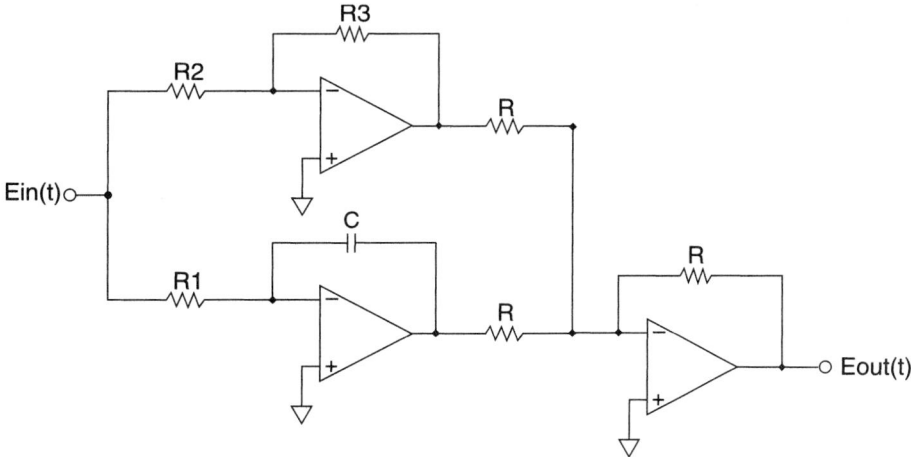

(figure 9-16)

Three-opamp PI controller

As before, the third opamp is both summing and inverting to keep the signal at the correct polarity. The transfer function is the sum of the integrator gain and the proportional gain.

$$K_P = \frac{R_3}{R_2} \qquad K_I = \frac{1}{sCR_1} \qquad |G_C(s)| = \frac{E_{out}(s)}{E_{in}(s)} = \frac{R_3}{R_2} + \frac{1}{sR_1C}$$

It should be clear from the above transfer function that it is in the form of Equation 9.8.

When Equation 9.8 is applied to the prototype second-order system, the system changes from type one to type two and has the form

$$G_C(s)G_P(s) = \frac{\omega_n^2(K_P s + K_I)}{s^2(s + 2\zeta\omega_n)} \qquad (9.10)$$

This system has no error for either a step or ramp input, but it does have an error for a parabolic input. Notice that a zero is placed at $s = -K_I/K_P$ in the s-plane. The system order has been increased to three, and it is possible for instability to occur if K_I and K_P are not chosen carefully.

Frequency Response Considerations in the PI Controller

In terms of frequency response, the PI controller has a corner frequency set by $\omega_I = K_I/K_P$, with a maximum phase of $-90°$. The frequency and phase plots of the integral controller are shown in Figure 9-17.

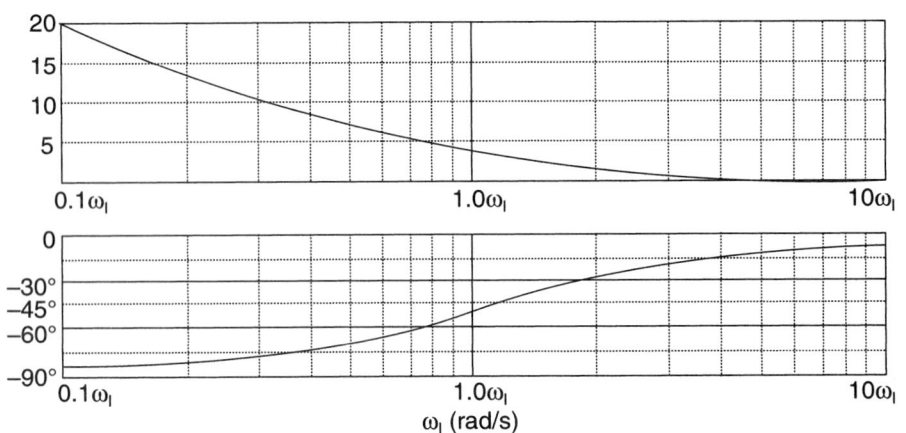

(figure 9-17)

Magnitude and phase of
integral controller

To meet specified gain and phase margins, the open-loop uncompensated Bode plot is used. First, determine the uncompensated gain and phase margin; then note the magnitude of the gain at the frequency where the phase margin is specified. This is a new cutoff frequency ω_c. The integral controller must provide the required attenuation at that point. The required attenuation is calculated from

$$|G_P(j\omega_C)| = -20\log_{10}(K_P) \ dB \qquad (9.11)$$

The magnitude of the required gain is found by rearranging the equation as

$$K_P = 10^{-|G_P(j\omega_c)|/20} \qquad (9.12)$$

Once K_P is determined, it is necessary only to determine the value of K_I to complete the design. Since the phase of the integral controller affects the phase of the original system, it is prudent to place the corner frequency K_I/K_P one to two decades below the new cutoff frequency.

To summarize, a properly designed PI controller will improve the damping ratio and overshoot, decrease the bandwidth, attenuate high-frequency noise, and improve the gain and phase margins of a system. An example of this type of control follows.

Example 9-4

A PI controller is used to convert a type zero system to a type one system as shown in Figure 9-18. The system specifications require that the system settle within 0.4 seconds for 2 percent error, and that the system have zero error for a step input.

Example 9-4 (continued)

Solution

By definition, the conversion to a type one system will provide zero error for a step input. In addition, the settling time is $T_{s(2\%)} = 0.4$ seconds. The open-loop transfer function of the system is

$$G_C(s)G_P(s) = \frac{20K_P\left(s + \dfrac{K_I}{K_P}\right)}{s(s + 5)}$$

If K_I/K_P is made equal to 5, a pole cancellation will occur. If you assume that the cancellation is exact, the closed-loop transfer function becomes

$$\frac{Y(s)}{R(s)} = \frac{20K_P}{s + 20K_P}$$

The time constant of the controller is the reciprocal of $20K_P$; and if that value is made equal to 0.1 seconds, the system will settle at less than 0.4 seconds. From the requirements, K_P must be

$$\frac{1}{20K_P} = 0.1 \qquad\qquad K_P = 0.5$$

The required value of K_I is then

$$\frac{K_I}{0.5} = 5 \qquad\qquad K_I = 0.5 * 5 = 2.5$$

The closed-loop transfer function of the system with the pole-zero cancellation is

$$\frac{Y(s)}{R(s)} = \frac{10}{s + 10}$$

The output of the system showing that the setline time is less than 0.4 seconds is shown in Figure 9-19.

(figure 9-18)

Conversion of type zero to type one system for Example 9-4

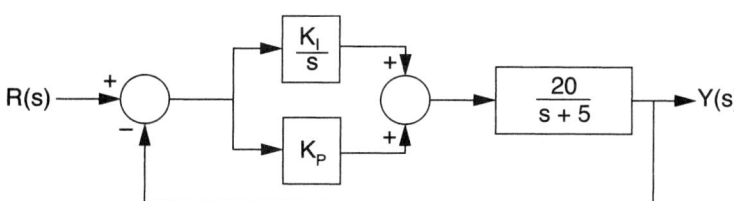

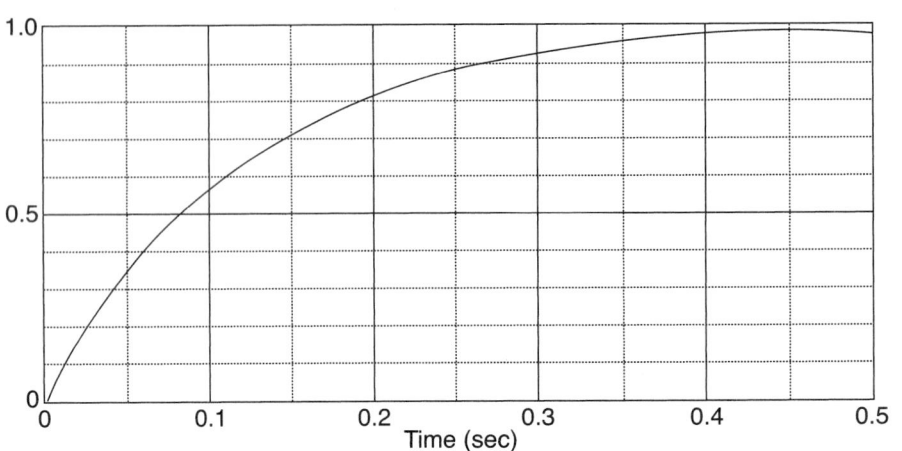

(figure 9-19)

Example 9-5 output for step input

Although you assumed an exact cancellation of the pole and zero, such a cancellation is not a practical situation due to system component variations. But near cancellations are possible, and small variations in pole-zero position do not produce wide variation in transient response.

Another use for the PI controller is when two or more dominant poles exist in a system. In that type of system, an increase from type zero to type one or from type one to type two will be affected. When the system type is changed, problems of instability arise and may place limitations on the response time of the system. Such a system is illustrated in Example 9-5 where the type one system is increased to type two.

Example 9-5

For the system shown in Figure 9-20, a PI controller is used to change the type number from type one to type two.
The specifications for the system are as follows:

    Ramp error = 0%
    Risetime $\leq$ 0.005 seconds
    Settling time (5%) $\leq$ 0.01 seconds
    Overshoot $\leq$ 5%

(figure 9-20)

Conversion of type one to type two system for Example 9-5

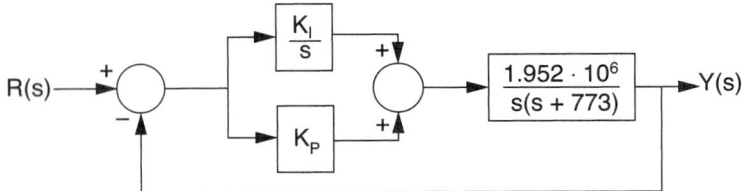

Example 9-5 (continued)

Solution

The forward path transfer function of the system with PI control is

$$G_P(s)G_C(s) = \frac{1.318 \cdot 10^6 K_P \left(s + \dfrac{K_I}{K_P} \right)}{s^2(s + 773)}$$

The error requirement is automatically met by changing the system from type one to type two. The overshoot requirement is met when the system exhibits a damping ratio of slightly more than 0.7. You will work with the damping ratio and overshoot first. Setting $K_I = 0$, the characteristic equation is

$$s^2 + 773s + 1.318 \cdot 10^6 K_P = 0$$

Assume that $\zeta = 0.707$; solving the characteristic equation for K_P gives a value of $K_P = 0.227$. Solving the second-order system for its time constant gives $\tau = 2.6$ msec.

A Routh test of the closed-loop system shows the system to be stable for $0 < K_I/K_P < 773$.

The placement of the zero is important in this design. If the zero is placed too far from the origin, the overshoot is excessive. It is best to place the zero so that it partially cancels the effect of one of the poles at the origin. If the zero is placed at $s = -10$, the integral gain is 2.27 and the closed-loop transfer function is

$$\frac{Y(s)}{R(s)} = \frac{2.992 \cdot 10^5(s + 10)}{s^3 + 773s^2 + 2.99 \cdot 10^5 s + 2.99 \cdot 10^6}$$

A **MATLAB** analysis of the system shows that the overshoot is 7 percent, the settling time is 10.6 msec, and the risetime is 4.12 msec. The only specification that is met is the risetime of the system. If you move the zero closer to the origin, the effect of one of the poles is negated and the complex poles of the second-order system dominate the response. Change the value of K_I from 2.27 to 1.135. The zero is moved to $s = -5$. Analysis of the system with the zero at -5 shows that the risetime and settling time is met, but the overshoot is 5.5 percent. As a last experimentation, the zero is moved to $s = -2.5$ with $K_I = 0.568$. Analysis of that system shows that all specifications are met. The responses of the system with K_P held constant and K_I allowed to vary are shown in Table 9-1. The step response of the system with $K_I = 0.568$ is shown in Figure 9-21.

(table 9-1)

PI control parameters

K_P	K_I	t_r	t_s	%OS
0.227	2.27	0.0041s	0.0106s	7.0%
0.227	1.135	0.0041s	0.0094s	5.5%
0.227	0.568	0.0039s	0.0083s	5%

Example 9-5 (continued)

Frequency Response Analysis of the PI Controller

The forward path transfer function of the uncompensated system is

$$G_P(s) = \frac{1.318 \cdot 10^6}{s(s + 773)}$$

Figure 9-22 shows the **MATLAB** Bode plot of the uncompensated system and the uncompensated phase margin and crossover frequency.

The uncompensated system has a phase margin of 37°. Let's specify the new phase margin to be 65° for the compensated system. From the uncompensated response, a phase of −115°, which gives the required margin, occurs at 354 rad/sec, with a magnitude of 12.8 dB. The required K_P to produce a −12.8 dB attenuation is calculated from Equation 9.12.

$$K_P = 10^{-12.8/20} = 0.229$$

If you place the zero at one decade below the new cutoff frequency, the calculated value of K_I is 8.1.

$$\frac{K_I}{K_P} = \frac{\omega_c}{10} \qquad K_I = \frac{0.229 \cdot 354}{10} = 8.1$$

That value turns out to be too large for the system, as it produces over 13 percent of overshoot and a phase margin of 59°. Let's change the zero to two decades below the new cutoff frequency, which yields $K_I = 0.81$. Analysis with that value provides a phase margin of 64.8° at 355 rad/sec. Finally, if you insert the values of K_I and K_P as used in the step response, you will find a good match for the frequency response and step response. The magnitude and phase curves for the system with the compensated values are shown in Figures 9-23 and 9-24. The curves for $K_I = 0.81$ and $K_I = 0.568$ are very close together, as are the phase curves for the values of K_I.

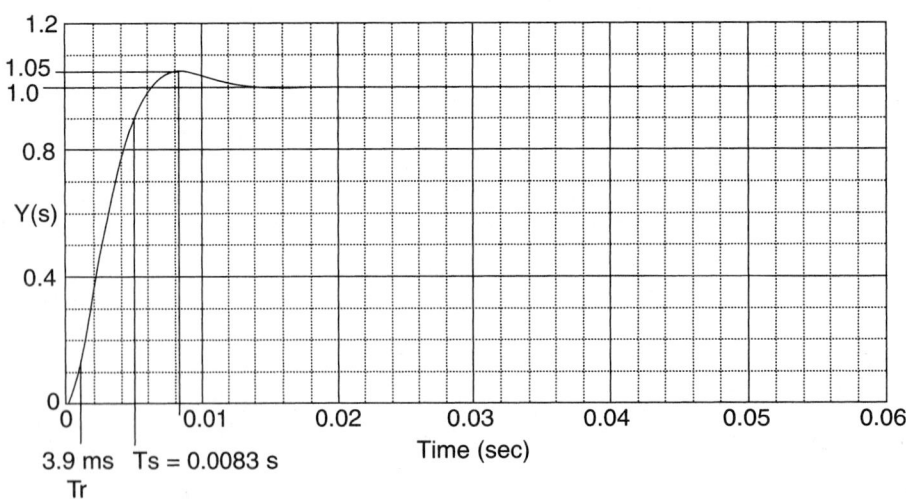

(figure 9-21)

Step response of Example 9-5 with $K_I = 0.568$

(figure 9-22)

Uncompensated frequency response of Example 9-5

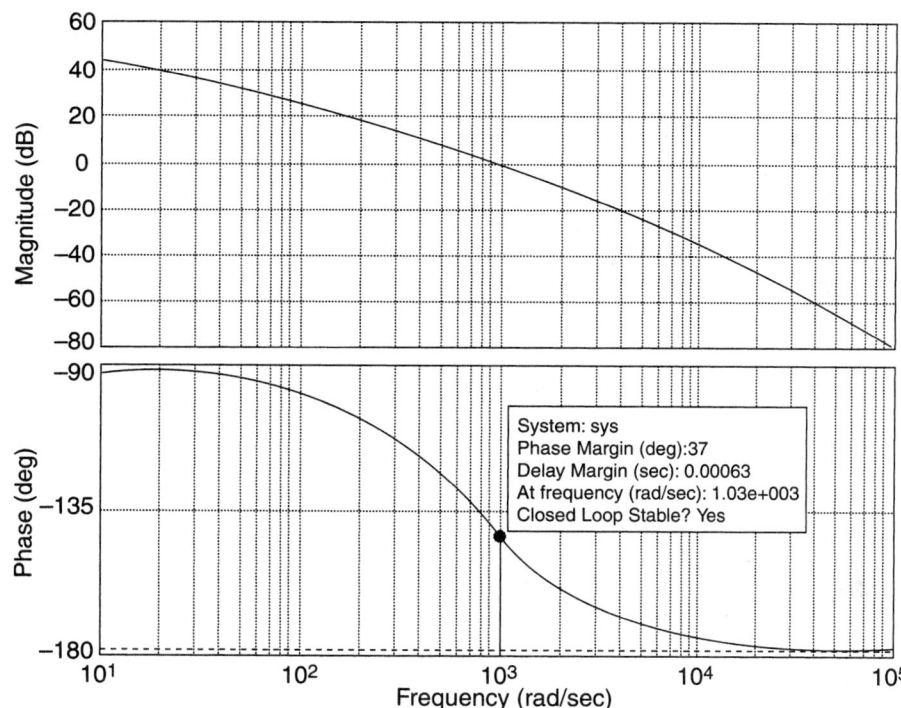

System: sys
Phase Margin (deg):37
Delay Margin (sec): 0.00063
At frequency (rad/sec): 1.03e+003
Closed Loop Stable? Yes

(figure 9-23)

Compensated magnitude responses of Example 9-5

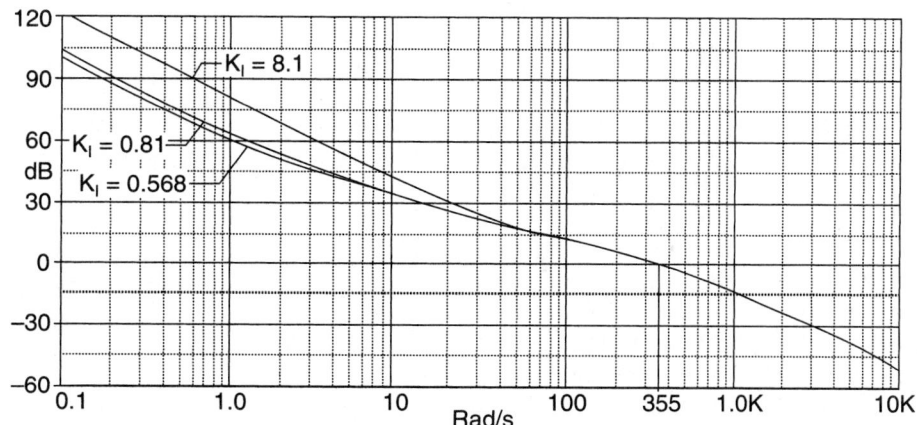

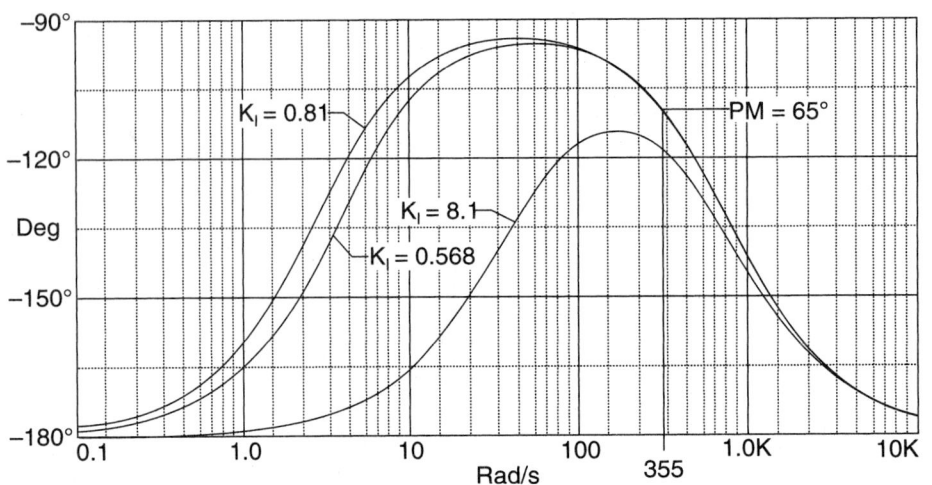

(figure 9-24)

*Phase margin for
Example 9-6*

9.6 Proportional-Integral-Derivative (PID) Control

From studying the PD controller, you found that you gained some control of rise- and settling times and that you could improve damping and widen the bandwidth of a system. From studying the PI controllers, you found that you could change the system type number and eliminate error in a system, but at the cost of a slower response and reduced bandwidth.

It is advantageous to apply the benefits of both types of controllers in a single system so you can obtain the benefits of both. The form of the PID controller is

$$G_P(s) = K_D s + K_P + \frac{K_I}{s} = \frac{K_D s^2 + K_P s + K_I}{s} \tag{9.13}$$

It is apparent that two zeros are added to a system. $GP(s)$ can be represented as a derivative controller term and an integrator term in the following manner:

$$G_C(s) = K_D s + K_P + \frac{K_I}{s} = (K_1 s + 1)\left(K_2 + \frac{K_3}{s}\right) \tag{9.14}$$

Placing the integral term over a common denominator and multiplying the factors provides the following values for K_D, K_P, and K_I:

$$G_C(s) = \frac{K_1 K_2 s^2 + (K_1 K_3 + K_2)s + K_3}{s} \tag{9.15}$$

$$K_D = K_1 / K_2$$

$$K_P = K_1 / K_3 + K_2$$

$$K_I = K_3$$

Either the integral or derivative controller can be designed first. The designs are based on the relative stability requirements for the system.

PID Controllers in the Time Domain

Example 9-6

Consider the second-order plant whose forward path transfer function is

$$G_P(s) = \frac{1.318 \cdot 10^6}{s(s + 400)}$$

The time domain specifications for the system are as follows:

Overshoot $\leq 5\%$

$t_r \leq 0.002\ s$

$T_s \leq 0.005\ s$ (2% criteria)

Solution

Following the concept of providing a near cancellation of one of the poles at the origin, you set $K_3/K_2 = -10$. That allows you to select values of K_1 such that the requirements are met. **MATLAB** will help you determine the proper value of K_1. Remember from the PD controller that the derivative zero is placed relatively far from the origin, which makes K_D relatively small. Allow K_2 to equal 1.0 and K_1 to equal 0.0005, 0.001, 0.0015, and 0.002.

To start, define the **MATLAB** transfer functions for the step response according to the gains specified.

$K_1 = 0.0005$.

n = 659 * [1 2010 2e4]; d = [1 1059 1.325e6 1.318e7];

$K_1 = 0.001$.

n = 1318 * [1 1010 1e4]; d = [1 1718 1.331e6 1.318e7];

$K_1 = 0.0015$.

n = 1977 * [1 677 6667]; d = [1 2377 1.338e6 1.318e7];

$K_1 = 0.002$.

n = 2636 * [1 510 5000]; d = [1 3036 1.344e6 1.318e7];

Only the **step(n,d)** command is needed to run the analyses for the responses. The overshoot, rise-time, and settling time vs. the derivative gain are shown in Table 9-2.

The table shows that either $K_1 = 0.0015$ or $K_1 = 0.002$ will work. The step response of the system for the K_1 gains specified is shown in Figure 9-25.

Example 9-6 (continued)

Frequency Domain Design of the PID Controller

To determine the frequency response of the system, you need to set some specifications for the phase margin, the bandwidth, and the maximum resonant peak of the system. From the time domain specifications of the problem, let's specify the following:

$BW \geq 1000$ rad/sec

$PM \geq 70°$

The magnitude and phase plots of the system are shown in Figure 9-26.

The figure clearly shows that the first system ($K_i = 0.0005$) does not meet the phase margin and peak requirements, the second system ($K_i = 0.001$) barely meets the phase margin requirements, and the third and fourth systems ($K_i = 0.0015$, $K_i = 0.002$) meet all requirements. Notice that the difference in bandwidth in the last two systems is almost 500 rad/sec. For that reason, the system whose $K_i = 0.0015$ should be selected. This will help minimize high-frequency noise problems. Table 9-3 shows the frequency domain performance of the systems vs. their rise- and settling times.

K_D	Tr	Ts	OS	
0.0005s	0.98 ms	6.3 ms	24.6%	**(table 9-2)**
0.001s	0.98 ms	4.49 ms	11.2%	*PID controller parameters*
0.0015 ms	0.88 ms	3.34 ms	4.2%	
0.002 ms	0.76 ms	1.16 ms	1.92%	

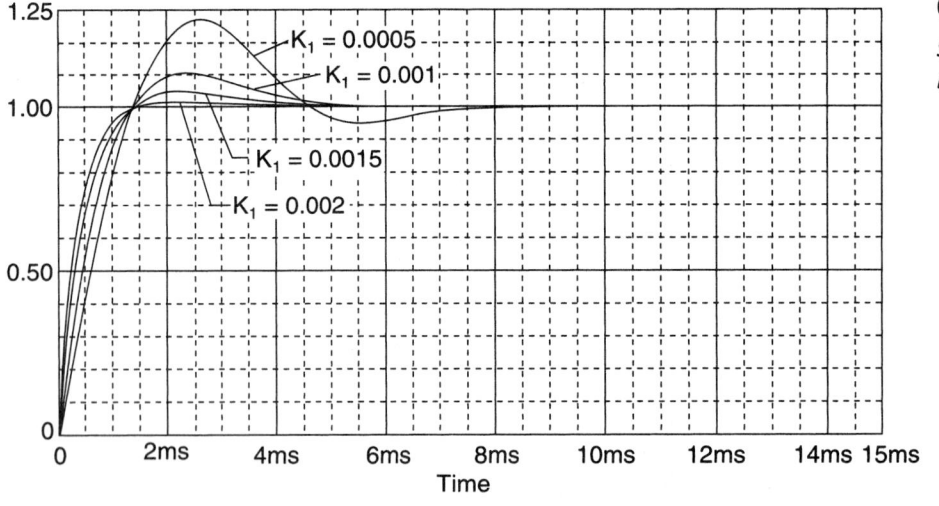

(figure 9-25)

Step responses of Example 9-6

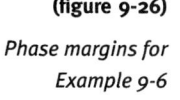

(figure 9-26)

Phase margins for Example 9-6

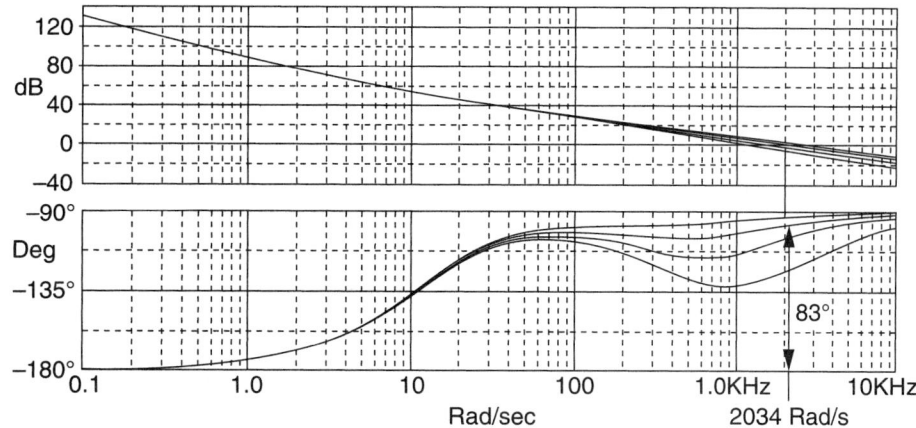

(table 9-3)

PID controller frequency response parameters

K₁	GM dB	PM Deg	Mᵣ	BW rad/sec	Tₛ msec	Tᵣ msec
0.0005	∞	49°	1.35	1730	6.73	1.08
0.001	∞	71°	1.11	1920	4.49	0.98
0.0015	∞	83°	1.04	2287	3.34	0.88
0.002	∞	88°	1.01	2764	1.16	0.76

Example 9-7

Consider a third-order attitude control system with the following open-loop transfer function:

$$G_P(s) = \frac{1.952 \cdot 10^9}{s(s + 773)(s + 2320)}$$

Let the time domain specifications be as follow:

Maximum overshoot ≤ 5%

$t_r \leq 0.005\ s$

$t_s \leq 0.01\ s$ (2% criteria)

Solution

Time Domain Design

In keeping with past practice, place the zero for the derivative controller relatively far from the origin; in this case, $s = -1000$. The zero for the integral part will be placed close enough to the origin to effect a partial cancellation of one of the poles at the origin—at $s = -10$. For this controller, you will allow K_2 to vary from a value of 1.0 to smaller values until all specifications are met.

Example 9-7 (continued)

Once again, **MATLAB** will be used to determine the peak response, the risetime, and the settling time. The values of K_2 are allowed to vary from 1.0 to 0.1 in 0.1 increments. The numerators and denominators for the transfer functions are as follows:

$n = 1.952e6 * [1\ 1010\ 1e4]; d = [1\ 3093\ 3.745e6\ 1.97e9\ 1.953e10];$	$(K_2 = 1.0)$
$n = 1.757e6 * [1\ 1010\ 1e4]; d = [1\ 3093\ 3.55e6\ 1.775e9\ 1.757e10];$	$(K_2 = 0.9)$
$n = 1.562e6 * [1\ 1010\ 1e4]; d = [1\ 3093\ 3.355e6\ 1.578e9\ 1.562e10];$	$(K_2 = 0.8)$
$n = 1.366e6 * [1\ 1010\ 1e4]; d = [1\ 3093\ 3.159e6\ 1.38e9\ 1.366e10];$	$(K_2 = 0.7)$
$n = 1.171e6 * [1\ 1010\ 1e4]; d = [1\ 3093\ 2.964e6\ 1.183e9\ 1.171e10];$	$(K_2 = 0.6)$
$n = 9.76e5 * [1\ 1010\ 1e4]; d = [1\ 3093\ 2.77e6\ 9.86e8\ 9.76e9];$	$(K_2 = 0.5)$
$n = 7.81e5 * [1\ 1010\ 1e4]; d = [1\ 3093\ 2.574e6\ 7.89e8\ 7.81e9];$	$(K_2 = 0.4)$
$n = 5.86e5 * [1\ 1010\ 1e4]; d = [1\ 3093\ 2.379e6\ 5.92e9\ 5.86e9];$	$(K_2 = 0.3)$
$n = 3.9e5 * [1\ 1010\ 1e4]; d = [1\ 3093\ 2.184e6\ 3.94e8\ 3.9e9];$	$(K_2 = 0.2)$
$n = 1.952e5 * [1\ 1010\ 1e4]; d = [1\ 3093\ 1.989e6\ 1.972e8\ 1.952e9];$	$(K_2 = 0.1)$

Table 9-4 shows the results of the step analysis.

From studying the table, it is evident that only $K_2 = 0.6$ causes the system to meet all specifications simultaneously. Also notice that as the value of K_2 is decreased below 0.4, the overshoot begins to increase and the settling time is increased dramatically. The values of K_P, K_D, and K_I are

$$K_I = K_2 K_3 = 0.6 \cdot 10 = 6.0$$

$$K_P = K_1 K_3 + K_2 = 0.001 \cdot 6 + .6 = 0.606$$

$$K_D = K_1 K_2 = 0.001 \cdot 0.6 = 0.0006$$

Notice that the PID design has resulted in a smaller K_D and a K_I large enough that the capacitors used in the design of the opamp controller will have reasonable values, meaning implementation of the design is easier.

Frequency Response Design

For this system, M_r will not be a consideration since it is less than 1.05 for all cases. The specifications for the frequency response of the system are

BW ≥ 500 rad/sec   PM ≥ 65°

The plot of the uncompensated system shows a crossover frequency of 741 rad/sec with a phase margin of 28.8°. To make the phase margin 65°, it is necessary to add 36.2° to the system. Measuring the magnitude for the additional phase shows that the magnitude is 7.76 dB, from which K_2 is calculated.

$$K_2 = 10^{\frac{-3.51}{20}} = 0.668$$

That is a little greater than the actual K_2 used, but would also result in an acceptable system. The plot of the compensated system with $K_2 = 0.6$ shows that the crossover frequency is 567 rad/sec and the phase margin is 68.1°, which fulfills the requirements. The plot of the uncompensated and the compensated systems is shown in Figure 9-27.

(table 9-4)

Step results for Example 9-7

K$_2$	%OS	T$_s$ msec	T$_r$ msec
1.0	9.07	5.49	1.55
0.9	7.79	5.93	1.68
0.8	6.64	6.56	1.86
0.7	5.54	7.35	2.09
0.6	4.41	8.49	2.44
0.5	3.39	10.4	2.94
0.4	2.72	18.6	3.72
0.3	2.81	49.0	5.05
0.2	3.85	92.3	7.88
0.1	6.71	166	15.5

(figure 9-27)

Frequency response of Example 9-7

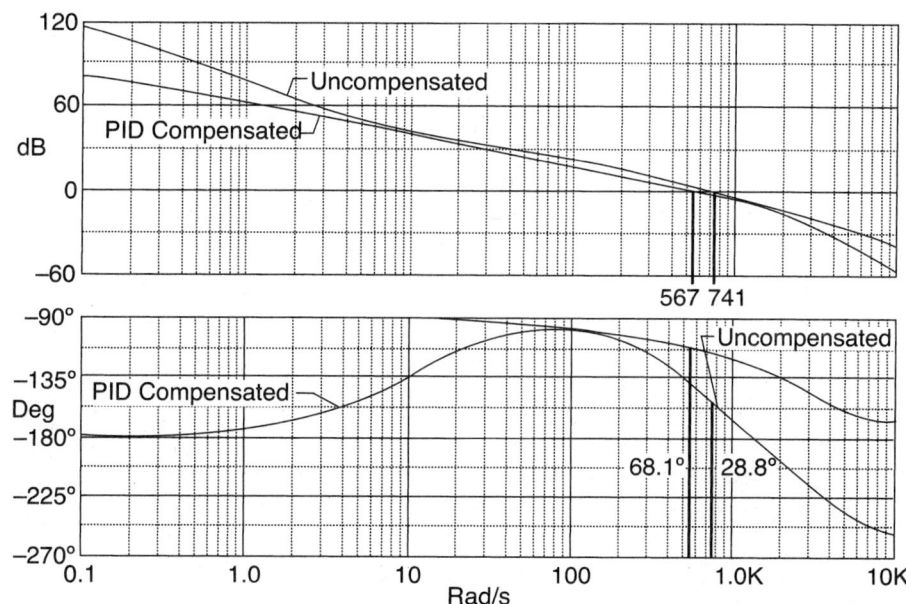

9.7 Loop Tuning

Several tuning methods have been designed to produce acceptable responses from control systems. In order to meet the steady-state and transient response specifications of a system, the engineer requires either a dynamic model or a detailed frequency response over a wide range of frequencies. Either of those may be difficult to obtain if the system already exists.

Further, those methods are used often in field tests of a process that may be difficult to determine or define. In that case, it may be less costly and more effective to adjust the system controller in the field. Electromechanical systems, such as servomechanisms that have more clearly defined dynamic models, can usually be defined well enough that field adjustment is unnecessary.

Tuning PID Controllers, Using the Ziegler-Nichols Method

By experimenting on the process itself, Callender et al. (1936) proposed a method of specifying satisfactory parameters for the PID controller based on measurement of plant parameters. Ziegler and Nichols (1942, 1943) extended the methodology by observing that a large number of process control systems exhibit a *process reaction curve* such as that shown in Figure 9-28.

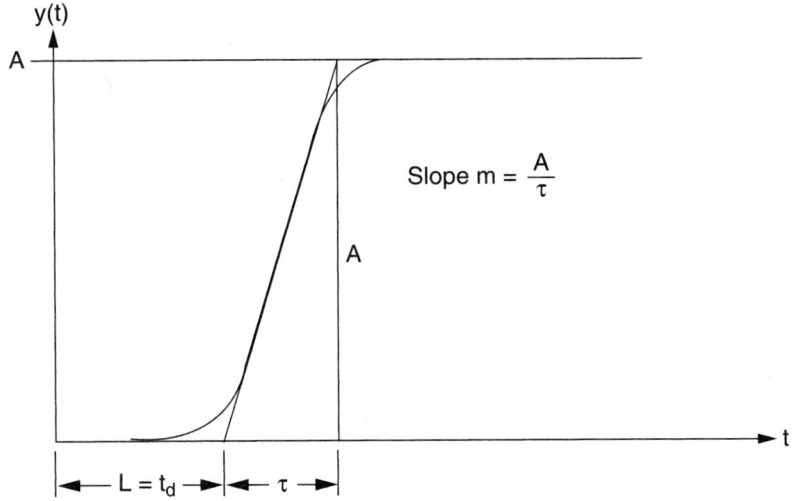

(figure 9-28)

Typical reaction curve of open-loop system

Note that the curve represents only the plant response and approximates a first-order response with a time delay t_d seconds. The response of the system can be approximated by

$$\frac{Y(s)}{R(s)} = \frac{Ae^{-sL}}{\tau s + 1} \tag{9.16}$$

The required constants can be determined by taking measurements from the unit step response of the process. If a tangent is drawn through the inflection point of the reaction curve, the slope of the line is $m = A/\tau$ and the intersection of the slope line with the time axis defines the time delay, $L = t_d$. J. B. Ziegler and N. B. Nichols proposed two methods of tuning the PID controller for that type of model.

The criteria for which of the two methods to use is determined by the presence of integrators in the open-loop systems. The first method uses a specific damping ratio to provide a quarter step in the damping of the system; the second method uses the ultimate gain and period of oscillation to provide the necessary system constants. When an integrator is in the open-loop transfer function, the first method cannot be used.

Ziegler-Nichols First Method

The first method measures the open-loop response of the plant; and the controller parameters are chosen such that the closed-loop step response decays to 0.25 greater than the final value at the second oscillation, as shown in Figure 9-29. The damping ratio of this decay results in $\zeta \approx 0.21$ and is called the *quarter-step method*.

(figure 9-29)

Quarter-step response

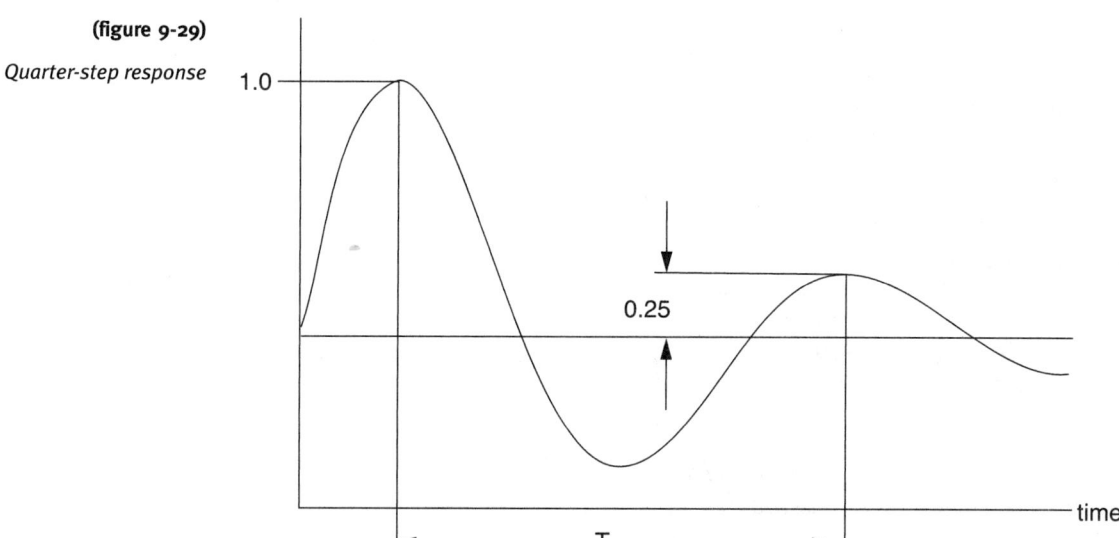

The system is opened between the controller and the plant. A small step is applied to the plant, and the plant output and feedback signal are measured. It is essential that none of the variables saturate. The test conditions are shown in Figure 9-30.

(figure 9-30)

Open-loop test for Ziegler Nichols first method

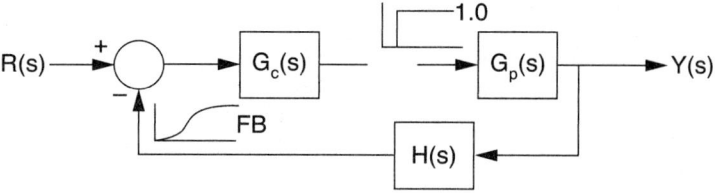

The regulator parameters suggested by Ziegler and Nichols are

$$G_C(s) = K_P\left(1 + \frac{1}{sT_I} + sT_D\right) \tag{9.17}$$

Where T_I is the integral time constant and T_D is the derivative time constant. The terms are given in Table 9-5. When using this method, it is essential that the measured variables not saturate.

Controller Type	Gain
Proportional	$K_P = 1/(m \cdot L)$
PI	$K_P = 0.9/(m \cdot L)$
	$T_I = L/0.3$
PID	$K_P = 1.2/(m \cdot L)$
	$T_I = 2L$
	$T_D = 0.5L$

(table 9-5)

Ziegler-Nichols first method

Example 9-8

An open-loop test of a temperature control system yields the reaction curve shown in Figure 9-31. The system open-loop transfer function is

$$G_P(s) = \frac{1}{(20s + 1)(50s + 1)}$$

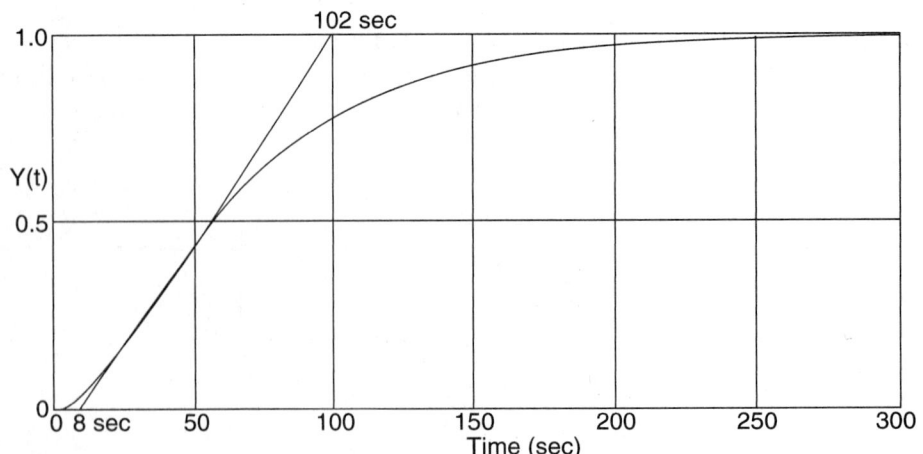

(figure 9-31)

Reaction curve for Example 9-8

Example 9-8 (continued)

Use the first Ziegler-Nichols method to determine K_P, K_D, and K_I for a quarter-step response PID control system.

Solution

Figure 9-31 shows that the delay time is $L \approx 8$ sec, the rise in temperature is from 0 to 100 percent, and the risetime is $\tau = 94$ sec, from which the slope of the line is

$$m = \frac{1}{\tau} = \frac{1}{94}$$

Using the expressions in Table 9-5

$$K_P = \frac{1.2 \cdot 94}{8} = 14.1$$

$$T_I = 2 \cdot 8 = 16 \text{ sec}$$

$$T_D = 0.5 \cdot 8 = 4 \text{ sec}$$

From Equation 9.17, the required $G_C(s)$ is

$$G_C(s) = \frac{56.4(s^2 + 0.25s + .0156)}{s}$$

Then the total transfer function for the PID system is

$$\frac{Y(s)}{R(s)} = \frac{0.0564s^2 + 0.0141s + 0.0009}{s^3 + 0.1264s^2 + 0.0154s + 0.0009} \cdot e^{-9s}$$

The response of the system for proportional gain only and PID control is shown in Figure 9-32. Keep in mind that those methods are a starting point for optimum tweaking of a system in the field and probably requires some refinement to obtain the best operation of the system.

(figure 9-32)

Closed-loop quarter-step response for Example 9-8

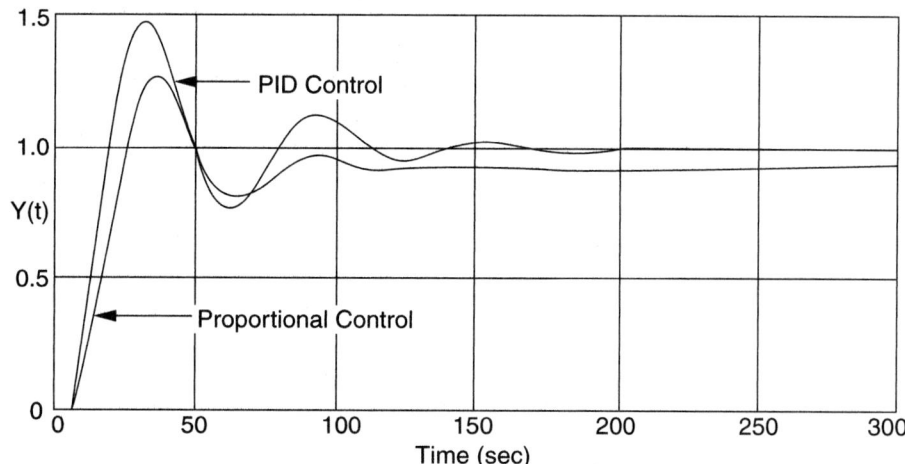

Ziegler-Nichols Second Method

The second method uses the closed-loop response of a system in which $T_I = \infty$, $T_D = 0$, and K_P is set to the value that first exhibits sustained oscillations. The value of K_P that produces sustained oscillation is K_{Cr}, and the period of the oscillations is P_{Cr}. When the system decays to a final value regardless of K_P, this method does not apply. The system should exhibit approximately 25 percent of overshoot, as in the first method. The critical gain and period are determined experimentally; values of K_P, T_I, and T_D are determined using the values in Table 9-6.

Controller Type	Gain	
Proportional	$K_P = 0.5 \cdot K_C$	
PI	$K_P = 0.45 \cdot K_C$ $T_I = P_{Cr}/1.2$	
PID	$K_P = 0.6 \cdot K_C$ $T_I = 0.5 \cdot P_C$ $T_D = 0.125 \cdot P_C$	

(table 9-6)

Ziegler-Nichols second method

Example 9-9

Consider the system shown in Figure 9-33 in which a PID controller is to be used to control the system. The PID controller has the form

$$G_C(s) = K_P\left(1 + \frac{1}{sT_I} + sT_D\right)$$

Use the second Ziegler-Nichols method to determine the response of the system. Obtain a step response curve of the controller.

Solution

Figure 9-33 shows that the transfer function of the system is

$$\frac{Y(s)}{R(s)} = \frac{K_P}{s^3 + 10s^2 + 16s + K_P}$$

To determine values of K_{Cr} and P_{Cr}, use a Routh table.

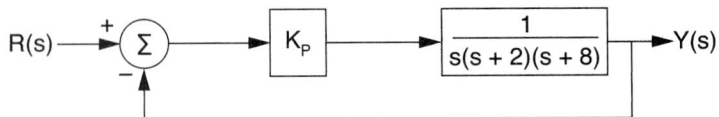

(figure 9-33)

Control system for Example 9-9

Example 9-9 (continued)

$$
\begin{array}{ccc}
s^3 & 1 & 16 \\[6pt]
s^2 & 10 & K_P \\[6pt]
s^1 & \dfrac{160 - K_P}{10} & \\[10pt]
s^0 & K_P &
\end{array}
$$

From which $K_{Cr} = 160$, $\omega = 4$, and $P_{Cr} = \dfrac{2\pi}{\omega} = \dfrac{2\pi}{4} = 1.57\,\text{s}$. Then calculating parameters from Table 9-6

$$
\begin{aligned}
K_P &= 0.6 \cdot K_{Cr} = 0.6 \cdot 160 = 96 \\
T_I &= 0.5 \cdot P_{Cr} = 0.5 \cdot 1.57 = 0.785\,\text{sec} \\
T_D &= 0.125 \cdot P_{Cr} = 0.125 \cdot 1.57 = 0.196\,\text{sec}
\end{aligned}
$$

Then

$$
G_C(s) = \frac{18.84\left(s^2 + 5.1s + 6.48\right)}{s}
$$

The transfer function of the system with the values calculated is

$$
\frac{Y(s)}{R(s)} = \frac{18.84s^2 + 96s + 122}{s^4 + 10s^3 + 34.84s^2 + 96s + 122}
$$

It is convenient to use **MATLAB** to produce the output curve of the system. The required commands are as follows:

```
sys = tf([18.84 96 122], [1 10 34.84 96 122])
step(sys)
```

The output of the system with the calculated parameters is shown in Figure 9-34.

Note that the overshoot for the parameters provided is over 60 percent and would be deemed unacceptable. However, a starting point has been established, and you can reduce the overshoot by increasing T_I since integral gain has a direct effect on overshoot.

As a test, you can double the integral time to 1.57 seconds, reducing the integral gain. Doing so results in the following new parameters:

$$
\begin{aligned}
K_P &= 96 \\
T_I &= 1.57\,\text{sec} \\
T_D &= 0.3927\,\text{sec}
\end{aligned}
$$

The new PID controller is

$$
G_C(s) = 37.7\left(\frac{s^2 + 2.547s + 1.611}{s}\right)
$$

Example 9-9 (continued)

And the new transfer function is

$$\frac{Y(s)}{R(s)} = \frac{37.7s^2 + 96s + 61.2}{s^4 + 10s^3 + 53.69s^2 + 96s + 61.2}$$

MATLAB is used to show the reduction in overshoot when the integral time is doubled. The commands are

    sys = tf([37.7 96 61.2], [1 10 53.7 96 61.2])
    step(sys)

And the output is shown in Figure 9-35.

While the overshoot is less than the 25 percent specified for the system corrections, the settling time is long, approximately 2 seconds. Since proportional gain has an effect on the risetime of a system, let's arbitrarily increase the proportional gain to $K_P = 160$. Note that that is the gain for oscillation of the uncompensated system. Then the transfer function becomes

$$\frac{Y(s)}{R(s)} = \frac{63s^2 + 160s + 102}{s^4 + 10s^3 + 79s^2 + 160s + 102}$$

The output of the corrected system is shown in Figure 9-36.

Note that the overshoot has increased slightly, but is within the range for the system. However, the settling time is decreased from 2 seconds to less than 1.4 seconds.

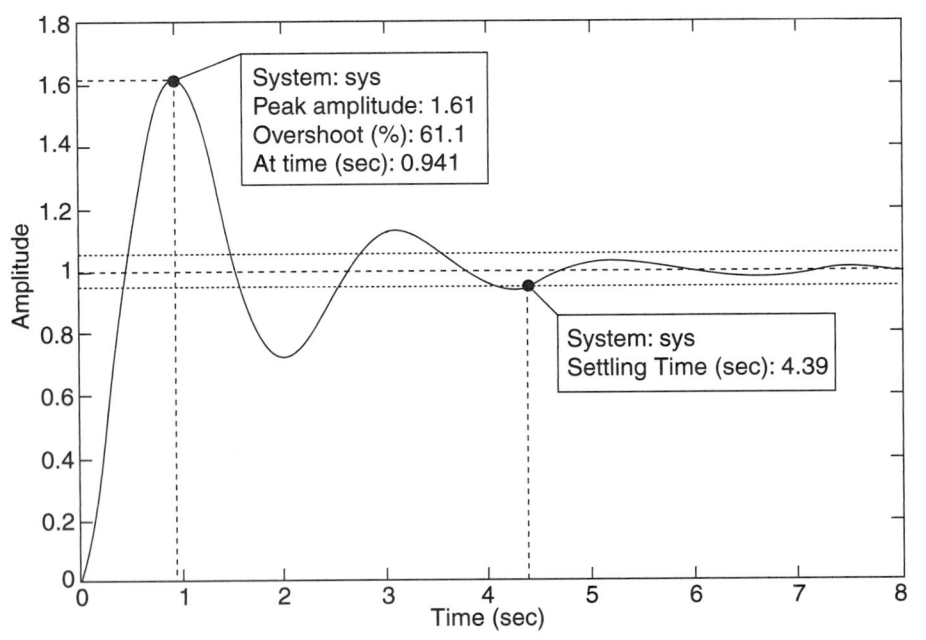

System: sys
Peak amplitude: 1.61
Overshoot (%): 61.1
At time (sec): 0.941

System: sys
Settling Time (sec): 4.39

(figure 9-34)

Output of Example 9-9 with calculated values

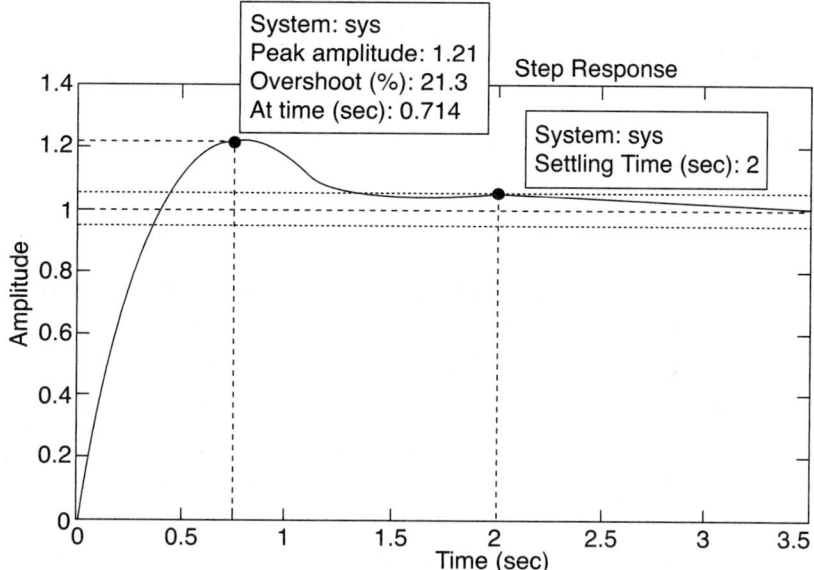

(figure 9-35)

Reduction of overshoot for Example 9-9

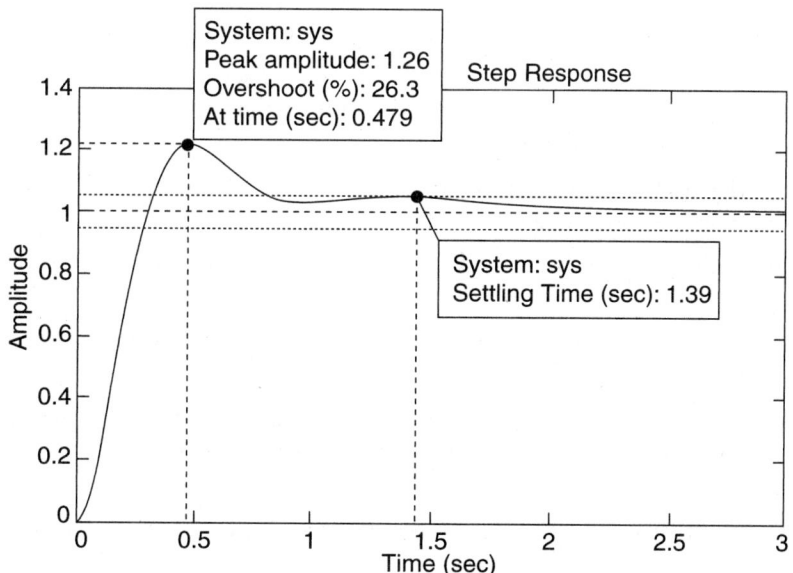

(figure 9-36)

Final correction for Example 9-9

Summary

➤ On-off, proportional, PD, PI, and PID controllers are used to control system parameters such as risetime, settling time, and error in well-defined dynamic models.

➤ The on-off controller is either full-on or full-off, and is not useable in high precision systems.

➤ The proportional controller primarily gives control of the damping ratio of a system at the cost of increasing overshoot.

➤ Compared to other forms of systems, proportional controllers may have an error for a step input and may have a larger overshoot to achieve a required error.

➤ The PD controller gives control of risetime and settling time and can add damping to a system to bring overshoot to an acceptable level and to control the phase margin of a system.

➤ The PD controller widens the bandwidth of a system, which makes it more susceptible to high-frequency noise than other forms of controllers.

➤ The PI controller is used to eliminate error for some forms of input signals and to improve the damping ratio of the system.

➤ The PI form of controller decreases the bandwidth of the system, making the system less susceptible to high-frequency noise.

➤ In general, the PI controller is slower than either the proportional or PD controller.

➤ The PID controller combines the best features of the proportional, PD, and PI controllers.

➤ The PID controller allows control of risetime, settling time, frequency response, gain/phase margins, and overshoot.

➤ The Ziegler-Nichols method of loop tuning was developed for the PID controller for adjusting a control system in the field.

➤ The first Ziegler-Nichols method is used when there are no integrations in the control system.

➤ The second Ziegler-Nichols method is used when there are integrations in the control system.

➤ MATLAB is used extensively in the design of controllers since, in some cases, it is necessary to iterate the design based on the variation of one or more of the gains in the system.

➤ MATLAB provides quick access to the system parameters, showing which values of gain fulfill the system requirements.

Problems

P9.1 For the following type zero open-loop transfer function

$$G_P(s) = \frac{100K_P}{s^2 + 50s + 100}$$

a. Calculate K_P such that the damping ratio of the closed-loop system is 0.707.

b. Use the computer program of your choice to plot the response to a step input and show that the error is less than 10 percent.

c. Calculate the settling time for the 5 percent and 2 percent criteria.

d. Use the damping ratio chart in Chapter 6 to determine the risetime of a 0.707 damping ratio. Is the risetime of this system approximately correct? Explain.

e. Plot the gain and phase of the system and determine the GM and PM of the system.

f. Calculate M_r for the system.

P9.2 For the following type one open-loop transfer function

$$G_P(s) = \frac{1.6 \cdot 10^6 K_P}{s(s + 3100)}$$

a. Calculate K_P such that the damping of the closed-loop system is 0.6. Calculate the overshoot of this system for a step input.

b. Calculate the time of the first peak of the system response to a step input.

c. Use the computer program of your choice to show the velocity error of this system. Calculate the value of K_V from the plot.

d. If K_V is doubled, what is the new value of K_P, the overshoot, and the damping ratio of the system?

e. Plot the gain and phase of the system for each of the two values of K_P. Show the difference in the margins and the change in M_r for the system.

P9.3 For the type one system of Figure P9-3

(figure P9-3)

System for Problem P9.3

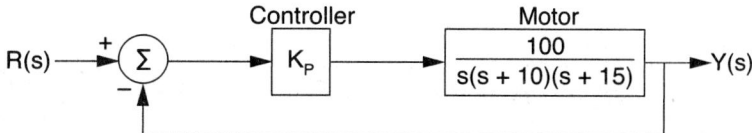

   a. Calculate K_p based on K_V equal to $1.5s^{-1}$. If a ramp input of $tu(t)$ is applied to the input, what is the error of the ramp in rad/sec?

   b. Using the second-order embedded system, determine the damping ratio and overshoot of the system.

   c. Provide a Bode plot of the gain and phase of the system. Show the gain and phase margins for the system.

P9.4   For the system shown in Figure P9-4, the following specifications are required:

     Risetime ≤ 0.1 sec

     Settling time ≤ 0.35 sec (5% criteria)

     Error for step input $\leq 5\%$

     Overshoot $\leq 10\%$

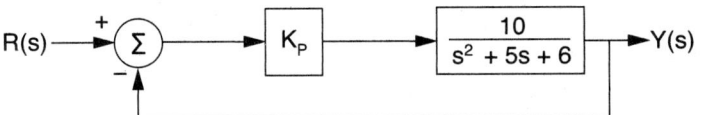

(figure P9-4)

System for Problem P9.4

     The overshoot and error present conflicting requirements for K_p. Use a PD controller to add damping to the system and make the system meet all specifications.

P9.5   For Problem P9.3, the specifications are changed as follows:

     Risetime ≤ 0.25 sec

     Settling time ≤ 0.6 sec (2% criteria)

     Overshoot $< 5\%$

     Using a pole-zero cancellation, design a PD controller to provide the necessary specifications. Assume a perfect pole-zero cancellation.

P9.6   For the following DC motor transfer function

$$G_P(s) = \frac{4.601 \cdot 10^6}{s(s + 1937)}$$

     The following specifications apply:

     Risetime ≤ 1 msec

     Settling time ≤ 2 msec (2% criteria)

     Overshoot $\leq 5\%$

     $K_V \leq 5000$ s^{-1}

The overshoot and K_V requirements cannot be obtained simultaneously with a pure proportional controller design. Design a PD controller that adds the required damping and meets the specifications.

P9.7   A control system with a PD controller is shown in Figure P9-7.

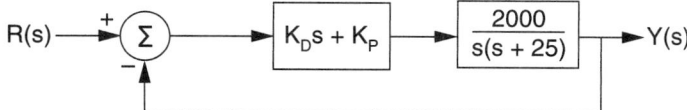

$$R(s) \longrightarrow \overset{+}{\underset{-}{\Sigma}} \longrightarrow \boxed{K_D s + K_P} \longrightarrow \boxed{\frac{2000}{s(s + 25)}} \longrightarrow Y(s)$$

a. Determine the values of K_P and K_D such that the damping ratio is 1.0 for $K_V = 80$.

b. Determine the values of K_P and K_D such that the damping ratio is 0.707 for $K_V = 80$.

P9.8   A system has an open-loop transfer function.

$$G_P(s) = \frac{5.538 \cdot 10^6}{s(s + 1937)}$$

Analysis of the system in the frequency domain reveals that the system has a crossover frequency of approximately 2 k·rad/sec, a phase margin of 44°, and an M_r of 1.334, with overshoot of 24 percent. The frequency response in terms of hertz shows that the system has response at 60 Hz and several of the harmonics of 60 Hz, which promotes noise problems. It is desired to decrease the crossover frequency by about two octaves and increase the phase margin to about 70°, with no overshoot.

The system is to have the following specifications:

> Overshoot = 0% for a step input
> Risetime ≤ 5msec
> Settling time ≤ 5 msec (5% criteria)
> Phase margin ≈ 70°

a. Use a PI controller to produce the required specifications. Start by calculating the value of K_P for a unity damping ratio. Then determine K_I for the rest of the specifications.

b. Provide a frequency response plot to show the new crossover frequency and phase margin. Provide both the uncompensated and compensated systems on one plot.

P9.9 A type zero system has three dominant poles as shown in Figure P9-9.

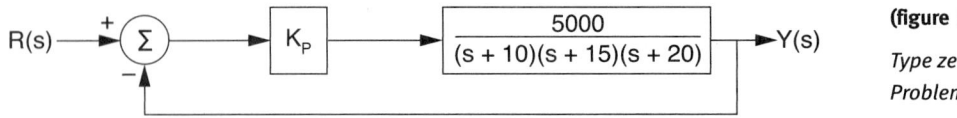

The set of specifications for the system requires that it have a step error of 0.01 of the input, a risetime less than 0.5 seconds, and a settling time of less than 1 second for the 2 percent criteria. The overshoot must not exceed 5 percent.

a. Use a Routh test to determine the limits of stable gain. Then calculate the value of gain required to meet the error requirement.

b. If the gain determined in Part a is too large, use a PI controller to cancel the pole at $s = -10$. Then use a root locus to determine the position of the poles to produce a damping ratio that is less than the required overshoot specification.

c. Plot the step response of the system and label response times.

P9.10 A control system with a type zero plant is shown in Figure P9-10.

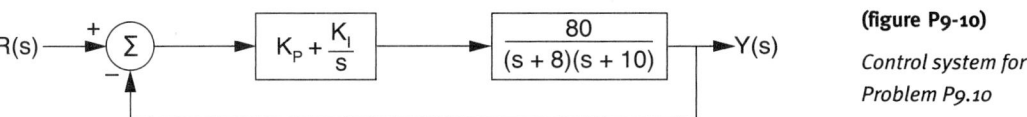

The system cannot follow a ramp input, so a PI controller is used to make the system type one.

a. Determine K_I such that $K_V = 10$.

b. Determine a suitable value for K_P such that the following specifications are met:

Overshoot < 10%

Risetime ≤ 0.15 sec

Settling time ≤ 0.5 sec

c. Plot the response of the system to a ramp input and show that the steady-state error is 0.1.

P9.11 One form of control assures that no single pole or pair of poles is dominant in a system. The general form of the characteristic equation of this type of system is $(s + \alpha)[(s + \alpha)^2 + \omega^2]$. For a PID controller, the general form of the characteristic equation is

$$s^3 + (2\zeta\omega_n + AK_D)s^2 + (\omega_n^2 + AK_P)s + AK_I = 0$$

Long division of the characteristic equation by $(s + a)$ yields the following first- and second-order expressions:

$$(s + \alpha)\left[s^2 + (2\zeta\omega_n + AK_D - \alpha)s + \frac{AK_I}{\alpha} \right]$$

From those equations, values for a and K_D can be determined, given the settling time required of the system. The defining equations for a and T_S are

$$\alpha = \frac{2\zeta\omega_n + AK_D}{3} \qquad T_s = \frac{10}{2\zeta\omega_n + AK_D - \alpha}$$

T_s is based on five time constants for full decay of the embedded second-order system.

A PID control system is shown in Figure P9-11.

(figure P9-11)

*PID system for
Problem P9.11*

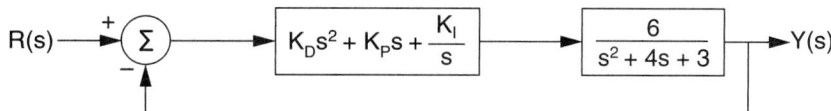

a. Determine a and K_D such that the real pole and the real parts of the complex poles are equal. Base the calculation on a settling time of 1.0 seconds and a damping ratio of 1.0.

b. Plot the system response to a unit step input. Determine the risetime, peak overshoot, and settling time for the system.

c. Plot the system response to a ramp input of $tu_s(t)$. Determine K_V from the plot.

d. Calculate K_V from the open-loop transfer function. Is the value approximately the same as the plot value? Explain.

P9.12   For the motor in Example 9-3, change the system to a PID controller. Maintain the integrator zero at $s = -10$ and determine a derivative gain such that the overshoot is approximately 5 percent, the settling time is less than 1.5 msec (5 percent criteria), and the risetime is less than 0.5 msec. The motor transfer function is shown as a convenience.

$$G_P(s) = \frac{1.952 \cdot 10^6}{s(s + 773)}$$

Hint: Set $K_P = 1$ and use the K_D calculated for Example 9-3 as a starting point.

P9.13   A type one control system has an open-loop transfer function given by

$$G_p(s) = \frac{1.6 \cdot 10^9}{s(s + 3100)(s + 200)}$$

Analysis of this system with PD control shows that the best overshoot occurs with $K_D = 0.0025$. All other parameters are met or exceeded. Design a PID controller such that it meets the following requirements:

Overshoot < 10%	BW > 200 rad/sec
Risetime ≤ 10 msec	PM ≥ 60°
Settling time ≤ 25 msec (5% criteria)	$M_r < 1.5$

a. Plot the response to a step function.

b. Plot the response to a unit ramp input.

c. Plot the open-loop magnitude and phase. Show the phase margin.

d. Plot the closed-loop magnitude and phase. Show that $M_r < 1.5$.

P9.14   For the type one system shown in Figure P9-14

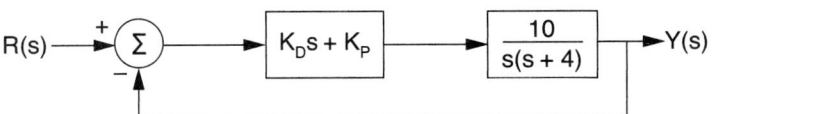

(figure P9-14)

PD system for Problem P9.14

Design a PD controller for a damping ratio of 0.707. Then design a PI controller to form a PID control system with the following specifications:

Overshoot ≤ 10%	Open-loop BW ≥ 5 rad/sec
Risetime ≤ 0.2 sec	PM ≥ 45°
Settling time (5%) ≤ 0.5 sec	

a. Plot the response to a unit step input.

b. Plot the response to a unit ramp input.

c. Plot the open-loop magnitude and phase of the system for the compensated and uncompensated systems. Mark all stability points and the bandwidth.

P9.15   For the following type one system, design a PID controller such that all of the poles of the controller are placed at $s = -9 + j0$. Also determine the position of the zeros of the closed-loop transfer function.

$$G_C(s) = \frac{10}{s(s + 20)}$$

P9.16   For the following type zero system, use a PID controller to convert to a type one system. The complex poles are to be positioned at $s = -25 \pm j10$, and the real pole is placed at $s = -100$.

$$G_P(s) = \frac{100}{s^2 + 20s + 200}$$

a. Plot the step response of the system and identify rise- and settling times and the maximum overshoot.

b. Plot the response to a unit ramp signal and determine K_V from the plot.

P9.17   For the following open-loop transfer function, use the Ziegler-Nichols first method to provide the necessary parameters for a quarter-step response.

$$G_P(s) = \frac{5}{(s + 2)(s + 3)}$$

Plot the plant response and determine the slope and delay time. Then plot the step response of the system and show that the second peak is approximately 25 percent of the first peak.

P9.18   A heat transfer system has the following transfer function:

$$G_P(s) = \frac{1}{(50s + 1)(100s + 1)}$$

Use the Ziegler-Nichols first method to provide a quarter-step response. Plot the plant response and determine the time delay. Show the output. Label the amount of settling time, including the fact that the second peak is 25 percent or less than the first peak.

P9.19   For the following control system plant that contains integrators, use the Ziegler-Nichols second method to determine K_P, T_I, and T_D for the system.

$$G_C(s) = \frac{K_P}{s(s + 2)(s + 4)}$$

Provide the response of the plant to a step input. Then minimize the overshoot, using the technique discussed.

P9.20   A control system is to use a PI controller. The start of the Routh table for this system is shown below. Determine the constants of the table. Then determine K_{Cr}, P_{Cr}, K_P, and T_I.

s^3	1	24
s^2	10	$5K_P$
s^1		
s^0		

Plot the output of the system and reduce the overshoot to 60 percent or less.

MATLAB Problems

M9.1   For Problems P9.1–P9.5, use **MATLAB** to provide all necessary plots of the system to verify your results.

M9.2   For the type one system shown, use **MATLAB** and the SISO Design Tool to determine the risetime and settling time, crossover frequency, and phase margin for the uncompensated system. Plot the uncompensated system closed-loop step response.

$$G_P(s) = \frac{5000}{s(s + 50)}$$

a. Add a real zero such that the damping ratio of the closed-loop system is approximately 0.8. Plot the closed-loop step response of the system, using the analysis menu. Then determine the value of K_D from the values given in the SISO Design Tool. (Assume that $K_P = 1$.)

b. From the methods outlined in the text, calculate a value for K_D, assuming $K_P = 1$. Plot the step response and show that the two agree.

c. Plot the magnitude and phase of the compensated system. What are the new crossover frequency and phase margin?

M9.3   A type one liquid level control system with unity feedback has the following open-loop transfer function:

$$G_C(s) = \frac{0.2}{s(s + 0.1)}$$

Design a PD controller such that the system has a 2 percent settling time of less than 10 seconds and a damping ratio of 0.707. Use **MATLAB** to confirm your design.

M9.4   A type zero system has the following plant transfer function:

$$G_P(s) = \frac{10}{s^2 + 12s + 10}$$

Provide a PI controller that causes all of the poles to have the same real parts, with a steady-state error of 0.078. Determine suitable values of K_p and K_I to provide the steady-state error. Use **MATLAB** to verify your results.

M9.5   A motor has the following parameters:

$K_T = 2.2$ oz-in/A	$L_a = 0.0016$ H
$K_b = 0.0156$ V/(rad/sec)	$J = 0.00027$ oz-in-s²
$R_a = 1.93$ ohms	$B = 0.0185$ oz-in-s

The specifications for the system are

$e_{ss} = 0.0015$ for a unit ramp input

Overshoot < 15%

Risetime < 5 msec

Settling time (5%) < 15 msec

a. Calculate the transfer function of the motor. Then use **MATLAB** to design a PI controller. (Hint: Use a real pole value of –577 for the evaluation of K_p.)

b. Factoring of the motor transfer function denominator shows that one of the poles is far from the origin and has a minimal effect on the system response. Remove the pole and recalculate the required K_p and K_I. Plot the new function ramp and step response, and compare the results of both analyses.

M9.6   An attitude control system has the following open-loop transfer function:

$$GC(s) = \frac{4500K}{s(s + 2100)}$$

For this system, the following specifications apply:

$K_V = 1000$	Closed-loop BW > 100 rad/sec
Overshoot < 1%	PM > 80°
Settling time < 2.5 msec	$M_r < 1.0$
Risetime < 2 msec	

a. Calculate a suitable value for K_V.

b. Use **MATLAB** and the SISO Design Tool to design a PID controller such that all specifications are met.

M9.7   For the following system, use **MATLAB** to plot the process reaction curve. Determine the slope and delay time. Then using the Ziegler-Nichols first method, design a PI system. Show that your design meets the quarter-step criteria.

$$G_P(s) = \frac{4}{(s + 1)(4s + 1)}$$

M9.8   For the following plant system, use the Ziegler-Nichols second method to determine the constants for a PID control system. Plot the time domain response of the system and reduce the overshoot. (Hint: Try doubling the integral time.)

$$G_P(s) = \frac{0.025}{s(s + .25)(s + .1)}$$

Chapter 10 | State-Space Design of Controllers

Chapter Objectives

After completing this chapter, you should be able to:

➤ Explain the reason state-space analysis has an advantage over the usual SISO design of control systems.

➤ Explain what a state variable is and how it is formulated.

➤ Explain the reason why a state-transition matrix is formulated.

➤ Use the state-transition matrix in the calculation of the time domain output of a control system.

➤ Derive a transfer function from the system matrixes in state-space.

➤ State the concept of state feedback and its application to control systems.

➤ Construct the state diagram of a control system.

➤ Construct state diagrams with state feedback.

➤ Explain what the eigenvalues of a matrix are.

➤ Explain the concept of controllability and observability.

10.1 Introduction

Previous chapters presented the concept of linear, time-invariant systems that have fixed components in the forward and feedback paths. Also developed were methods of exploring the stability of the systems, using root locus and frequency response techniques. Using the Laplace transform, you converted the system differential equations into algebraic equations in the s-plane. Those methods proved to be very powerful and useful techniques.

As control systems have become more complex and may have more than one input and output, the transfer function method does not apply since it is based on a single input, output, and linear time invariance. The type of analysis used when a system has multiple inputs and outputs is the **state variables** of the system, and it is called **state-variable analysis**.

This chapter will introduce the concepts of state variables and define the state-variable equations. To predict stability in the time domain, frequency response techniques can be used to extract the characteristic equations of the systems. In addition, two concepts used in modern control theory will be introduced: *controllability* and *observability*. This type of analysis also lends itself to adaptive and optimal control design. The SFGs of Chapter 4 are used extensively in this type of design.

10.2 What Are State Variables?

Some systems that are defined by one or more nth-order linear, time-invariant equations are more conveniently solved by breaking up the equation into n first-order functions. The n first-order equations are the state variables of the system.

As an example of determining state variables, consider the following nth-order, linear, constant coefficient differential equation:

$$\sum_{i=0}^{n} a_i \frac{d^i y}{dt^i} = u$$

That equation can be replaced by n first-order equations of the form

$$\frac{dx_1}{dt} = x_2$$

$$\frac{dx_2}{dt} = x_3$$

$$\vdots$$

$$\frac{dx_{n-1}}{dt} = x_n$$

From which the general form is

$$\frac{dx_n}{dt} = -\frac{1}{a_n}\left[\sum_{i=0}^{n-1} a_i x_{i+1}\right] + \frac{u}{a_n} \tag{10.1}$$

Where $x_1 \approx y$. You can write the general form as a *vector-matrix* form, which is

$$\frac{dx}{dt} = Ax + Bu \tag{10.2}$$

Further, if the system is single input/output, the matrix forms are developed as they are for single input/output systems.

$$
\begin{bmatrix} \dot{x}_1 \\ \dot{x}_2 \\ \vdots \\ \dot{x}_n \end{bmatrix}
=
\overset{\textstyle A}{
\begin{bmatrix}
0 & 1 & 0 & \cdots & 0 \\
0 & 0 & 1 & \cdots & 0 \\
\vdots & \vdots & \vdots & \ddots & \vdots \\
-\dfrac{a_0}{a_n} & -\dfrac{a_1}{a_n} & -\dfrac{a_2}{a_n} & \cdots & -\dfrac{a_{n-1}}{a_n}
\end{bmatrix}}
\overset{\textstyle x(t)}{
\begin{bmatrix} x_1 \\ x_2 \\ \vdots \\ x_n \end{bmatrix}}
+
\overset{\textstyle Bu(t)}{
\begin{bmatrix} 0 \\ 0 \\ \vdots \\ \dfrac{1}{a_n} \end{bmatrix}} u
\tag{10.3}
$$

Where $x_1, x_2, ..., x_n$ are the state variables and $u(t)$ is the scalar input to the system.

If the system is multiple input/output, the matrix expands to

$$
\begin{bmatrix} \dot{x}_1 \\ \dot{x}_2 \\ \vdots \\ \dot{x}_n \end{bmatrix}
=
\overset{\textstyle A}{
\begin{bmatrix}
0 & 1 & 0 & \cdots & 0 \\
0 & 0 & 1 & \cdots & 0 \\
\vdots & \vdots & \vdots & \ddots & \vdots \\
-\dfrac{a_0}{a_n} & -\dfrac{a_1}{a_n} & -\dfrac{a_2}{a_n} & \cdots & -\dfrac{a_{n-1}}{a_n}
\end{bmatrix}}
\overset{\textstyle x(t)}{
\begin{bmatrix} x_1 \\ x_2 \\ \vdots \\ x_n \end{bmatrix}}
+
\overset{\textstyle B}{
\begin{bmatrix}
b_{11} & b_{12} & \cdots & b_{1r} \\
b_{21} & b_{22} & \cdots & b_{2r} \\
\vdots & \vdots & \ddots & \vdots \\
b_{n1} & b_{n2} & \cdots & b_{nr}
\end{bmatrix}}
\overset{\textstyle u(t)}{
\begin{bmatrix} u_1 \\ u_2 \\ \vdots \\ u_r \end{bmatrix}}
\tag{10.4}
$$

Whose vector-matrix form is

$$\frac{dx}{dt} = Ax + Bu \tag{10.5}$$

Where $x \approx x(t)$, A is the $n \times n$ matrix of constants a_{ij}, B is the $n \times r$ matrix of constants b_{ij}, and u is an r-vector of input functions. The addition of output vectors makes the state-variable forms complete. The output vectors have the form

$$Y = Cx + Du$$

$$
\begin{bmatrix} y_1 \\ y_2 \\ \vdots \\ y_n \end{bmatrix}
=
\overset{\mathbf{C}}{\begin{bmatrix} c_{11} & c_{12} & \cdots & c_{1n} \\ c_{21} & c_{22} & \cdots & c_{2n} \\ \vdots & \vdots & \ddots & \vdots \\ c_{m1} & c_{m2} & \vdots & c_{mn} \end{bmatrix}}
\overset{x(t)}{\begin{bmatrix} x_1 \\ x_2 \\ \vdots \\ x_n \end{bmatrix}}
+
\overset{\mathbf{D}}{\begin{bmatrix} d_{11} & d_{12} & \cdots & d_{1n} \\ d_{21} & d_{22} & \cdots & d_{12} \\ \vdots & \vdots & \ddots & \vdots \\ d_{m1} & d_{m2} & \cdots & d_{mn} \end{bmatrix}}
\overset{u(t)}{\begin{bmatrix} u_1 \\ u_2 \\ \vdots \\ u_n \end{bmatrix}}
\tag{10.6}
$$

10.3 State-Transition Matrix

The state-transition matrix is important to the time response of linear systems in that it determines the effect of initial conditions on the systems. This section will discuss procedures that relate to and define state-space and transform approaches that define the state-transition matrix. In addition, the matrix will be defined and some of its properties identified.

If you consider Equation 10.2 in the s-plane, including any initial conditions, you have

$$sX(s) - x(0) = AX(s) + BU(s)$$

$$X(s)[sI - A] = x(0) + BU(s)$$

The identity matrix I multiplies s to make the s and A matrices the same size for subtraction. Rearrangement of the equation yields

$$X(s) = \frac{x(0)}{[sI - A]} + \frac{B}{[sI - A]}U(s) = [sI - A]^{-1} \cdot x(0) + [sI - A]^{-1} \cdot BU(s) \tag{10.7}$$

You can simplify the expression by substituting $\Phi(s)$ for $[sI - A]^{-1}$ to yield

$$X(s) = \Phi(s) \cdot x(0) + \Phi(s) \cdot BU(s) \tag{10.8}$$

And taking the inverse Laplace transform of $[sI - A]^{-1}$ yields

$$\phi(t) = \mathcal{L}^{-1}\{[sI - A]^{-1}\} \tag{10.9}$$

$\phi(t)$ is called the **state-transition matrix**. A complication arises in that the second term in Equation 10.8 is a product in the s-plane, which may be solved using the convolution integral to transform to the general form in the time domain.

$$x(t) = \phi(t)x(0) + \int_0^t \phi(t - \tau) \cdot Bu(\tau)d\tau \tag{10.10}$$

The first term in the equation is transformed to e^{At}, using the fact that the identity matrix has a determinant of 1 and the term reduces to $\Phi(s) = \dfrac{1}{s - A} \cdot e^{At}$ can be defined as

$$e^{At} = I + At + \frac{(At)^2}{2!} + \frac{(At)^3}{3!} + \cdots \tag{10.11}$$

From the series in Equation 10.11, $\phi(0) = I$. In addition, Equation 10.10 can be revised to become

$$x(t) = e^{At}x(0) + \int_0^t e^{A(t-\tau)} Bu(\tau)d\tau \tag{10.12}$$

It is important to understand that the state-transition matrix is useful only when the initial time is defined to be at $t = 0$, as it is the response of the system that is excited only by the initial conditions. With Equation 10.10, you can accurately calculate the transition matrix, using classical methods. An example will illustrate the method.

Example 10-1

Consider the state equations

$$\begin{bmatrix} \dot{x}_1 \\ \dot{x}_2 \end{bmatrix} = \overset{A}{\begin{bmatrix} -3 & 1 \\ -2 & 0 \end{bmatrix}} \begin{bmatrix} x_1 \\ x_2 \end{bmatrix} + \overset{B}{\begin{bmatrix} 4 \\ -5 \end{bmatrix}} r(t)$$

$$y(t) = [1 \quad -1] \begin{bmatrix} x_1 \\ x_2 \end{bmatrix}$$

Solution

Starting with $[sI - A]$, you have

$$[sI - A] = \begin{bmatrix} s & 0 \\ 0 & s \end{bmatrix} - \begin{bmatrix} -3 & 1 \\ -2 & 0 \end{bmatrix} = \begin{bmatrix} s + 3 & -1 \\ 2 & s \end{bmatrix}$$

From which $[sI - A]^{-1}$ is given by

$$[sI - A]^{-1} = \frac{\text{Adj}\,[sI - A]^T}{\det\,[sI - A]} = \frac{1}{s^2 + 3s + 2} \cdot \begin{bmatrix} s & 1 \\ -2 & s + 3 \end{bmatrix}$$

$$= \begin{bmatrix} \dfrac{s}{s^2 + 3s + 2} & \dfrac{1}{s^2 + 3s + 2} \\[3mm] \dfrac{-2}{s^2 + 3s + 2} & \dfrac{s + 3}{s^2 + 3s + 2} \end{bmatrix}$$

Example 10-1 (continued)

Taking the inverse Laplace transform of [$sI - A$] gives

$$\phi(t) = \mathcal{L}^{-1}[sI - A]^{-1} = \begin{bmatrix} \left(-e^{-t} + 2e^{-2t}\right) & \left(e^{-t} - e^{-2t}\right) \\ \left(-2e^{-t} + 2e^{-2t}\right) & \left(2e^{-t} - e^{-2t}\right) \end{bmatrix}$$

The total unforced and forced responses are then

$$x(t) = \begin{bmatrix} \left(-e^{-t} + 2e^{-2t}\right) & \left(e^{-t} - e^{-2t}\right) \\ \left(-2e^{-t} + 2e^{-2t}\right) & \left(2e^{-t} - e^{-2t}\right) \end{bmatrix}$$

$$x(0) + \int_0^t \begin{bmatrix} \left(-e^{-(t-\tau)} + 2e^{-2(t-\tau)}\right) & \left(e^{-(t-\tau)} - e^{-2(t-\tau)}\right) \\ \left(-2e^{-(t-\tau)} + 2e^{-2(t-\tau)}\right) & \left(2e^{-(t-\tau)} - e^{-2(t-\tau)}\right) \end{bmatrix} \begin{bmatrix} 4 \\ -5 \end{bmatrix} d\tau$$

or

$$x(t) = \begin{bmatrix} \left(-e^{-t} + 2e^{-2t}\right) & \left(e^{-t} - e^{-2t}\right) \\ \left(-2e^{-t} + 2e^{-2t}\right) & \left(2e^{-t} - e^{-2t}\right) \end{bmatrix} x(0) + \begin{bmatrix} 9e^{-t} - 6.5e^{-2t} - 2.5 \\ 18e^{-t} - 6.5e^{-2t} - 11.5 \end{bmatrix} \quad t \geq 0$$

Alternatively, you can determine the second term by taking the inverse Laplace transform of [$sI - A$]$^{-1} \cdot BU(s)$, as follows:

$$\mathcal{L}^{-1}[sI - A]^{-1}BU(s) = \mathcal{L}^{-1}\left(\begin{bmatrix} \dfrac{s}{s^2 + 3s + 2} & \dfrac{1}{s^2 + 3s + 2} \\ \dfrac{-2}{s^2 + 3s + 2} & \dfrac{s + 3}{s^2 + 3s + 2} \end{bmatrix} \begin{bmatrix} 4 \\ -5 \end{bmatrix} \dfrac{1}{s}\right)$$

$$= \begin{bmatrix} 9e^{-t} - 6.5e^{-2t} - 2.5 \\ 18e^{-t} - 6.5e^{-2t} - 11.5 \end{bmatrix} \quad t \geq 0$$

And $y(t)$ is given by

$$y(t) = x_1(t) - x_2(t)$$
$$y(t) = 9e^{-t} - 6.5e^{-2t} - 2.5 - 18e^{-t} + 6.5e^{-2t} + 11.5 = -9e^{-t} + 9$$
$$= 9(1 - e^{-t})$$

10.4 Transfer Functions in State-Space

In your study of frequency response, you found the transfer function of a single input/output system to be useful in predicting the stability of the system. The characteristic equation developed in the transfer function gives the location of the system poles and allows you to predict, to some degree, the response of the system to an input.

The output of a system is given by

$$Y(s) = CX(s) + DU(s) \qquad (10.13)$$

And the transfer function requires that all initial cditions be set to zero. $X(s)$ is given by

$$X(s) = [sI - A]^{-1} \cdot BU(s)$$

From which

$$Y(s) = \{C[sI - A]^{-1} \cdot B + D\}U(s) \qquad (10.14)$$

$Y(s)/U(s)$ is the transfer function matrix, an $m \times n$ matrix $P(s)$ with the following form:

$$P_{mn}(s) = \begin{bmatrix} P_{11}(s) & P_{12}(s) & \cdots & P_{1n}(s) \\ \vdots & \vdots & \ddots & \vdots \\ \vdots & \vdots & \ddots & \vdots \\ P_{m1}(s) & P_{m2}(s) & \cdots & P_{mn}(s) \end{bmatrix} = \frac{Y_m(s)}{U_n(s)} \qquad (10.15)$$

The determinant of that matrix gives the denominator the transfer function. The characteristic equation is then

$$\det[sI - A] = 0 \qquad (10.16)$$

Let's use the system from Example 10-1 to show that the s output equation is obtained using this method.

Example 10-2

Already having $[sI - A]^{-1}$ from Example 10-1, you substitute it in the output equation for $P(s)$.

Solution

$$P(s) = \begin{bmatrix} 1 & -1 \end{bmatrix} \begin{bmatrix} \dfrac{s}{s^2 + 3s + 2} & \dfrac{1}{s^2 + 3s + 2} \\ \dfrac{-2}{s^2 + 3s + 2} & \dfrac{s + 3}{s^2 + 3s + 2} \end{bmatrix} \begin{bmatrix} 4 \\ -5 \end{bmatrix}$$

Example 10-2 (continued)

From which

$$P(s) = \begin{bmatrix} 1 & -1 \end{bmatrix} \begin{bmatrix} \dfrac{4s-5}{s^2+3s+2} \\[4mm] \dfrac{-5s-23}{s^2+3s+2} \end{bmatrix} = \dfrac{4s-5}{s^2+3s+2} + \dfrac{5s+23}{s^2+3s+2} = \dfrac{9(s+2)}{(s+1)(s+2)}$$

$$= \dfrac{9}{s+1}$$

Using $U(s) = 1/s$ as an input

$$Y(s) = \dfrac{9}{s(s+1)}$$

And

$$y(t) = 9(1 - e^{-t})$$

The previous examples generated only a simple transfer function; it would be beneficial to investigate a more complex system.

Example 10-3

The following state equations define a control system. Determine the characteristic equation and transfer function matrix of the system.

$$\begin{bmatrix} \dot{x}_1 \\ \dot{x}_2 \\ \dot{x}_3 \end{bmatrix} = \begin{bmatrix} -4 & 0 & 2 \\ -1 & -1 & 0 \\ 3 & 0 & -3 \end{bmatrix} \begin{bmatrix} x_1 \\ x_2 \\ x_3 \end{bmatrix} + \begin{bmatrix} -3 & -2 \\ 4 & 1 \\ 0 & 0 \end{bmatrix} \begin{bmatrix} u_1 \\ u_2 \end{bmatrix}$$

$$\begin{bmatrix} y_1 \\ y_2 \end{bmatrix} = \begin{bmatrix} 1 & 1 & 1 \\ -1 & 0 & 1 \end{bmatrix} \begin{bmatrix} x_1 \\ x_2 \\ x_3 \end{bmatrix}$$

Solution

The $[sI - A]$ matrix is

$$[sI - A] = \begin{bmatrix} s & 0 & 0 \\ 0 & s & 0 \\ 0 & 0 & s \end{bmatrix} - \begin{bmatrix} -4 & 0 & 2 \\ -1 & -1 & 0 \\ 3 & 0 & -3 \end{bmatrix} = \begin{bmatrix} s+4 & 0 & -2 \\ 1 & s+1 & 0 \\ -3 & 0 & s+3 \end{bmatrix}$$

Example 10-3 (continued)

And the inverse is

$$[sI - A]^{-1} = \begin{bmatrix} \dfrac{(s+1)(s+3)}{(s+1)(s^2+7s+6)} & 0 & \dfrac{2(s+1)}{(s+1)(s^2+7s+6)} \\ \dfrac{-(s+3)}{(s+1)(s^2+7s+6)} & \dfrac{(s+4)(s+3)-6}{(s+1)(s^2+7s+6)} & \dfrac{-2}{(s+1)(s^2+7s+6)} \\ \dfrac{-3(s+1)}{(s+1)(s^2+7s+6)} & 0 & \dfrac{(s+4)(s+1)}{(s+1)(s^2+7s+6)} \end{bmatrix}$$

$$= \begin{bmatrix} \dfrac{(s+3)}{(s^2+7s+6)} & 0 & \dfrac{2}{(s^2+7s+6)} \\ \dfrac{-(s+3)}{(s+1)(s^2+7s+6)} & \dfrac{1}{(s+1)} & \dfrac{-2}{(s+1)(s^2+7s+6)} \\ \dfrac{-3}{(s^2+7s+6)} & 0 & \dfrac{(s+4)}{(s^2+7s+6)} \end{bmatrix}$$

Then, using Equation 10.14, you have

$$P(s) = \begin{bmatrix} 1 & 1 & 1 \\ -1 & 0 & 1 \end{bmatrix} \begin{bmatrix} \dfrac{(s+3)}{(s^2+7s+6)} & 0 & \dfrac{2}{(s^2+7s+6)} \\ \dfrac{-(s+3)}{(s+1)(s^2+7s+6)} & \dfrac{1}{(s+1)} & \dfrac{-2}{(s+1)(s^2+7s+6)} \\ \dfrac{-3}{(s^2+7s+6)} & 0 & \dfrac{(s+4)}{(s^2+7s+6)} \end{bmatrix} \begin{bmatrix} -3 & -2 \\ 4 & 1 \\ 0 & 0 \end{bmatrix}$$

Performing the required operations, $P(s)$ is reduced to a single 2×2 matrix with the form

$$P(s) = \begin{bmatrix} \dfrac{s^2+10s+15}{(s+1)(s^2+7s+6)} & \dfrac{-s(s+5)}{(s+1)(s^2+7s+6)} \\ \dfrac{3s}{s^2+7s+6} & \dfrac{2s}{s^2+7s+6} \end{bmatrix}$$

From which the characteristic equation is

$$s^3 + 8s^2 + 13s + 6 = 0$$

State Feedback

State feedback is very useful in the placement of a system's poles. The concept is that the state variables are fed back to the input summing junction multiplied by a linear gain factor $k_1, k_2, \ldots, k_n$. The states are then fed back as $k_1 x_1 + k_2 x_2 + \ldots + k_n x_n$.

$$Kx(t) = \begin{bmatrix} k_1 & k_2 & k_n \end{bmatrix} \begin{bmatrix} x_1(t) \\ x_2(t) \\ \vdots \\ x_n(t) \end{bmatrix}$$

The state model is modified by the use of these gain factors from its original form. The A, B, C, and D matrices are initially applied to the plant, and the overall state model is then completed by augmenting the plant as required to include the feedback model.

Figure 10-1 shows a SISO system with state feedback.

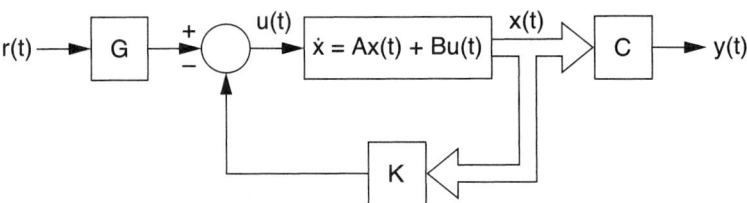

(figure 10-1)

SISO system with state feedback

From looking at the figure, it is clear that the controller input, $u(t)$, is given by

$$u(t) = gr(t) - Kx(t)$$

That changes the $\dot{x}(t)$ equality to

$$\dot{x}(t) = Ax(t) + B[gr(t) - Kx(t)]$$

And K is a row matrix multiplied by B. The determinant, $\det(sI - A + BK) = 0$, is the characteristic equation of the system.

Example 10-4

For the following state equations, determine the characteristic equation and the values of K such that the poles of the system are placed at $s = -5 \pm j5$.

$$\begin{bmatrix} \dot{x}_1 \\ \dot{x}_2 \end{bmatrix} = \begin{bmatrix} 0 & 2 \\ -1 & -3 \end{bmatrix} \begin{bmatrix} x_1 \\ x_2 \end{bmatrix} + \begin{bmatrix} 0 \\ 1 \end{bmatrix} [k_1 \quad k_2]u(t)$$

The characteristic equation of the system with poles as required is $s^2 + 10s + 50 = 0$. Since the determinant $(sI - A + BK)$ is the characteristic equation, the solution proceeds as follows.

Solution

$$(sI - A + BK) = \begin{bmatrix} s & -2 \\ (1 + k_1) & (3s + 3k_2) \end{bmatrix} + \begin{bmatrix} 0 \\ 1 \end{bmatrix}[k_1 \quad k_2]$$

Example 10-4 (continued)

And the characteristic equation is

$$\det(sI - A + BK) = s^2 + (3 + 3k_2)s + (2 + 2k_1) = 0$$

Solving the resulting two equations in two unknowns produces

$$3 + k_2 = 10, \quad k_2 = 10 - 3 = 7$$

$$2 + 2k_1 = 50, \quad k_1 = \frac{50 - 2}{2} = 24$$

The poles are now placed at $s = -5 \pm j5$.

MATLAB can also be used to generate the values of k_1 and k_2 directly from the **A** and **B** matrices as follows:

    A = [0 2;−1 −3]; B = [0 1]';
    p = [−5 + 5i −5 − 5i];
    k = place (A, B, p)

The place (A, B, p) command calculates the values of k_1 and k_2. The result of the calculation is $k_1 = 24$, $k_2 = 7$; the values are reversed from the characteristic equation.

10.5 Construction of the State Diagram of a System

Chapter 4 covered the concepts of the SFG and Mason's gain formula, but it did not include initial states in that development. The use of the SFG or block diagram along with Mason's gain formula, including initial states, is convenient in the solution of state systems.

The fundamental form of an SFG with an initial state is shown in Figure 10-2.

(figure 10-2)

Graphical representation of Equation 10.17

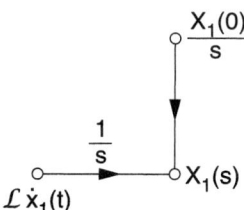

Consider a state variable x_1, which is represented as above.

$$x_1(t) = \int_0^t \dot{x}_1(t)\, dt + x_1(0) \tag{10.17}$$

Whose Laplace transform is

$$X_1(s) = \frac{1}{s}\left[\mathcal{L}\dot{x}_1(t)\right] + \frac{x_1(0)}{s} \qquad (10.18)$$

Addition of the necessary number of similar structures and the feedback loops for each state variable of a differential equation yields the state diagram of the system. Let's clarify the method with an example.

Example 10-5

Consider the following state model:

$$\begin{bmatrix} \dot{x}_1 \\ \dot{x}_2 \end{bmatrix} = \begin{bmatrix} 0 & 1 \\ -4 & -3 \end{bmatrix}\begin{bmatrix} x_1 \\ x_2 \end{bmatrix} + \begin{bmatrix} 0 \\ 5 \end{bmatrix}u \text{ and } x(0) = \begin{bmatrix} 1 \\ 0 \end{bmatrix}$$

The system input is $u(t) = 2tu_s(t)$. Construct the state diagram of the system and determine the time domain response for $X_1(s)$ and $X_2(s)$.

Solution

The state diagram of the system is shown in Figure 10-3.
Applying Mason's gain formula for $X_1(s)$

$$\Delta = 1 - \left(\frac{-4}{s}\right) - \left(\frac{-3}{s^2}\right) = \frac{s^2 + 4s + 3}{s^2}$$

$$\Delta P_1 = \frac{1}{s}\cdot\left(1 + \frac{3}{s}\right) = \frac{s+3}{s^2} \qquad \Delta P_2 = \frac{5}{s^2}\cdot\frac{2}{s^2} = \frac{10}{s^4}$$

$$\Delta P_1 + \Delta P_2 = \frac{s^3 + 3s^2 + 10}{s^4}$$

$$X_1(s) = \frac{\dfrac{s^3 + 3s^2 + 10}{s^4}}{\dfrac{s^2 + 4s + 3}{s^2}} = \frac{s^3 + 3s^2 + 10}{s^2(s^2 + 4s + 3)}$$

Example 10-5 (continued)

A similar analysis for X_2 (s) yields

$$X_2(s) = \frac{\dfrac{-4s + 8}{s^3}}{\dfrac{s^2 + 4s + 3}{s^2}} = \frac{-4s + 8}{s(s^2 + 4s + 3)}$$

To determine the time domain response of the X_1 (s) and X_2 (s), partial fraction expansion of the two equations is required. You can use MATLAB for that purpose. Define the numerator and denominator; then use the residual command for the poles and residues, as follows:

```
n = [1 3 0 10]; d = [1 4 3 0 0];
[r, k] = residue(n,d)
```

This results in the residues of 3.333, –4.444, 6, and –0.556 and poles at 0, 0, –3, and –1. That gives the following s-plane function for X_1 (s) and x_1 (t).

$$X_1(s) = \frac{3.333}{s^2} - \frac{4.444}{s} + \frac{6}{s + 1} - \frac{0.556}{s + 3}$$

$$x_1(t) = [3t - 4.444]u_s(t) + [6e^{-t} - 0.556e^{-3t}]u_s(t)$$

For X_2 (s), the results using a similar decomposition are residues of 2.667, –6, and 3.333; poles at 0, –1, and –3; and X_2 (s), x_2 (t).

$$X_2(s) = \frac{2.667}{s} - \frac{6}{s + 1} + \frac{3.333}{s + 3}$$

$$x_2(t) = [2.667]u_s(t) - [6e^{-t} + 3.333e^{-3t}]u_s(t)$$

(figure 10-3)

State diagram for Example 10-5

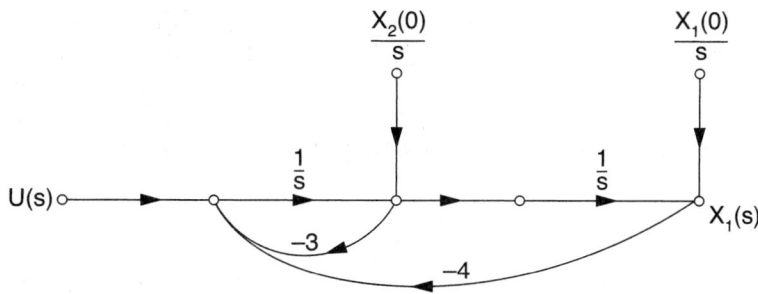

State Diagrams with State Feedback

The state diagram of a system with state feedback is found as in the previous section. The state feedback from the outputs $X_1(s)$ and $X_2(s)$ is returned to the summing junction at $U(s)$. The state diagram for Example 10-4 is shown in Figure 10-4.

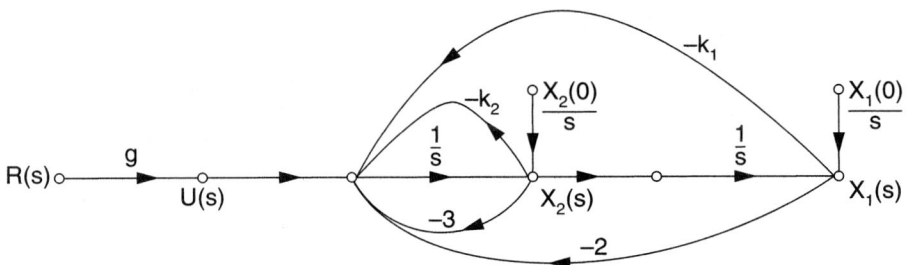

(figure 10-4)

State diagram for Example 10-4

Mason's gain formula can be used to determine the transfer function of the system for simulation.

10.6 Controllability

Much of modern control system design is carried out in the time domain rather than in the frequency domain by the application of state variables. The characteristic equation allows you to predict the time response and stability of a system, but it is also important to know if you have direct input to all of the state variables. This property is known as the *controllability* of a system. The definition of controllability is as follows:

> *A system is controllable if every state can be driven to a required output with a control input.*

To observe whether a system is controllable, it is necessary to provide a decoupled system diagram. One method is to diagonalize the A matrix and construct a parallel form of the state diagram. Diagonalization assures that the rows of the A matrix are linearly independent and that the form of the parallel diagram is readily constructed using the matrix values.

Before proceeding, you need to understand the concept of eigenvalues.

Eigenvalues

The roots of the characteristic equation of a system are called the **eigenvalues** of the system if the coefficients of the A matrix are all real, its eigenvalues are real, or they occur in complex conjugate pairs. To find the eigenvalues of A (i.e., the values of λ) the following method is used:

$$det(\lambda I - A) = 0 \qquad (10.19)$$

Where eigenvalues of A are the values, λ_i, that satisfy Equation 10.19.

An example of determining the eigenvalues of a matrix is shown in the following example.

Example 10-6

Determine the eigenvalue matrix for the following **A** matrix.

Solution

$$A = \begin{bmatrix} 0 & 1 & 0 \\ 0 & 0 & 1 \\ -24 & -26 & -9 \end{bmatrix}$$

$$(\lambda I - A) = \begin{bmatrix} \lambda & 0 & 0 \\ 0 & \lambda & 0 \\ 0 & 0 & \lambda \end{bmatrix} - \begin{bmatrix} 0 & 1 & 0 \\ 0 & 0 & 1 \\ -24 & -26 & -9 \end{bmatrix} = \begin{bmatrix} \lambda & -1 & 0 \\ 0 & \lambda & -1 \\ 24 & 26 & (\lambda + 9) \end{bmatrix}$$

The characteristic equation of the system is

$$\det(\lambda I - A) = \lambda^3 + 9\lambda^2 + 26\lambda + 24$$
$$= (\lambda + 2)(\lambda + 3)(\lambda + 4)$$

The eigenvalues of **A** are the roots of the characteristic equation.

$$\lambda_1 = -2, \ \lambda_2 = -3, \ \lambda_3 = -4$$

MATLAB can be used to generate the eigenvalues of the system, using the **eig** command as follows:

$$A = \begin{bmatrix} 0 & 1 & 0; & 0 & 0 & 1; -24 & -26 & 9 \end{bmatrix};$$
$$eig(A)$$

That generates the same values, $\lambda_1 = -2$, $\lambda_2 = -3$, and $\lambda_3 = -4$.

From the eigenvalues, you can construct a Vandermonde matrix with which to diagonalize the A matrix. The Vandermonde matrix is a matrix whose top row is all 1's and whose succeeding rows are powers of the eigenvalues, as shown.

$$V = \begin{bmatrix} 1 & 1 & 1 & \cdots & 1 \\ \lambda_1 & \lambda_2 & \lambda_3 & \cdots & \lambda_n \\ \lambda_1^2 & \lambda_2^2 & \lambda_3^2 & \cdots & \lambda_n^2 \\ \vdots & \vdots & \vdots & \ddots & \vdots \\ \lambda_1^{n-1} & \lambda_2^{n-1} & \vdots & \ddots & \lambda_m^{n-1} \end{bmatrix} \tag{10.20}$$

The Vandermonde matrix for the matrix of Example 10-6 is

$$V = \begin{bmatrix} 1 & 1 & 1 \\ -2 & -3 & -4 \\ 4 & 9 & 16 \end{bmatrix}$$

Diagonalization of a matrix is done by using the following equality:

$$\bar{A} = V^{-1}AV \tag{10.21}$$

Diagonalization of the A matrix of Example 10-6 proceeds as shown in the following example.

Example 10-7

Diagonalize the matrix of Example 10-6, using Equation 10.21.

Solution

Use MATLAB to obtain the inverse matrix of V.

$$V^{-1} = \begin{bmatrix} 6 & 3.5 & 0.5 \\ -8 & -6 & -1 \\ 3 & 2.5 & 0.5 \end{bmatrix}$$

$$\bar{A} = \begin{bmatrix} 6 & 3.5 & 0.5 \\ -8 & -6 & -1 \\ 3 & 2.5 & 0.5 \end{bmatrix} \begin{bmatrix} 0 & 1 & 0 \\ 0 & 0 & 1 \\ -24 & -26 & -9 \end{bmatrix} \begin{bmatrix} 1 & 1 & 1 \\ -2 & -3 & -4 \\ 4 & 9 & 16 \end{bmatrix}$$

$$= \begin{bmatrix} -2 & 0 & 0 \\ 0 & -3 & 0 \\ 0 & 0 & -4 \end{bmatrix}$$

MATLAB can be used to generate the diagonalized matrix directly with the following command:

$$[V, A] = \text{eig}(A)$$

When that command is run, two matrixes are generated: one is a matrix of eigenvectors, which have not been discussed, and the other is the diagonalized $\bar{A}$ matrix.

For a system to be completely controllable, a necessary and sufficient condition is that the system have an $n \times nr$ controllability matrix, where the matrix has a rank of n. The general form of the controllability matrix is

$$M_C = \begin{bmatrix} B & AB & \ldots & A^{n-1}B \end{bmatrix} \tag{10.22}$$

Let's investigate a system, using the A matrix of Example 10-6.

Example 10-8

The transfer function of a system is given as

$$\frac{Y(s)}{U(s)} = \frac{3s^2 + 18s + 26}{s^3 + 9s^2 + 26s + 24} = \frac{1}{s+2} + \frac{1}{s+3} + \frac{1}{s+4}$$

Solution

The system matrices derived from the transfer function are

$$\begin{bmatrix} \dot{x}_1 \\ \dot{x}_2 \\ \dot{x}_3 \end{bmatrix} = \begin{bmatrix} -2 & 0 & 0 \\ 0 & -3 & 0 \\ 0 & 0 & -4 \end{bmatrix} \begin{bmatrix} x_1 \\ x_2 \\ x_3 \end{bmatrix} + \begin{bmatrix} 0 \\ 1 \\ 1 \end{bmatrix} u(t)$$

$$y(t) = \begin{bmatrix} 1 & 1 & 1 \end{bmatrix} \begin{bmatrix} x_1 \\ x_2 \\ x_3 \end{bmatrix}$$

From which the state diagram is obtained, as shown in Figure 10-5.
From the general formula for a controllability matrix, the matrix for this system is

$$M_C = \begin{bmatrix} B & AB & A^2B \end{bmatrix} = \begin{bmatrix} 0 & 0 & 0 \\ 1 & -3 & 9 \\ 1 & -4 & 16 \end{bmatrix}$$

Since a row of all zeros occurs in M_c, it is singular and this system is not controllable.

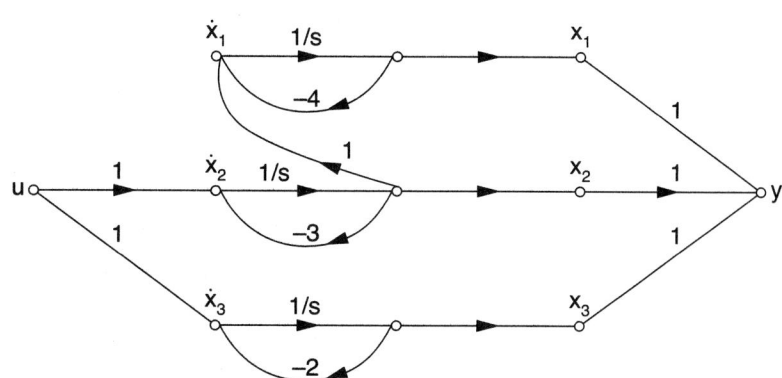

(figure 10-5)

State diagram for Example 10-8

Example 10-8 (continued)

Once again, **MATLAB** can be used to generate the controllability directly. Keep in mind that each matrix manipulation produces a row vector for each manipulation. The **MATLAB** commands for this are

$A = [-2\ 0\ 0;\ 0\ -3\ 0;\ 0\ 0\ -4];\ B = [0\ 1\ 1]';\ C = [1\ 1\ 1];$
Mc = ctrb((A, B)
rank(Mc)

The rank of *Mc* is 2, and the system has at least one uncontrollable state, which is a state for which you have no access to its input.

While a system may be uncontrollable, it still may be observable. That is, you may inspect the effect of every state variable on at least one of the outputs of the system.

10.7 Observability

Often, it is desirable to obtain information regarding the state variables by inspecting the inputs and outputs of the system. If any of the states cannot be observed at an output, the state is said to be unobservable. In most cases, the system is simply said to be unobservable.

A system is observable if every state output can be measured at any of the system outputs.

For a system to be observable, it is a necessary and sufficient condition that the following $n \times np$ observability matrix have rank n.

$$M_O = \left[\begin{array}{c} C \\ \hline CA \\ \hline \vdots \\ \hline CA^{n-1} \end{array} \right] \tag{10.23}$$

Example 10-9

Using the same system as in Example 10-7, the observability matrix is generated from

Solution

$$C = \begin{bmatrix} 1 & 1 & 1 \end{bmatrix} \quad A = \begin{bmatrix} -2 & 0 & 0 \\ 0 & -3 & 0 \\ 0 & 0 & -4 \end{bmatrix}$$

Example 10-9 (continued)

$$M_O = \begin{bmatrix} C \\ CA \\ CA^2 \end{bmatrix} = \begin{bmatrix} 1 & 1 & 1 \\ -2 & -3 & -4 \\ 4 & 9 & 16 \end{bmatrix}$$

The observability matrix has a determinant of –24 and is, therefore, nonsingular and of rank 3, indicating that the system is completely observable.

MATLAB also can be used to generate the observability matrix directly. The form of the command in this case requires the use of semicolons to separate each matrix manipulation since each manipulation produces a column vector of the solution.

$$M_O = obsv(A,C)$$

The rank of this matrix is 3, and the system is completely observable.

Since the same system is found to be uncontrollable but observable, some conditions on controllability need to be mentioned.

1. A system may be controllable and observable.
2. A system may be controllable but unobservable.
3. A system may be uncontrollable but observable.
4. A system may be uncontrollable and unobservable.

A final word on controllability and observability: It should be clear from Figure 10-5 that the outputs of x_1, x_2, and x_3 are all observable since they connect to a single output. However, since the input is connected only to x_2 and x_3, those states are controllable. However, x_1 is not connected since there is no direct input from $u(t)$ to its input, which fulfills condition 3 above.

Summary

➤ The state-variable approach can be applied to the solution of control system requirements.

➤ Single input/output systems, or large systems with multiple inputs/outputs, and initial conditions can be accommodated using state-variable analysis.

➤ The state equations developed were found to be coupled first-order differential equations.

➤ The time solution of the equations was made possible by the use of exponential matrices and the use of the state-transition matrix.

➤ The state matrices provided solutions that were both numerical and mathematical.

➤ Determination of the transfer function matrix and the characteristic equation of a system uses the determinant of $[sI - A]$ matrix.

➤ The eigenvalues and diagonalization of matrices provide a method of decoupling the differential equations of a system to produce another form of state diagram.

➤ State variables give an insight to the concepts of controllability and observability.

➤ The controllability matrices give insight into whether the individual input states of a control system are accessible.

➤ The observability matrices give insight into whether the individual output states of a control system are accessible.

➤ For a system to be controllable, all input states must be measurable.

➤ To be observable, all output states must be measurable.

➤ Calculating by hand the necessary controllability and observability matrices is a very tedious and time-consuming effort.

➤ MATLAB functions for producing the controllability and observability matrices decrease the tedium of that task.

Problems

P10.1   The state equations of a linear time invariant system are given by

$$\frac{dx(t)}{dt} = Ax(t) + Bu(t)$$

Find the state-transition matrix $\phi(t)$, eigenvalues, and characteristic equation for the following matrices:

a. $A = \begin{bmatrix} 0 & 1 \\ -2 & -3 \end{bmatrix}$ $B = \begin{bmatrix} 0 \\ 1 \end{bmatrix}$

b. $A = \begin{bmatrix} 0 & -1 \\ 1 & -3 \end{bmatrix}$ $B = \begin{bmatrix} 0 \\ 1 \end{bmatrix}$

c. $A = \begin{bmatrix} 0 & -3 \\ -1 & 0 \end{bmatrix}$ $B = \begin{bmatrix} 0 \\ 1 \end{bmatrix}$

d. $A = \begin{bmatrix} -2 & 0 \\ 0 & -3 \end{bmatrix}$ $B = \begin{bmatrix} 0 \\ 1 \end{bmatrix}$

e. $A = \begin{bmatrix} 0 & 1 \\ -2 & -4 \end{bmatrix}$ $B = \begin{bmatrix} 0 \\ 1 \end{bmatrix}$

P10.2   For the following state equations, determine the eigenvalues and diagonalize the matrix.

a. $A = \begin{bmatrix} 0 & 1 \\ -2 & -8 \end{bmatrix}$ $B = \begin{bmatrix} 0 \\ 3 \end{bmatrix}$

b. $A = \begin{bmatrix} 0 & 1 & 0 \\ 0 & 0 & 1 \\ -6 & -11 & -6 \end{bmatrix}$ $B = \begin{bmatrix} 0 \\ 1 \\ 1 \end{bmatrix}$

P10.3   For the following state equations, determine the values for k_1, k_2, and k_3 such that the closed-loop poles are located at $(s + 10 \pm j10)$ and $(s + 50)$.

$$\begin{bmatrix} \dot{x}_1 \\ \dot{x}_2 \\ \dot{x}_3 \end{bmatrix} = \begin{bmatrix} 0 & 2 & 0 \\ 0 & -1 & 20 \\ 0 & -5 & -50 \end{bmatrix} \begin{bmatrix} x_1 \\ x_2 \\ x_3 \end{bmatrix} + \begin{bmatrix} 0 \\ 0 \\ 10 \end{bmatrix} u(t)$$

$$y(t) = \begin{bmatrix} 1 & 0 & 0 \end{bmatrix} \begin{bmatrix} x_1 \\ x_2 \\ x_3 \end{bmatrix}$$

P10.4   For the following state equations, determine the transfer function between $Y(s)$ and $U(s)$ according to the formula

$$\frac{Y(s)}{U(s)} = \left(C\big[(sI - A)\big]^{-1} B + D \right)$$

a. $A = \begin{bmatrix} 0 & 3 \\ -2 & -3 \end{bmatrix}$ $B = \begin{bmatrix} 0 \\ 3 \end{bmatrix}$ $C = \begin{bmatrix} 1 & 0 \end{bmatrix}$ $D = 1$

b. $A = \begin{bmatrix} 0 & 1 & 0 \\ 00 & 1 \\ -4 & -8 & -12 \end{bmatrix}$ $B = \begin{bmatrix} 0 \\ 0 \\ 4 \end{bmatrix}$

$C = \begin{bmatrix} 1 & 1 & 0 \end{bmatrix}$ $D = 1$

P10.5   For the following state equations with initial conditions, draw the state diagram and determine the time domain output for $X_1(s)$ and $X_2(s)$. The system input is $tu_s(t)$.

$$\begin{bmatrix} \dot{x}_1 \\ \dot{x}_2 \end{bmatrix} = \begin{bmatrix} 0 & 1 \\ -6 & -5 \end{bmatrix} \begin{bmatrix} x_1 \\ x_2 \end{bmatrix} + \begin{bmatrix} 0 \\ 5 \end{bmatrix} u(t) \quad x(0) = \begin{bmatrix} 1 \\ 0 \end{bmatrix}$$

P10.6   For the following state equations with initial conditions, draw the state diagram and determine the time domain output for $X_1(s)$, $X_2(s)$, and $X_3(s)$. The system input is $u(t) = 2tu_s(t)$. Use

$$X(s) = [sI - A]^{-1} x(0) + [sI - A]^{-1} BU(s)$$

$$\begin{bmatrix} \dot{x}_1 \\ \dot{x}_2 \\ \dot{x}_3 \end{bmatrix} = \begin{bmatrix} 0 & 1 & 0 \\ 0 & 0 & 1 \\ -6 & -11 & -6 \end{bmatrix} \begin{bmatrix} x_1 \\ x_2 \\ x_3 \end{bmatrix} + \begin{bmatrix} 0 \\ 0 \\ 3 \end{bmatrix} u(t) \qquad x(0) = \begin{bmatrix} 1 \\ 0 \\ 0 \end{bmatrix}$$

P10.7   For the motor equivalent shown in Figure P10-7, select a set of state equations and determine a state model for the motor. Use only two state variables.

(figure P10-7)

Motor analogy

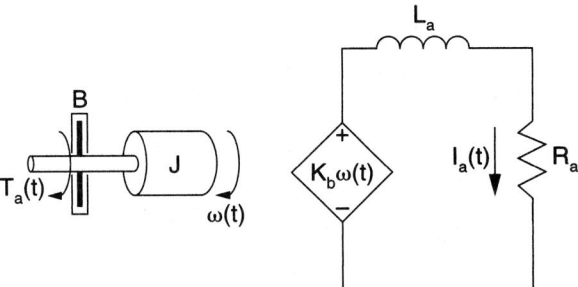

P10.8   For the following motor constants, develop the transfer function and provide the state equations for the motor.

$$R_a = 2\ \Omega \qquad\qquad J_m = 0.0001\ \text{oz-in-s}^2$$
$$L_a = 4\ \text{mH} \qquad\qquad B_m = 0.2\ \text{oz-in}$$
$$K_T = 5\ \text{oz-in/A} \qquad K_b = 0.02\ \text{V/(rad/s)}$$

P10.9   For the following state equations, diagonalize the matrix and calculate the characteristic equation. Use

$$\bar{A} = V^{-1} A V$$

$$\begin{bmatrix} \dot{x}_1 \\ \dot{x}_2 \\ \dot{x}_3 \end{bmatrix} = \begin{bmatrix} 0 & 1 & 0 \\ 0 & 0 & 1 \\ -15 & -23 & -9 \end{bmatrix} \begin{bmatrix} x_1 \\ x_2 \\ x_3 \end{bmatrix} + \begin{bmatrix} 0 \\ 0 \\ 1 \end{bmatrix} u(t)$$

$$y(t) = \begin{bmatrix} 1 & 0 & 0 \end{bmatrix} \begin{bmatrix} x_1 \\ x_2 \\ x_3 \end{bmatrix}$$

P10.10   For the following state equations, determine the time domain output of the system for a unit step input.

$$\begin{bmatrix} \dot{x}_1 \\ \dot{x}_2 \end{bmatrix} = \begin{bmatrix} -8 & 6 \\ -6 & 4 \end{bmatrix} \begin{bmatrix} x_1 \\ x_2 \end{bmatrix} + \begin{bmatrix} 4 \\ -6 \end{bmatrix} u(t)$$

$$y(t) = \begin{bmatrix} 1 & -1 \end{bmatrix} \begin{bmatrix} x_1 \\ x_2 \end{bmatrix}$$

P10.11   For the state diagram shown in Figure P10-11, determine the transfer function of the system and determine if the system is controllable and observable. Provide the controllability and observability matrices of the system.

(figure P10-11)

State diagram

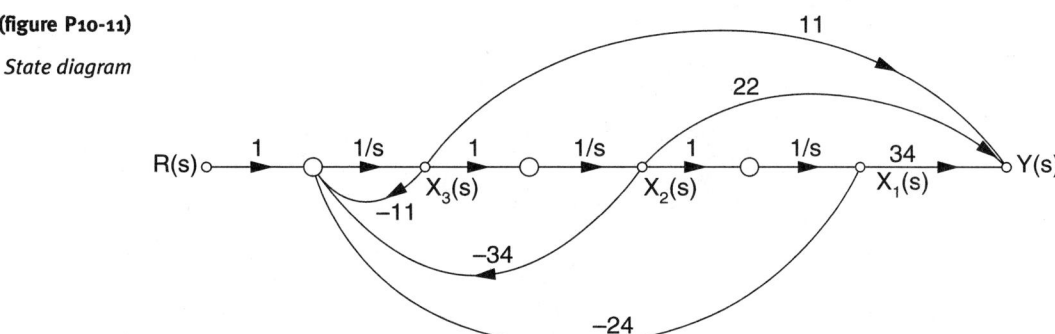

P10.12   For the following state equations, determine if the systems are controllable and observable.

$$\begin{bmatrix} \dot{x}_1 \\ \dot{x}_2 \\ \dot{x}_3 \end{bmatrix} = \begin{bmatrix} 0 & 1 & 0 \\ 0 & 2 & 0 \\ -4 & -6 & -4 \end{bmatrix} \begin{bmatrix} x_1 \\ x_2 \\ x_3 \end{bmatrix} + \begin{bmatrix} 1 \\ 0 \\ 0 \end{bmatrix} u(t)$$

a.

$$y(t) = \begin{bmatrix} 0 & 0 & 1 \end{bmatrix} \begin{bmatrix} x_1 \\ x_2 \\ x_3 \end{bmatrix}$$

$$\begin{bmatrix} \dot{x}_1 \\ \dot{x}_2 \\ \dot{x}_3 \end{bmatrix} = \begin{bmatrix} 0 & 1 & 0 \\ 1 & 0 & 0 \\ -6 & -11 & -6 \end{bmatrix} \begin{bmatrix} x_1 \\ x_2 \\ x_3 \end{bmatrix} + \begin{bmatrix} 0 \\ 2 \\ 1 \end{bmatrix} u(t)$$

b.

$$y(t) = \begin{bmatrix} 1 & 0 & 0 \end{bmatrix} \begin{bmatrix} x_1 \\ x_2 \\ x_3 \end{bmatrix}$$

$$\begin{bmatrix} \dot{x}_1 \\ \dot{x}_2 \\ \dot{x}_3 \end{bmatrix} = \begin{bmatrix} 0 & 1 & 1 \\ 0 & 0 & 1 \\ -12 & -24 & -60 \end{bmatrix} \begin{bmatrix} x_1 \\ x_2 \\ x_3 \end{bmatrix} + \begin{bmatrix} 0 \\ 0 \\ 1 \end{bmatrix} u(t)$$

c.

$$y(t) = [1 \ 0 \ 0] \begin{bmatrix} x_1 \\ x_2 \\ x_3 \end{bmatrix}$$

P10.13 For the following plant and controller, set the poles to be at $s = -5.0$ and draw a minimum phase diagram. Write the A, B, C, and D matrices and determine the time domain response. Is the system controllable and observable? Why or why not?

$$G_P(s) = \frac{8}{s^2 + 6s + 8}, \qquad G_C(s) = \frac{K_D s^2 + K_P s + K_I}{s}, \qquad x(0) = 0$$

P10.14 For the following state equations, determine the transfer function of the system and draw the state diagram. Determine the damping ratio of the embedded second-order system and provide a root locus for the system.

$$\begin{bmatrix} \dot{x}_1 \\ \dot{x}_2 \\ \dot{x}_3 \end{bmatrix} = \begin{bmatrix} 0 & 1 & 0 \\ -80 & -48 & 64 \\ -1 & 0 & 0 \end{bmatrix} \begin{bmatrix} x_1 \\ x_2 \\ x_3 \end{bmatrix} + \begin{bmatrix} 0 \\ 4 \\ 1 \end{bmatrix} u(t)$$

$$y(t) = [1 \ 0 \ 0] \begin{bmatrix} x_1 \\ x_2 \\ x_3 \end{bmatrix}$$

P10.15 A control system has the following state equations:

$$\begin{bmatrix} \dot{x}_1 \\ \dot{x}_2 \\ \dot{x}_3 \end{bmatrix} = \begin{bmatrix} 0 & 2 & 0 \\ 0 & 0 & 2 \\ 0 & 0 & 0 \end{bmatrix} \begin{bmatrix} x_1 \\ x_2 \\ x_3 \end{bmatrix} + \begin{bmatrix} 0 \\ 0 \\ 1 \end{bmatrix}$$

$$y(t) = [1 \ 0 \ 0] \begin{bmatrix} x_1 \\ x_2 \\ x_3 \end{bmatrix}$$

Determine the following conditions:

a. Is the system asymptotically stable?

b. Can state feedback be used to place the poles of the system without restriction? Explain your answer.

c. If the poles can be placed using state feedback, place the poles such that they are at $s = -5$ and $s = -5 \pm j5$.

d. Provide the state diagram of the system with state feedback.

P10.16   For the following state equations, determine the transfer function and determine if the system is controllable and observable. Provide a state diagram for the system.

$$\begin{bmatrix} \dot{x}_1 \\ \dot{x}_2 \\ \dot{x}_3 \end{bmatrix} = \begin{bmatrix} -1 & 1 & 0 \\ 0 & 0 & 2 \\ -15 & -23 & -9 \end{bmatrix} \begin{bmatrix} x_1 \\ x_2 \\ x_3 \end{bmatrix} + \begin{bmatrix} 0 \\ 1 \\ 1 \end{bmatrix} u(t)$$

$$y = \begin{bmatrix} 1 & 0 & 0 \end{bmatrix} \begin{bmatrix} x_1 \\ x_2 \\ x_3 \end{bmatrix}$$

MATLAB Problems

M10.1   Use **MATLAB** to verify your calculations for Problems P10.2 and P10.4.

M10.2   For the following matrices, use **MATLAB** to place the poles of the system at $s = -8$ and $s = -4 \pm j6$. Produce the time domain response of the system.

$$A = \begin{bmatrix} 0 & 1 & 0 \\ 0 & 0 & 10 \\ -6 & -11 & -6 \end{bmatrix}, \qquad B = \begin{bmatrix} 0 \\ 0 \\ 1 \end{bmatrix}, \qquad C = \begin{bmatrix} 41.6 & 0 & 0 \end{bmatrix}, \qquad D = 0$$

M10.3   For the following state equations, use **MATLAB** to determine the K matrix and place the poles at $s = -10$ and $s = -5 \pm j2$. Use Simulink to provide the output for a $2tu_s(t)$ input.

$$\begin{bmatrix} \dot{x}_1 \\ \dot{x}_2 \\ \dot{x}_3 \end{bmatrix} = \begin{bmatrix} 0 & 1 & 0 \\ 0 & 0 & 10 \\ 0 & 0 & -15 \end{bmatrix} \begin{bmatrix} x_1 \\ x_2 \\ x_3 \end{bmatrix} + \begin{bmatrix} 0 \\ 0 \\ 1 \end{bmatrix}$$

$$y(t) = \begin{bmatrix} 29 & 0 & 0 \end{bmatrix} \begin{bmatrix} x_1 \\ x_2 \\ x_3 \end{bmatrix} \qquad D = 0$$

M10.4   For Problem P10.8, add an integrator to the system and define the transfer function of the system. Then calculate the error for a $tu_s(t)$ input. Provide the response of the system to the input and confirm the error function. (Use Simulink to plot the response, then label the error.)

M10.5   For Problem P10.15, show that the system is both controllable and observable with the **K** matrix as developed. What value of gain factor must be used to make the system exhibit unity gain? Use **Simulink** to show the gain factor placement for the system. Show the error for a $2tu_s(t)$ input.

M10.6   Use **MATLAB** to determine if the following systems are controllable and observable.

a. $A = \begin{bmatrix} -2 & 0 & 1 \\ 0 & -2 & 1 \\ 1 & 0 & -2 \end{bmatrix}$, $B = \begin{bmatrix} 1 \\ 1 \\ 1 \end{bmatrix}$, $C = \begin{bmatrix} 0 & 1 & 0 \end{bmatrix}$

b. $A = \begin{bmatrix} -2 & 0 & 1 \\ 0 & -2 & 1 \\ 1 & 0 & -2 \end{bmatrix}$, $B = \begin{bmatrix} 1 \\ 0 \\ 1 \end{bmatrix}$, $C = \begin{bmatrix} 0 & 1 & 0 \end{bmatrix}$

c. $A = \begin{bmatrix} -2 & 0 & 1 \\ 0 & -2 & 1 \\ 1 & 0 & -2 \end{bmatrix}$, $B = \begin{bmatrix} 0 \\ 1 \\ 1 \end{bmatrix}$, $C = \begin{bmatrix} 0 & 1 & 0 \end{bmatrix}$

M10.7   Convert the block diagram of Figure M10-7 into an SFG and write the matrices. Use **MATLAB** to determine the transfer function. Then provide the step response of the system. Is the system controllable and observable? Why or why not?

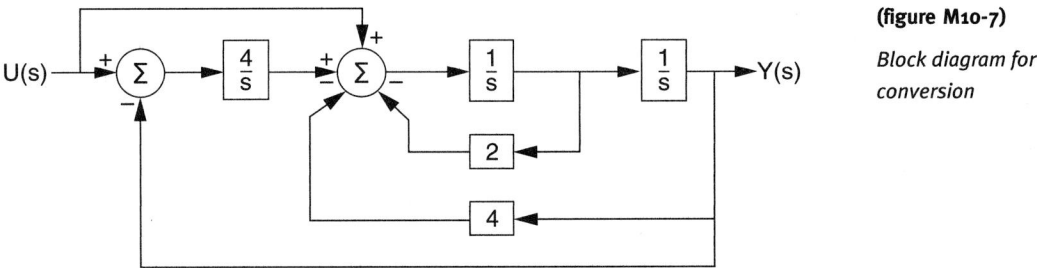

(figure M10-7)

Block diagram for conversion

Part 2

Chapter 11

Discrete Systems

Chapter Objectives

After completing this chapter, you should be able to:

➤ Explain the role of the zero-order hold in analog-to-digital conversion.

➤ Explain what is meant by the resolution of an A/D converter.

➤ Explain how the conversion time of an A/D converter is calculated.

➤ Explain how the accuracy of an A/D converter is calculated.

➤ Explain the methods used to increase the apparent frequency response of an A/D converter.

➤ Explain the methodology used to convert the difference equation to a transfer function.

➤ Show the methods used to convert the transfer function to a difference equation.

➤ Show by calculation that the z-plane is a unit circle and that all stable system poles exist inside the unit circle.

➤ Calculate the proper sampling time to convert the continuous time transfer function into a discrete transfer function.

➤ Explain the methods used to calculate the frequency response of a discrete system.

11.1 Introduction

A continuous system can be approximated by sampling a finite number of points at discrete times. In terms of the accuracy of the approximation, as the number of points increases toward infinity, the more accurate the approximation of the system becomes. That concept should be familiar to you from developing the derivative and integral in calculus. The implication is that all continuous systems start as a finite set of points, and using the concepts of limits allows you to represent the system in closed form.

In the past, computers and microprocessors were not commonplace and engineers were forced to represent systems as closed or continuous forms. As digital tools have evolved, the approximation of a continuous system by digital methods has become very practical. That method is more an idea whose time has come, rather than a completely new method. Based on that fact, this chapter will discuss the application of discrete methods to the approximation of control systems.

11.2 Sample and Hold, Zero-Order Hold

Figure 11-1 shows the fundamental elements of a digital control system.

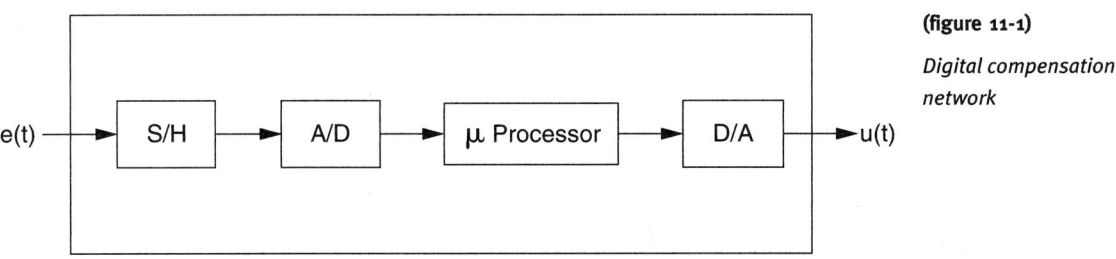

(figure 11-1)

Digital compensation network

The sample-and-hold block of the figure presents a practical impulse sample of the applied signal and provides an accurate sample of frequencies involved in the signal.

The technique often involves charging a capacitor to the value of the applied signal at discrete times and holding the charge until the A/D conversion is performed. That is called a *zero-order hold process*. Once the sample is digitized, it is presented to the microprocessor and the iterative difference equation is implemented to generate the present control action. To accommodate the continuous nature of a practical system, the D/A converter is utilized to convert the digital output of the microprocessor to an analog signal.

A/D Conversion

One method to convert an analog signal to digital form is successive approximation. For this method, the number of bits that can be converted is important, and the more bits, the more accurate the conversion of the analog signal.

Several parameters of the A/D converter need to be discussed. They are as follows:

Resolution. The smallest detectable change at the output. In this case, the resolution is given by

$$RES = \frac{V}{2^{n-1}} \tag{11.1}$$

Where n is the number of bits of the converter.

Conversion time (T_s). Also known as **aperture time** (T_a). This represents the time required for the conversion to occur. This parameter is supplied by the manufacturer and is indicated by an end-of-conversion (EOC) pulse. (Note: The input should not change by more than one-half of the least significant bit (LSB) during this time, or an erroneous conversion will result. For that reason, one or two of the LSBs are not used to prevent erroneous conversion.)

Accuracy. This is affected by zero drift, power supply drift, gain error, nonlinearity errors, and noise. In most cases, the accuracy will be input signal-related and not device-related.

To gain an appreciation of the terms just defined, you will examine a typical A/D converter.

Example 11-1

The specifications of a successive approximation A/D converter are as follows:

$T_s = 10\ \mu s$
$V = \pm 5\ V$
$N = 16\ bits$

Calculate:
  a. Resolution of the converter.
  b. Bandwidth of input to guarantee accurate conversion of a sample.

Solution

  a. Using Equation 10.1

$$RES = \frac{5}{2^{16-1}} = 152.6\ \mu V$$

Example 11-1 (continued)

b. This presents more of a problem. The answer to this part defines the need for the sample-and-hold block. Since the input can change only by, at most, one-half of the LSB during the conversion time, you need to define some parameters for the determination of the frequency response. Let's define the input voltage as $v(t) = V \sin(\omega t)$ and the change at the input of $\Delta v(t)$. Refer to Figure 11-2 for those parameters.

Remembering that the rate of change of a curve is found by differentiating the equation of the curve, the maximum rate of change of the sinewave occurs where the slope passes through zero. Differentiating the sine wave yields

$$\frac{d}{dt}v(t) = V\omega \cos(\omega t) \tag{11.2}$$

And that implies a maximum rate when $\cos(\omega t) = 1$. From which

$$\frac{d}{dt}v(t) = V\omega$$

From which the approximation of the change is given by

$$\frac{\Delta v(t)}{T_c} = V\omega \tag{11.3}$$

Since the maximum change is one-half the resolution

$$V\omega = \frac{76.3\mu V}{10\mu s} = 7.63 \text{ V/sec}$$

$\omega = 2\pi f$ and $V = 5$, and the sampling frequency is approximated by

$$f = \frac{7.63}{5 \cdot 2\pi} = 0.243 \text{ Hz}$$

The implication here is that the bandwidth of the signal input must be less than 0.24 Hz for an accurate conversion.

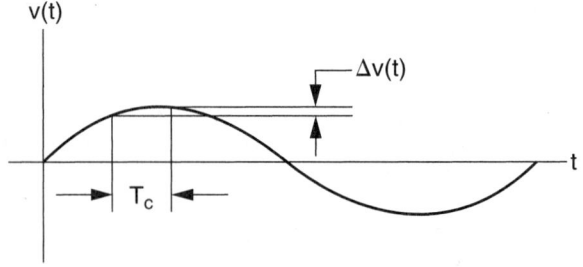

(figure 11-2)

Sinusoidal input signal for conversion

It is clear from the example that the bandwidth is less than impressive. If used alone, the A/D converter can perform 100,000 conversions with a very limited bandwidth. It would appear that if you wanted to increase the bandwidth, you would need to purchase a converter with a shorter conversion time, which may be cost prohibitive. The question is whether there is another way to increase the effective bandwidth of the converter. If you were to take a fast sample of the signal and hold that value during the conversion time, you would effectively decrease the conversion time and increase the bandwidth.

Sample-and-Hold Block

To accomplish the goal of increasing the bandwidth, a simplified circuit that performs the function of sample and hold is shown in Figure 11-3.

(figure 11-3)

Sample-and-hold circuit

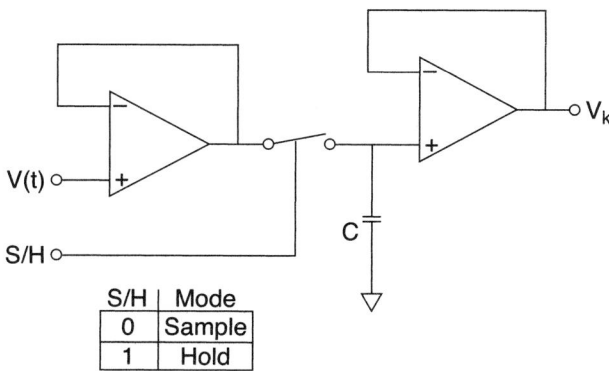

S/H	Mode
0	Sample
1	Hold

With reference to Figure 11-3, the circuit provides an analog memory of the input signal by using the capacitor, C. The capacitor charges to $v(t)$ while the switch is closed (logical 0) and holds the voltage while the switch is open (logical 1). The length of time the switch is closed is considered the aperture time (T_a) of the circuit. A typical range of conversion times for the S/H function can be from 50 nanoseconds to 100 picoseconds. Let's examine the result of digitizing the output of the S/H circuit V_k.

If you consider the analog-to-digital converter of Example 11-1 to have a sample-and-hold circuit with an aperture time of 200 picoseconds, you can use Equation 11.3 to calculate the new bandwidth.

$$f = \frac{76.3 \cdot 10^{-6}}{2\pi \cdot 200 \cdot 10^{-12} \cdot 5} = 12.143 \text{ KHz}$$

With the sample-and-hold circuit, you can sample sinusoidal voltages up to about 12 KHz, which is more practical than the bandwidth of the A/D converter alone.

While you can sample faster signals because of the reduced aperture time of the S/H circuit, the conversion time of the A/D converter still controls how many samples per second can be taken, in this case, 100,000.

The methods of converting a continuous signal into a digitized signal for computerization of a control system have been provided here. What about the required information for the computer to process the data presented? You will now investigate the methods used to extract the necessary information from the continuous system to form the digital system.

11.3 Transfer Functions of Discrete Systems

The definition of the transfer function remains the same as for the continuous time system. This chapter will introduce the z-plane and the location of poles and zeros in the z-plane. You will construct transfer functions from difference equations and vice versa. You will compare the pole-zero locations in the z-plane to the pole-zero locations in the s-plane. Frequency response of discrete systems will also be investigated. Further, some of the devices required to perform the digital-to-analog and analog-to-digital conversions will be discussed.

Transfer Function from Difference Equation

You will consider difference equations containing past samples only. That implies the use of the advance theorem discussed in Chapter 2. Three steps are required to form the transfer function:

1. Take z-transforms of both sides of the difference equation, using the advance theorem.
2. Solve for the output/input relationship in terms of z.
3. Multiply the numerator and denominator by the highest positive order of z.

An example of the application of the process is provided in Example 11-2.

Example 11-2

For the following difference equation, determine $U(z)/E(z)$.

$$u_k = u_{k-1} + 0.8e_k - 0.2e_{k-1}$$

Solution

From the difference equation, the following identities are found:

$$u_k = U(z) \quad e_k = E(z) \quad u_{k-1} = U(z)z^{-1} \quad e_{k-1} = E(z)z^{-1}$$

Rewriting the difference equation, using the identities and rearranging, yields

$$U(z)\left(1 - z^{-1}\right) = E(z)\left(0.8 - 0.2z^{-1}\right)$$

$$\frac{U(z)}{E(z)} = \frac{\left(0.8 - 0.2z^{-1}\right)}{\left(1 - z^{-1}\right)}$$

Then multiplying numerator and denominator by the highest positive order of z

$$\frac{U(z)}{E(z)} = \frac{\left(0.8 - 0.2z^{-1}\right)}{\left(1 - z^{-1}\right)} \cdot \frac{z}{z} = \frac{0.8z - 0.2}{z - 1}$$

The discrete transfer function allows you to do analysis and design in the discrete frequency domain. Implementation of the transfer function by computer requires that you have the difference equation from the discrete transfer function.

Difference Equation from Transfer Function

The methodology of producing the difference equation from the transfer function also involves three steps:

1. Multiply the numerator and denominator of the transfer function by the highest negative order of z.
2. Cross-multiply and take inverse z-transforms.
3. Solve for the present output in terms of past samples of the output and input.

An example of this transform follows.

Example 11-3

For the following transfer function, provide a difference equation to allow computer simulation.

$$\frac{U(z)}{E(z)} = \frac{z^2 + z}{z^2 - 2z + 1}$$

Solution

The inverse transforms for this transfer function are

$$u_k = U(z) \quad u_{k-1} = z^{-1} \quad u_{k-2} = z^{-2} \quad e_k = E(z) \quad e_{k-1} = E(z)z^{-1}$$

First, multiply the numerator by z^{-2}; then cross-multiply to form

$$\frac{U(z)}{E(z)} = \frac{z^2 + z}{z^2 - 2z + 1} \cdot \frac{z^{-2}}{z^{-2}} = \frac{1 + z^{-1}}{1 - 2z^{-1} + z^{-2}}$$

$$U(z)\left(1 - 2z^{-1} + z^{-2}\right) = E(z)\left(1 + z^{-2}\right)$$

From which the difference equation, using the identities above, is

$$U(z) - 2U(z)z^{-1} + U(z)z^{-2} = E(z) + E(z)z^{-1}$$

$$u_k = e_k + e_{k-1} + 2u_{k-1} - u_{k-2}$$

It is obvious that the implementation of the difference equation via a computer program is easily constructed using almost any programming language.

z-Plane and Pole Locations

In the s-plane, the poles are located in the left half of the plane for a stable system, on the $j\omega$ axis for a marginally stable system, or in the right-half plane for an unstable system.

Similar conditions exist for the poles of a discrete system, but the positions require the definition of a new plane, called the z-plane.

By referring to Tables 2-1 and 11-1, you can provide the pole locations in the s-plane and z-plane.

(table 11-1)

Relationship of s-plane and z-plane poles

Pole Locations		
	s-plane	**z-plane**
Step	$s = 0$	$z = 1$
Damped exponential	$s = -a$	$z = e^{-aT}$
Sine, cosine	$s_{1,2} = \pm j\omega$	$z_{1,2} = \cos(\omega T) \pm j\sin(\omega t)$
Damped sine, cosine	$s = -a \pm j\omega$	$z_{1,2} = e^{-aT}\left[\cos(\omega T \pm j\sin(\omega t)\right]$

From Table 11-1, it appears that the relationship between the s- and z-planes is

$$z = e^{sT} \tag{11.4}$$

The location of zeros in the z-plane is not necessarily correct with respect to Equation 11.1, but the location of the poles dictates system stability, and the equation is sufficient.

From your knowledge of the conditions on stability in the s-plane, you find that the limit of stability occurs when $s = \pm j\omega$. You can use the same conditions to determine the limit of stability in the z-plane. Consider the following:

$$z_{1,2} = \cos(\omega T) \pm j\sin(\omega T) \tag{11.5}$$

Which in polar form is

$$z_{1,2} = 1 \angle \pm \omega T \tag{11.6}$$

Equation 11.5 describes a unit circle, which defines the limit of stable gain. The unit circle is equivalent to the $j\omega$ axis on the s-plane. From Table 11-1, it is evident that all systems with poles inside the unit circle are stable systems. This leaves the area outside the unit circle for unstable systems. Figure 11-4 shows the regions of the z-plane.

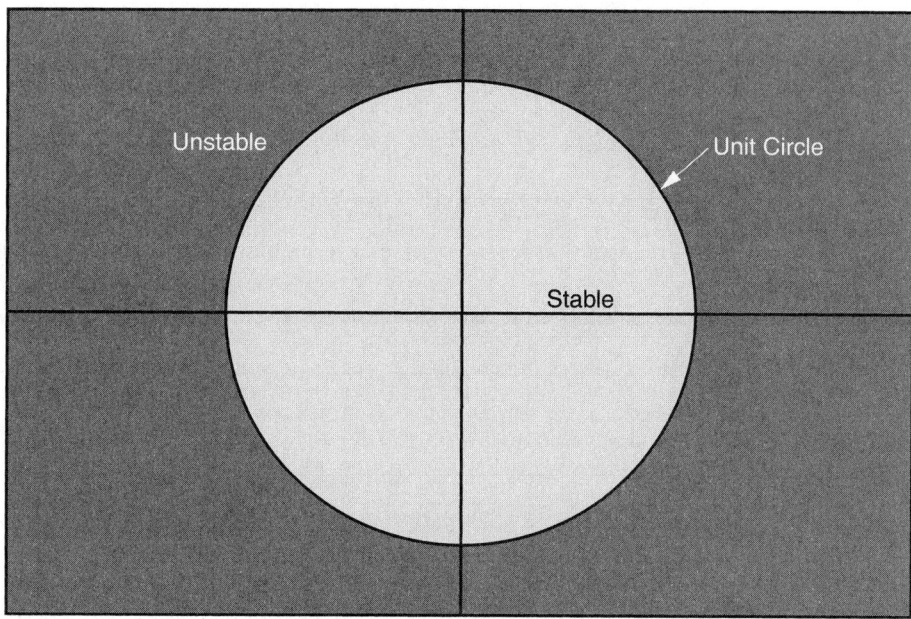

(figure 11-4)

Stability on the z-plane

At this point, you will investigate the implications of the sampling time on the pole positions and the maximum sampling time allowed.

Sampling Time

You can see from the description of the poles of discrete system that the sampling time, T, is contained in the description. From Equation 11.3, the poles are unique if

$$\omega T \leq \pi \tag{11.7}$$

Now

$$T = \frac{2\pi}{\omega_s} \qquad \omega_s \rightarrow \text{sampling frequency} \tag{11.8}$$

Which yields

$$\omega_s = 2 \cdot \omega \tag{11.9}$$

Equation 11.9 shows that the sampling time should be at least two times the frequency of the sampled function. In more formal terms,

If a signal is to be reconstructed from samples, it should be sampled at a frequency that is at least 2 times the highest frequency component of the signal.

This provides a guideline for the proper selection of sampling frequency.

Practically, a sampling frequency at least 10 times the highest frequency component would be chosen.

Example 11-4

A continuous system has the following closed-loop transfer function. The system is computer-controlled. What is the required sampling time, T_s?

$$P(s) = \frac{s + 29}{s^2 + 10s + 29}$$

Solution

If you rewrite the transfer function in the following form:

$$P(s) = \frac{s + 29}{(s + 5)^2 + 4}$$

It is evident that the highest frequency component is 2 rad/sec and that twice the sampling frequency would be a minimum of 4 rad/sec. Practically, the sampling frequency should be about 40 rad/sec. Using Equation 11.5, the sampling time is

$$T = \frac{2\pi}{40} = 0.157 \text{ seconds}$$

Any sampling time less than that value is acceptable since it approximates the continuous system more accurately.

Many computer systems are required to multitask. It is important to sample at the proper rate and not to oversample, since doing so requires microprocessor time, time that could be used to perform other tasks. Consider the following example.

Example 11-5

A microprocessor-based control system is multitasking and can handle up to 10 tasks in real time. A discrete lead controller that requires 15 milliseconds of execution time is to be used for each of the 10 identical plants. What is the minimum sampling time required for each task?

Solution

Assuming the 15-millisecond time contains the A/D conversion, solution of the difference equation, and D/A conversion for each task, the 10 tasks will require 0.15 seconds. This figure assumes 100 percent computer usage, which is not practical since processes are known to have variations in their execution time and there is no room for error with this level of usage. In this case, a maximum

Example 11-5 (continued)

of 75 percent utilization will be specified for the computer, which requires 0.15/0.75, or 0.2 seconds, to allow for process variations.

Understand that the processes were identical in this example. If the processes are different, the required sampling time calculation is more involved. In some cases, manufacturers may supply an approximation of computer usage for their devices.

These two examples provide a glimpse into the complexity of designing components for discrete systems with multitasking capabilities. In some high-speed systems, the computer, which is fast, cannot close the loop, resulting in errors and shutdowns. In those cases, a dedicated microprocessor is one solution. Another problem that is incurred in sampling systems becomes apparent in the frequency response of a system. You will now turn your attention to the area of frequency response.

Frequency Response of Discrete Systems

All of the frequency response methods previously discussed can be extended to the discrete form of control system. In the continuous system, the response is obtained by replacing the Laplace operator by $j\omega$. The magnitude and phase of the system vs. frequency was then plotted.

For the discrete transfer function, $P(z)$, the frequency response, considering Equation 11.4, is given by replacing the Laplace operator, s, with $j\omega$ then, $P(z) = P(e^{j\omega t})$ and the DC gain is $P(1)$. The magnitude and phase are determined by setting z to be $e^{j\omega t}$. Keeping in mind that

$$e^{j\omega T} = \cos(\omega T) + j\sin(\omega T) \tag{11.10}$$

The magnitude and phase can be determined. An example will help clarify the process.

Example 11-6

For the following z-plane transfer function, determine the expression for the magnitude and phase of the system. Assume the sampling time to be 1 second.

$$P(z) = \frac{0.165(z + 0.085)}{z^2 - 1.189z + 0.368}$$

Solution

The DC gain of the system is 1.0, and $T = 1.0$ second. The first step is to convert the transfer function to the $e^{j\omega t}$ form by replacing z with $e^{j\omega}$.

$$P(e^{j\omega T})_{T=1} = \frac{0.165(e^{j\omega} + 0.085)}{e^{j2\omega} - 1.189e^{j\omega} + 0.368}$$

Example 11-7 (continued)

Using Equation 11.10, the transfer function in magnitude and phase for the system is

$$\left|P(e_{j\omega T})_{T=1}\right| = \frac{\sqrt{[0.1646 \cdot \cos(\omega) + 0.014]^2 + [0.1646 \cdot \sin(\omega)]^2}}{\sqrt{[\cos(2\omega) - 1.189 \cdot \cos(\omega) + 0.819]^2 + [\sin(2\omega) - 1.189 \cdot \sin(\omega)]^2}}$$

$$\angle P(e^{j\omega T})_{T=1} = \tan^{-1}\left(\frac{0.1646\sin(\omega)}{0.1646 \cdot \cos(\omega) + 0.0144}\right) - \tan^{-1}\left(\frac{\sin(2\omega) - 1.189 \cdot \sin(\omega)}{\cos(2\omega) - 1.189 \cdot \cos(\omega) + 0.819}\right)$$

A plot of the frequency response of the system would show it to approximate a low-pass filter. And the bandwidth is repeated for integral multiples of the sampling frequency. That indicates that higher frequencies will be included in the low-pass system. To circumvent that situation, you can add a prefilter to the system at the input to filter out the higher frequencies. The order of the filter is dictated by the sampling rate.

Those examples have pointed out some of the difficulties in implementing the discrete system.

Summary

➤ In this chapter, you have investigated the A/D converter.

➤ The resolution, sampling rate, and frequency response of the A/D converter have been determined and calculated.

➤ To increase the apparent frequency response, a sample-and-hold system and an A/D converter with a narrow aperture time is used.

➤ Although the frequency response is increased, the number of samples per second remains the same, as it is set by the conversion time of the A/D converter.

➤ We have produced the transfer function from difference equations.

➤ We have also produced the difference equation from the transfer function.

➤ The z-plane is produced by mapping the s-plane poles onto a unit circle.

➤ The unit circle is the boundary between stable and unstable systems.

➤ The minimum sampling time is to be sampled at a rate of 4 to 10 times the minimum rate calculated.

➤ The frequency response of discrete systems is similar to that of continuous systems with the exception that the sampling time, T, replaces the continuous time, t, in the continuous system.

➤ The magnitude response repeats itself at integer intervals of the sampling rate. To circumvent that, a prefilter whose order is dictated by the sample rate is required.

➤ If the sampling frequency is increased, the order of the filter can be decreased, another positive tradeoff.

Problems

P11.1   A certain A/D converter has the following specifications:

$$Tc = 20\mu s, \ N = 16, \ V = 10 \text{ V}$$

a. Using the upper 12 bits of the converter, calculate the resolution and frequency response with no sample and hold.

b. Add a sample-and-hold system with an aperture time of 500 ps. Then recalculate the frequency response of the system.

P11.2   A tachometer generator provides 5.0 V at 4000 rpm. Can the A/D converter of Problem P11.1 provide 0.1 percent regulation of the system? Explain.

P11.3   Provide transfer functions for the following difference equations:

a. $u_k = 2e_k + u_{k-1}$
b. $u_k = 0.5e_k + 3u_{k-1} - u_{k-2}$
c. $u_k = e_{k-1} + e_{k-2} + u_{k-1}$
d. $u_k = e_{k-1} + 2e_{k-2} + 1.8u_{k-1} - 1.9u_{k-2} + 3u_{k-3}$

P11.4   For the following transfer functions, provide the difference equations for computer simulation.

a. $\dfrac{U(z)}{E(z)} = \dfrac{z}{z^2 - 0.7z + .1}$

b. $\dfrac{U(z)}{E(z)} = \dfrac{z^2 + z}{z^2 - 1.2z + .5}$

c. $\dfrac{U(z)}{E(z)} = \dfrac{z^2 + 3z}{z^3 - 1.6z^2 + 1.8z - 2.4}$

d. $\dfrac{U(z)}{E(z)} = \dfrac{z^2 + 2z}{z^3 - 1.5z^2 + 0.66z - 0.08}$

P11.5   Show that the z-transform for the unit ramp function is

$$F(z) = \frac{Tz^{-1}}{\left(1 - z^{-1}\right)^2}$$

P11.6   Consider the control system shown in Figure P11-6. This system represents a control system with a lag compensator and a second-order plant. Generate the difference equations for the controller and plant. Let sampling time T be a variable.

(figure P11-6)

Control system for Problem P11.6

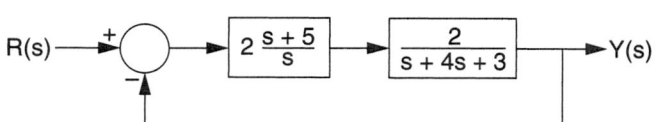

P11.7   For the system in Problem P11.6, repeat the problem but replace the lag
         compensator with a lead compensator of the form

$$G_C(s) = \frac{2s}{(s + 4)}$$

P11.8   For the following two systems, determine the sampling time from the sam-
         pling theorem.

   a. $P(s) = \dfrac{8}{s^2 + 4s + 16}$

   b. $P(s) = \dfrac{s + 4}{s^2 + 4s + 10}$

P11.9   For Problem P11.8, generate the difference equations for the sampling time
         calculated. Then generate the difference equations for a sampling time of
         1 second and compare the results. What conclusion do you draw about the
         effect of sampling times on the pole locations of the systems?

P11.10   For the following z-plane transfer functions, determine the steady-state DC
         gain of the systems.

   a. $Y(z) = \dfrac{z^2 + 2}{(1 - z^{-1})(z^2 - 1.5z + 0.7)}$

   b. $Y(z) = \dfrac{z^2 + 2z}{(1 - z^{-1})(z - 0.6)(z - 0.3)(z - .1)}$

P11.11   For the following difference equations, determine the z-transform and plot
         the output for a 0.1 second sampling time.

   a. $u_k = 0.1e_{k-1} + 2u_{k-1} - u_{k-2}$
   b. $u_k = 0.1e_{k-1} + 0.1e_{k-2} + u_{k-1}$

P11.12   Using Table 2-3, $T = 0.1$ second, find the z-transforms for the following dis-
         crete time functions:

   a. $f_k = e^{-5kT}$
   b. $f_k = 3e^{-5kT}\cos(5kT)$
   c. $f_k = 6\cos(6kT)$
   d. $f_k = 5e^{-5kT}\left[\cos(4kT) + 2\sin(4kT)\right]$

P11.13   For Problem P11.12, find the inverse z-transform. Are the equations the
         same? Why or why not?

P11.14   For the following s-plane transfer function, convert the equation in the z-plane. (The sampling time is 0.1 second.) Determine the magnitude and phase equations and plot them to 30 rad/sec. Show that the DC gain of the discrete system is unity.

$$\frac{Y(s)}{U(s)} = \frac{8}{s^2 + 6s + 8}$$

P11.15   For the following z-plane transfer function, determine the magnitude and phase equations of the system. Plot them to 80 radians.

$$\frac{Y(z)}{U(z)} = \frac{0.081z - 0.027}{z^2 - 1.622z + 0.676}$$

Chapter 12 | Discrete Control Systems

Chapter Objectives

After completing this chapter, you should be able to:

➤ Explain the use of the ZOH in determining the open-loop and closed-loop transfer function of a discrete system.

➤ Show both the manual method and the use of **MATLAB** to convert a continuous system to a discrete system.

➤ Explain the six different methods of forming the transfer function using **MATLAB** to form the closed-loop transfer function.

➤ Show the methods used to form the closed-loop transfer function by hand and with **MATLAB**.

➤ Explain the concepts of stability in the z-plane, relative to the s-plane.

➤ Explain how to use bilinear transformation from the z-plane to the r-plane which allows us to provide a Routh table for a discrete system.

➤ Explaine the rules of the discrete root locus and its construction on the z-plane.

➤ Show the use of **MATLAB** for plotting the discrete root locus and determining the parameters of a discrete control system as poles and zeros are moved.

12.1 Introduction

In order to analyze the performance of a digital compensating network, it is necessary to obtain the response of the continuous plant relative to a discrete control action. The most common method is by the use of the ZOH process. This process controls the generation of $G(z)$, which allows analysis of the closed-loop system in the discrete domain. The closed-loop transfer function and characteristic equation can be developed, the pole locations can be plotted on the z-plane, and the discrete root locus can be formed.

Chapter 2 developed the concept of approximating the differential equation of a continuous system as a difference equation and then transforming the function to the z-plane. You will begin this chapter using the ZOH as a controlling factor in determining the specific transfer function of a system.

12.2 *G(z)* from a Continuous Plant Function with Zero-Order Hold

In Chapter 11, you learned that the output of the discrete compensation network is obtained from a D/A converter through a sample-and-hold system. The control action u_k is passed to the D/A converter, which provides a continuous staircase signal, $u(t)$.

The process is shown in Figure 12-1.

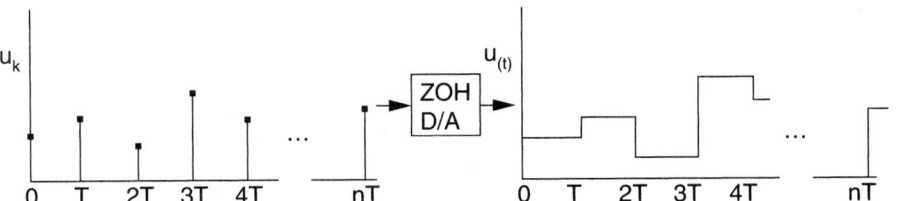

(figure 12-1)

Zero-order hold process

Figure 12-1 shows the output of the ZOH to be the sum of an infinite number of rectangular pulses. If you consider the pulses and take the nth pulse, you will find it to be represented as

$$u_n \cdot \mathcal{L}^{-1}\left[\frac{G(s)}{s}\right] \tag{12.1}$$

Relative to a step input, u_n/s. The discrete output would then be

$$u_n Z \cdot \left\{\mathcal{L}^{-1}\left[\frac{G(s)}{s}\right]\right\} \tag{12.2}$$

Considering two pulses delayed by n and $(n + 1)$ samples, respectively, the nth rectangular pulse output is

$$Y_n(z) = u_n z^{-n} Z\left\{\mathcal{L}^{-1}\left[\frac{G(s)}{s}\right]\right\} - u_n z^{-(n+1)} Z\left\{\mathcal{L}^{-1}\left[\frac{G(s)}{s}\right]\right\} \qquad (12.3)$$

Simplifying that equation yields

$$Y_n = u_n z^{-n}(1 - z^{-1}) \cdot Z\left\{\mathcal{L}^{-1}\left[\frac{G(s)}{s}\right]\right\} \qquad (12.4)$$

And the total system output is given by

$$Y(z) = \sum_{n=0}^{\infty} Y_n(z) \qquad (12.5)$$

You can then replace $u_n z^{-n}$ with Equation 12.5, which yields

$$Y(z) = U(z) \cdot \left(\sum_{n=0}^{\infty} u_n z^{-n}\right)(1 - z^{-1}) Z\left\{\mathcal{L}^{-1}\left[\frac{G(s)}{s}\right]\right\} \qquad (12.6)$$

Where u_n is represented by the summation in Equation 12.6, which then yields

$$\frac{Y(z)}{U(z)} = G_C(z) = \left(1 - z^{-1}\right) Z\left\{\mathcal{L}^{-1}\left[\frac{G(s)}{s}\right]\right\} \qquad (12.7)$$

In the derivations above, you relied on a continuous plant, and it should be clear that the method applies to any continuous system. The implication here is that the compensating networks of continuous systems can be transferred to the discrete domain.

The general form of the ZOH in the z-plane is $(1 - z^{-1})$, and the application of the ZOH to a continuous compensating system is given as

$$G_C(z) = \left(1 - z^{-1}\right) Z\left\{\mathcal{L}^{-1}\left[\frac{G_C(s)}{s}\right]\right\}$$

Which shows the conversion of $G_C(s)$ to $G_C(z)$. To show the application, you will begin by considering a simple single-pole compensator.

Example 12-1

Consider the following compensation scheme, using a sampling time of 0.1 seconds.

$$G_C(s) = \frac{1}{s + 5}$$

Provide the transfer function of the compensator in the z-plane.

Solution

The inverse Laplacian of that function is

$$\mathcal{L}^{-1}\left[\frac{G_C(s)}{s}\right] = \mathcal{L}^{-1}\left[\frac{1}{s(s + 5)}\right] = 0.2 - 0.2e^{-5t}$$

And

$$Z\left\{\mathcal{L}^{-1}\left[\frac{G_C(s)}{s}\right]\right\} = \frac{0.2}{1 - z^{-1}} - \frac{0.2}{1 - 0.6065z^{-1}}$$

Multiplying the function by $(1 - z^{-1})$ and simplifying, you obtain

$$(1 - z^{-1})\left\{\mathcal{L}^{-1}\left[\frac{G_C(s)}{s}\right]\right\} = \frac{0.2(1 - z^{-1})}{1 - z^{-1}} - \frac{0.2(1 - z^{-1})}{1 - 0.6065z^{-1}}$$

$$G_C(z) = \frac{0.0787z^{-1}(1 - z^{-1})}{(1 - z^{-1})(1 - 0.6965z^{-1})} \cdot \frac{z}{z} = \frac{0.0787}{z - 0.6065}$$

This can be a tedious process for systems higher than first-order. You also need to keep in mind that this is an open-loop form of function. Fortunately, **MATLAB** methods can be used to produce the discrete function directly from the s-plane equations. Let's investigate the application of **MATLAB** to a second-order plant.

Example 12-2

Consider the following second-order plant. Calculate a proper sampling time 5 times the minimum, provide $G(z)$ using ZOH, and provide the output of the system to a step input.

$$G(s) = \frac{3}{s^2 + 4s + 3}$$

It should be clear that this system has two real poles at $s = -1$ and $s = -3$. Analysis shows that the minimum sampling frequency is 1 rad/sec. So you will use a sampling frequency of 0.2 seconds.

Solution

MATLAB has a function called *c2d*, which converts a continuous time transfer function to its discrete form. The format of the **c2d** command is

$$\text{Sysd} = \text{c2d(Sysc, T}_s\text{, 'Method')}$$

Example 12-2 (continued)

Where *Sysc* is the continuous transfer function, T_s is the sampling time, and 'Method' is the method of conversion. The default is the ZOH. There are a total of six methods, which will be discussed shortly.

The first thing you must do is determine the transfer function of the continuous system using the **tf** command.

$n = 3; d = [1\ 4\ 3];$ % definition of the numerator and denominator

Sysc = tf(n,d);   % convert to the transfer function

Next, you convert the continuous system to the discrete transfer function, using the **c2d** command.

Sysd = c2d(Sysc, 0.2, 'zoh')   % convert to discrete system

Which results in

Transfer function:

   $0.04631z + 0.03548$

   \-

$z\char`^2 - 1.368z + 0.4493$

Sampling time: 0.2

In providing the response of the system to a step input, the command is the same as for a continuous except that you provide the discrete system as the function to convert.

step(Sysd);   % provide the response to a step input

As stated previously, there are six methods of conversion for the **c2d** command. They are:

1. 'zoh'—uses the zero-order hold as the conversion method.
2. 'foh'—uses a first-order-hold approximation to discrete (triangular approximation).
3. 'imp'—impulse invariant discretization.
4. 'tustin'—bilinear approximation.
5. 'prewarp'—Tustin approximation with frequency prewarping. The critical frequency ω_c (in rad/sec) is specified as a fourth input by

   Sysd = c2d(Sysc,'prewarp',ω_c)

6. 'matched'—matched pole-zero method (for SISO systems only).

If the 'method' is not specified in the **c2d** command, the system produces the ZOH model.

Discrete Closed-Loop Transfer Functions

Discrete closed-loop transfer functions are arrived at using the same rationale as for continuous systems. A block diagram or SFG will provide the closed-loop transfer function. There are some differences, though, that need to be discussed. One of the differences is that systems that were stable for all values of gain in a continuous system may very well not be stable in the z-plane.

The reason is that as the system gain is increased, the poles of the discrete system migrate toward the outer edge of the unit circle where the system becomes marginally stable. The general form of a discrete control system is shown in Figure 12-2.

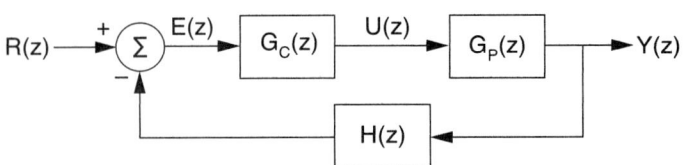

(figure 12-2)

Discrete canonical form

The transfer function is arrived at using the same technique as for the continuous form.

$$\frac{Y(z)}{R(z)} = \frac{G_C(z)G_P(z)}{1 + G_C(z)G_P(z)H(z)} \tag{12.8}$$

And the error is

$$E(z) = \frac{R(z)}{1 + G_C(z)G_P(z)H(z)} \tag{12.9}$$

The closed-loop characteristic equation is

$$1 + G_C(z)G_P(z)H(z) = 0 \tag{12.10}$$

An example will clarify this method.

Example 12-3

Consider the discrete control system of Example 12-1. Calculate the following parameters:
 a. Closed-loop transfer function
 b. Steady-state error for a unit step input
 c. Location of the system pole on the z-plane as the gain varies from zero to infinity
The system is shown in Figure 12-3.

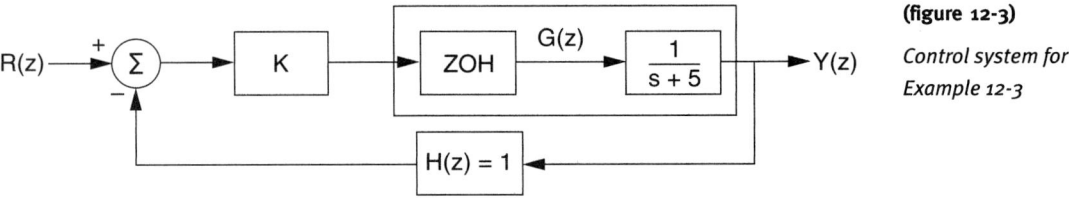

(figure 12-3)

Control system for
Example 12-3

Example 12-3 (continued)

Solution

a. To obtain $G(z)$, you need to apply the ZOH. Referring to Example 12-1, to close the loop, $0.0787K$ must be added to the denominator of the transfer function, which yields

$$\frac{Y(z)}{R(z)} = \frac{0.0787K}{z - 0.6065 + 0.0787K}$$

b. The error function is given by Equation 12.9 and is

$$E(z) = \frac{z - 0.6065}{(1 - z^{-1})(z - 0.6065 + 0.0787K)}$$

The steady-state error $e(\infty)$ is found using the final value theorem and is

$$e(\infty) = \lim_{z \to 1}(1-z^{-1}) \cdot E(z) = \frac{0.3935}{(0.3035 + 0.0787K)}$$

Examination of that equation shows that as K increases, the error decreases, which is the same as for the continuous system.

c. The characteristic equation gives the location of the system pole on the z-plane.

$$z - 0.6065 + 0.0787K = 0$$

When $K = 0$, the pole is located on the positive real axis at $z = 0.6065$. It should be clear that as K increases, the pole moves toward -1.0 on the real axis. When the pole is located at -1.0, the system becomes unstable; that occurs when the gain is equal to 20.4. The position is shown in Figure 12-4.

(figure 12-4)

Migration of system pole
for Example 12-3

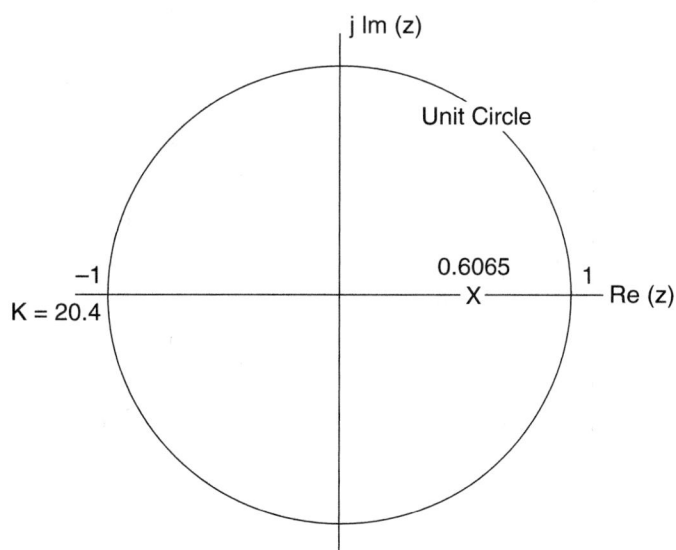

The last example gave a glimpse into the concept of the root locus in the z-plane. However, the computation of the root locus for system orders higher than two is more complex. Generally, design engineers will design the compensator in the s-plane and then transfer them into the z-plane, using the ZOH process. An excursion into the root locus in the z-plane should serve to convince you of the complexity of that method.

12.4 Stability in the z-Plane

As in the s-plane, stability is paramount in the z-plane, but there is an important difference between the s-plane and z-plane. In the s-plane, a first-order system can never become oscillatory, but in the z-plane the first-order system can be an oscillator if the parameters used are incorrect.

Discrete System Routh Tables

In previous examples, it was shown that all roots of the discrete transfer function must lie within the unit circle for the system to be stable. You determined that most of the forms of analysis that you learned about can be transferred to the z-plane. There is one exception to the rules: the Routh-Hurwitz criterion, which is restricted to the imaginary axis of the s-plane as the limit of stability of the system. It can be applied only to continuous systems.

The Routh criterion can still be used if you have a transform that maps the unit circle in the z-plane to the imaginary axis of another complex plane.

One method of doing that is to use a bilinear transformation of the form

$$z = \frac{ar + b}{cr + d} \tag{12.11}$$

Where a, b, c, and d are all real constants and r is a complex variable that maps circles in the z-plane onto straight lines in the r-plane. The simplest form of transformation is

$$z = \frac{1 + r}{1 - r} \tag{12.12}$$

That is referred to as the *r-transformation*. Once the characteristic equation in z is transformed to the r-plane, the Routh table can be applied as usual. Some examples will help you understand the process.

Example 12-4

Consider the following characteristic equation of a discrete control system:

$$z^3 - 1.198z^2 + 0.441z - 0.0498 = 0$$

Convert that characteristic equation to its r-plane and use the Routh table to determine the stability of the system.

Solution

First, use the equivalent transform on the characteristic equation.

$$\left(\frac{1+r}{1-r}\right)^3 - 1.198\left(\frac{1+r}{1-r}\right)^2 + 0.441\left(\frac{1+r}{1-r}\right) - 0.0498$$

Simplifying the equation and eliminating the common denominator yields

$$2.68r^3 + 3.608r^2 + 1.51r + 0.193 = 0$$

Placing the characteristic equation into a Routh table, you find that the system has no sign changes in the first column and that the system is stable.

r^3	2.689	1.510
r^2	3.608	0.193
r^1	1.366	
r^0	0.193	

Let's consider another z-plane characteristic equation, only this time with a gain factor K.

Example 12-5

Consider the following characteristic equation of a discrete control system:

$$z^3 + 0.2z^2 + .5z + K = 0$$

Convert that characteristic equation to its r-plane equivalent and use the Routh table to determine the limits of stable gain of the system.

Solution

The procedure starts the same way as Example 12-4. The main difference is that the function depending on K must be added to the system to form the table. Doing so yields

$$(1.3 - K)r^3 + (2.7 + 3K)r^2 + (2.7 - 3K)r + (1.7 + K) = 0$$

Example 12-5 (continued)

When the characteristic equation is placed in the Routh table, a quadratic function for the gain is produced. The same rule as the one for continuous systems applies here. Also, the gain must be greater than zero. The Routh table is

$$
\begin{array}{ccc}
r^3 & (1.3 - K) & (2.7 - 3K) \\
r^2 & (2.7 + 3K) & (1.7 + K) \\
r^1 & \dfrac{5.08 - 0.4K - 8K^2}{(2.7 + 3K)} & \\
r^0 & (1.7 + K) &
\end{array}
$$

Solution of the quadratic for the gain yields

$$K_1 = 0.8223 \qquad\qquad K_2 = -0.7723 \qquad\qquad 0 < K < 0.8223$$

Substituting $K = 0.8223$ into the r-plane equation and factoring to its roots provides two identical roots on the imaginary axis, proving that maximum gain provides a marginally stable system.

Clearly, the method for using the Routh table is more tedious in the z-plane than in the Laplace domain, but it is a required method. The process for finding the root locus for system orders greater than one is shown in the next section.

Root Locus on the z-Plane

Example 12-3 showed the location of the single pole to be on the real axis of the z-plane, which is typical for first-order systems. For the example, the sampling time was 0.1 seconds and the pole was located at 0.6065. If the sampling time was doubled, the sampling time would be 0.2 seconds and the pole would move closer to the origin of the z-plane, indicating that the gain of the system had decreased. The locus crosses the unit circle when $z = -1.0$. Substitution of this value in the characteristic equation allows you to calculate the maximum gain of the circuit.

The determination of the root locus for second-order systems is more complex than that for the first-order system. To investigate the second-order system, let's start by reexamining some of the conditions for continuous second-order plants or systems.

Consider the following characteristic equation:

$$s^2 + 2\zeta\omega_n s + \omega_n^2 = 0 \tag{12.13}$$

Considering complex roots, you obtain

$$s_{1,2} = -\zeta\omega_n \pm j\omega_n\sqrt{1 - \zeta^2} \tag{12.14}$$

When the poles are mapped into the z-plane using

$$z_{1,2} = e^{s_{1,2} T}$$ (12.15)

the poles are located at:

$$z_{1,2} = e^{-\zeta \omega_n T \pm j \omega_n \sqrt{1-\zeta^2} T}$$ (12.16)

Since the z-plane exhibits symmetry about the real axis, just as in the s-plane root locus, you can examine only one of the poles to determine its position in the z-plane.

$$z_1 = e^{-\zeta \omega_n T} \left[\cos\left(\omega_n \sqrt{1 - \zeta^2}\, T \right) + j \sin\left(\omega_n \sqrt{1 - \zeta^2}\, T \right) \right]$$ (12.17)

Figure 12.5 shows the location of this pole on the z-plane.

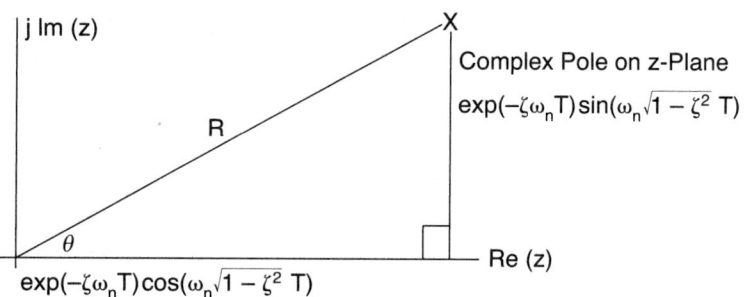

It is clear from Figure 12-5 that the R is given by

$$R = e^{-\zeta \omega_n T}$$ (12.18)

And

$$\cos\theta = \cos\left(\omega_n \sqrt{1 - \zeta^2}\, T \right)$$ (12.19)

That equation reduces to

$$\theta = \omega_n \sqrt{1 - \zeta^2}\, T$$ (12.20)

Substituting Equation 12.20 into Equation 12.18 gives

$$R = e^{-\frac{\zeta\theta}{\sqrt{1-\zeta^2}}}$$

(12.21)

Equation 12.21 defines the loci of constant damping ratios on the z-plane. The equation will assist in the design of discrete systems with the same characteristics as a continuous system. Figure 12-6 shows the unit circle and the constant damping ratio lines on the z-plane.

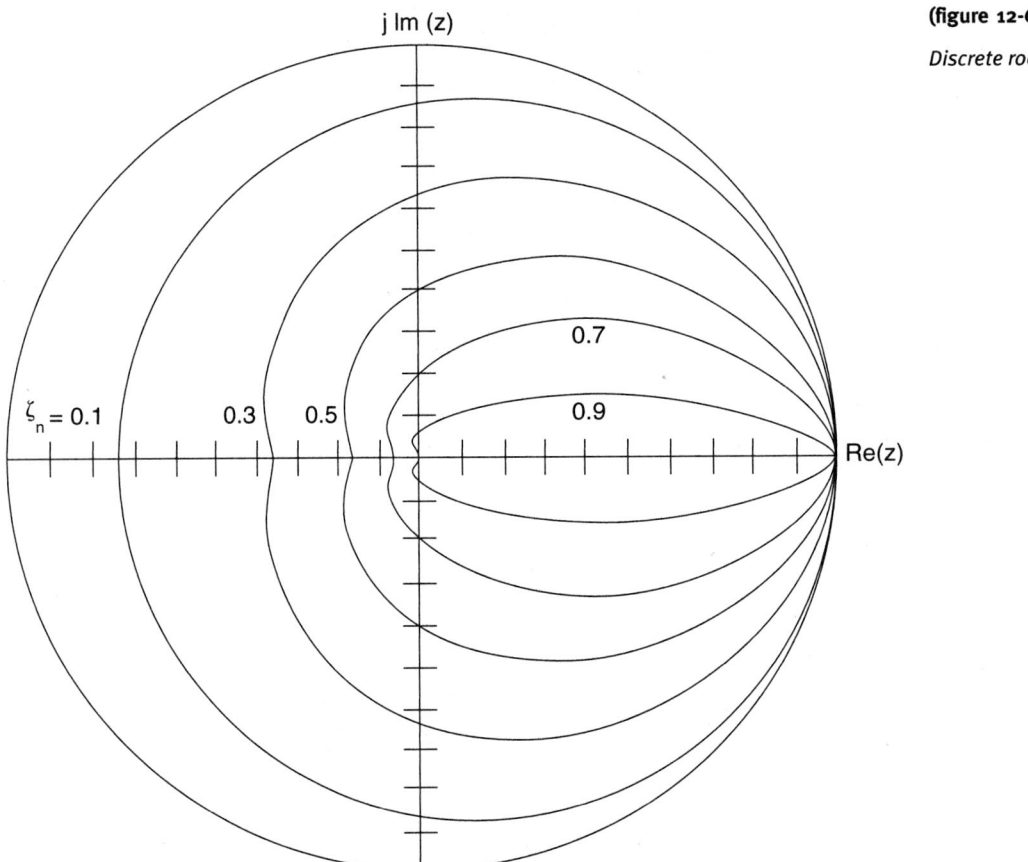

(figure 12-6)

Discrete root locus

The general form for the second-order discrete system is

$$z^2 - bz + c = 0$$

(12.22)

When the roots of the characteristic equation are imaginary, the roots are given by

$$z_{1,2} = \frac{b \pm j\sqrt{4c - b^2}}{2} \qquad (12.23)$$

Referring to Figure 12-5 and replacing the exponential functions with their equivalents from Equation 12.22, some algebraic manipulation shows values of R and θ to be:

$$R = \sqrt{c} \qquad (12.24)$$

$$\theta = \cos^{-1}\left(\frac{b}{2\sqrt{c}}\right) \qquad (12.25)$$

It should be clear that Equations 12.21 and 12.24 are equivalent. Substituting Equation 12.25 into Equation 12.21 and simplifying, the result is

$$\cos\left[-\frac{\sqrt{1 - \zeta^2}}{2\zeta}\ln c\right] - \frac{b}{2\sqrt{c}} = 0 \qquad (12.26)$$

Equation 12.26 provides a method for calculating the gain required for a specific damping ratio. Let's investigate the steps involved in providing a root locus on the z-plane for a second-order plant.

Three steps are required in plotting the root locus on the z-plane.

Step 1. The open-loop discrete transfer function provides the location of the poles on the z-plane. This transfer function must remain second-order. With $z = 1.0$ as a reference, the root locus lies to the left of an odd-numbered pole or zero. This should be familiar from the continuous root locus.

Step 2. If a breakaway point exists, the locus is a circle with its center at the zero and a radius equal to the distance from the zero to the breakaway point. The breakaway point is calculated by setting the discriminant equal to zero.

$$b^2 - 4c = 0 \qquad (12.27)$$

That generates a gain value, and solution of the characteristic equation gives the breakaway point.

Step 3. It is important to know the maximum gain possible for the system as well as the gain required to produce a specific damping ratio. If the locus leaves the unit circle at $z = \pm 1.0$, the maximum gain is found by substituting that value into the characteristic equation. If the locus leaves at any other point on the unit circle, the radius R is unity and $c = 1.0$. If the locus crosses the specified damping ratio curve, then Equation 12.26 can be used to find the gain by iteration.

Understand that the open-loop poles tend to the open-loop zeros as the gain approaches infinity, similarly to the Laplace domain. An example will help clarify that fact.

Example 12-6

Consider the following second-order open-loop transfer function:

$$G_C(z)G_P(z) = K\frac{0.0132z + 0.0115}{z^2 - 1.646z + 0.6703}$$

The system is to operate with a damping ratio of 0.707. What is the required gain? What is the maximum gain allowed? Provide a root locus of the system in the z-domain.

Solution

The zero is located at -0.871, and the poles are located at $z_1 = 0.907$ and $z_2 = 0.739$. The characteristic equation is

$$z^2 - (1.646 - 0.0132K)z + (0.6703 + 0.0115K) = 0$$

The damping ratio of the system is greater than 1, so the discriminant is $(b^2 - 4c)$.

$$(1.646 - 0.0132K)^2 - 4(0.6703 + 0.0115K)$$

When the discriminant is solved for the values of K, the result is $K_1 = 0.313$ and $K_2 = 513.48$. If you substitute the values of K into the characteristic equation and solve for the roots, the breakaway and break-in points are found.

$$K = 0.313$$
$$z^2 - 1.642z + 0.674 = 0$$
$$\text{double root at } z \approx 0.815$$
$$K = 513.48$$
$$z^2 + 5.132z + 6.575 = 0$$
$$\text{double root at } z \approx -2.5$$

Since the root locus does not leave the unit circle at -1.0 as evidenced by the breakaway and break-in points, you can use the identity that $c = 1$ to solve for the maximum gain for stability.

$$0.6703 + 0.0115K = 1$$
$$K_{max} = \frac{1 - .6703}{0.0115} = 28.7$$

Example 12-6 (continued)

This system is to exhibit a damping ratio of 0.707, and you need to use Equation 12.26 to iterate a gain for the damping ratio. That is a tedious process, and selection of the gain values is important. The equation for this system is

$$\cos\left[-\frac{\sqrt{1-0.707^2}}{2\cdot 0.797}ln(0.6703+0.0115\cdot K)\right] - \frac{1.646-0.0132\cdot K}{2\sqrt{0.6703+0.0115\cdot K}} = 0$$

By iteration, the gain required for a damping ratio of 0.707 is 1.42.
The root locus is shown in Figure 12-7.

(figure 12-7)

Root locus for Example 12-6

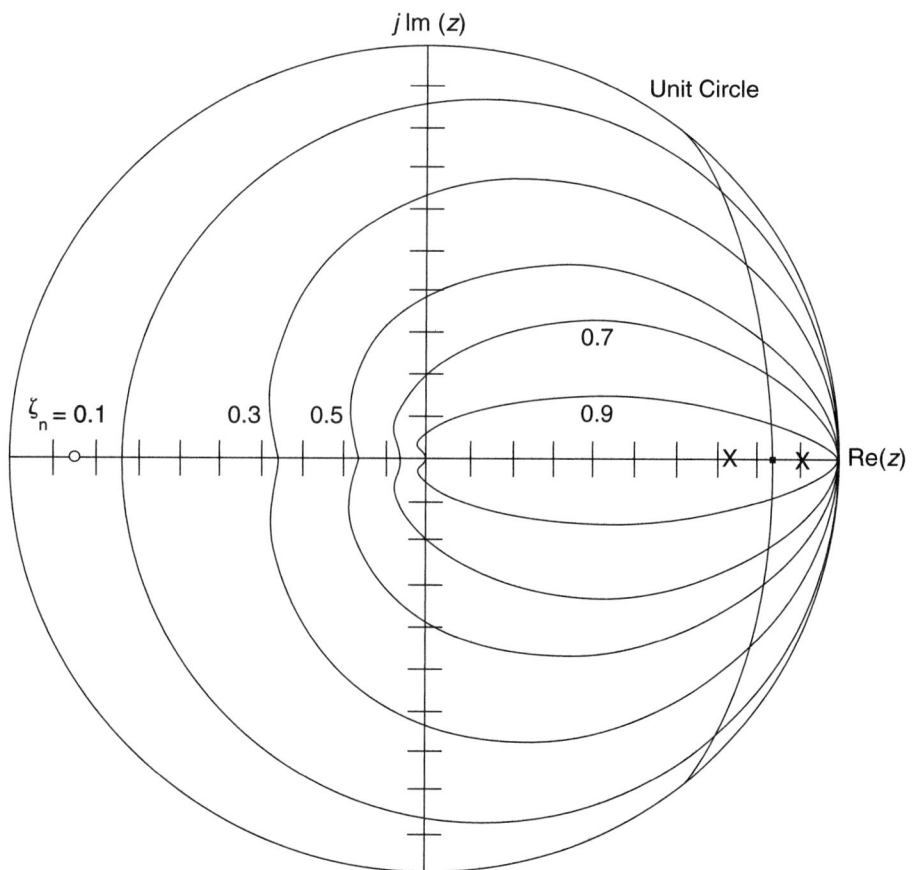

12.5 MATLAB for the z-Plane Root Locus

MATLAB has functions that have already been used in the continuous systems in previous chapters. The functions are *rlocus(name of functions)* and *step(name of functions)*. The **MATLAB** program for providing the functions of Example 12-6 is as follows:

$n = 3; d = [1\ 4\ 3];$	% Continuous system numerator and denominator
$k = 0:.005:30;$	% Set gain for calculation of the root locus.
sys = tf(n,d);	% Transfer function of continuous system.
sysd = c2d(sys, 0.1)	% Convert continuous system to discrete system.
step(sysd)	% Generate the step response of the discrete system for $T =$ % 0.1 sec.
rlocus(sysd, k)	% Generate the root locus of the discrete system for $T =$ % 0.1 sec and gains to 30.
axis 'equal'	% Set up the axes to produce a real circle.

The step response and root locus are shown in Figures 12-8 and 12-9, respectively.

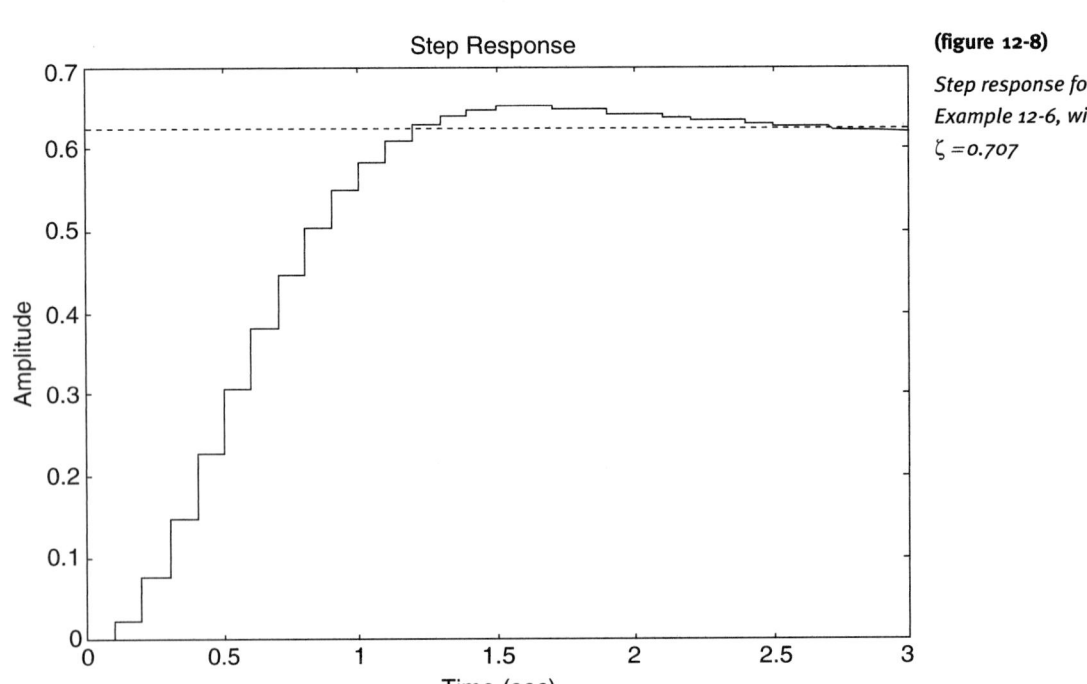

(figure 12-8)

Step response for Example 12-6, with $\zeta = 0.707$

(figure 12-9)

Root locus for Example 12-6

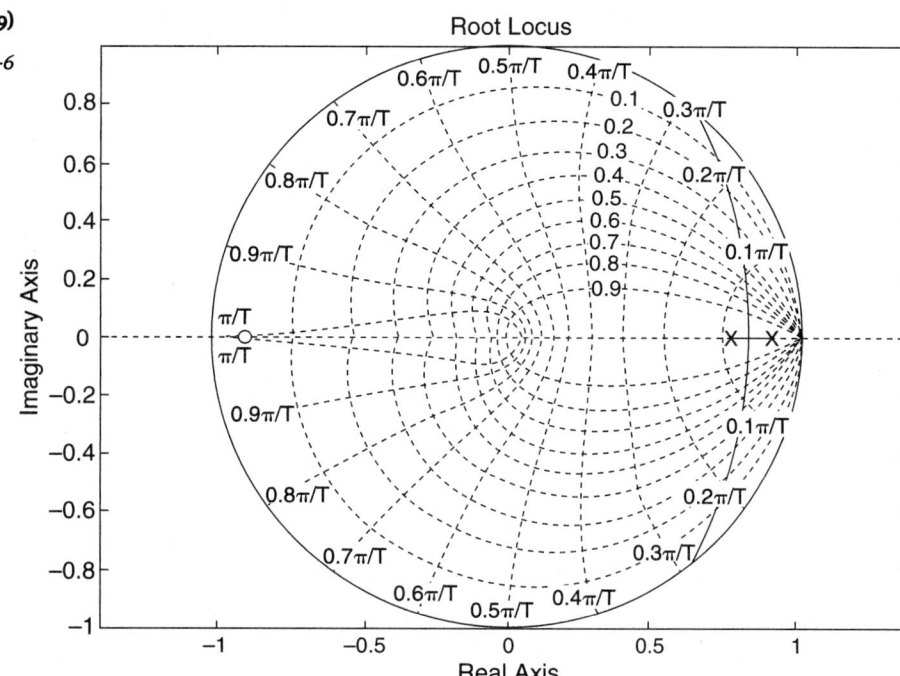

The gain to be calculated was chosen as the maximum gain for stability since plotting the entire root locus would cause the unit circle to be quite small and difficult to manipulate.

The grid for the root locus is turned on by right-clicking inside the plot and selecting *grid*. Values of gain at the point where the root locus crosses the unit circle and the 0.7 damping ratio line are found by placing the cursor at the point of interest and clicking. The gains shown by that procedure are approximately 29 and 1.4, which bear out the calculations for the system.

Summary

➤ The ZOH is used to derive the transfer function of the discrete control system. While other methods can be used, the ZOH is the most common.

➤ There is more than one method of producing the discrete transfer function from a continuous system.

➤ The Routh table is not directly applicable to discrete control systems. The method of bilinear transformation to the r-plane is used to apply the transformation to the Routh table for discrete systems.

➤ The maximum gain for stability in the z-plane is determined using the Routh table generated by the bilinear transformation.

➤ The discrete control system poles are constrained to be inside a unit circle if the system is to be stable.

➤ Even though a single pole system is unconditionally stable in a continuous system, the same is not necessarily true in the discrete domain.

➤ The distribution of the poles on the z-plane is found the same way as in the continuous system and is plotted in a similar manner.

➤ If any of the poles are outside the unit circle, the system is unstable.

➤ The root locus of the discrete system is plotted as a circle whose radius is the distance from the zero to the breakaway point on the real axis or as a circle whose radius is the distance from the zero to a pole on the z-plane.

➤ MATLAB is useful in the calculation of the root locus and step response of discrete systems.

➤ The discrete transfer function is readily found using **MATLAB**; there are several selectable methods of obtaining the transfer function, with the default being the ZOH method.

➤ The step response in the time domain and the root locus in the z-plane can be obtained using **MATLAB**.

Problems

P12.1   For the following continuous second-order plants, determine the discrete versions, using ZOH.

      a. $G(s) = \dfrac{1}{s^2 + 4s + 1}$

      b. $G(s) = \dfrac{20}{s^2 + 6s + 25}$

      c. $G(s) = \dfrac{s + 4}{s^2 + 4s + 8}$

P12.2   For the following closed-loop continuous system, design K for a damping ratio of 0.707. Then develop the discrete transfer function using ZOH. What is the final value of the output?

$$G(s) = \frac{6K}{s^2 + 13s + (36 + 6K)}$$

P12.3   For the following motor constants, generate the motor transfer function in the s-plane. Then choose a proper sampling time and convert the system to a discrete system, using ZOH. (Use 5 times the minimum sampling frequency.)

$La = 1$ mH $Kt = 5$ oz-in/A
$Ra = 5\,\Omega$ $Kd = 0.125$ oz-in/rad/sec
$Ke = 0.014$ V/rad/sec $J = 0.001$ oz-in-s^2

P12.4   If the motor of Problem P12.3 is placed in a proportional control system with a gain K of 2.5, what must the new sampling frequency be? What is the damping ratio of the new system? Produce the transfer function of the new system, using ZOH. Assume $H(z) = 1.0$.

P12.5   Derive an expression for the damped natural frequency ω_d in the discrete time domain.

P12.6   Show that the settling time of a second-order discrete control system is given by

$$Ts = \frac{5T}{\ln\sqrt{c}}$$

P12.7   For the following discrete characteristic equation, provide a Routh table, convert the equation to the r-plane, and determine if the system is stable.

$$z^3 + 1.2z^2 + 2z - 3 = 0$$

P12.8   For Problem P12.2, use the Routh-Hurwitz criterion to determine the limits of asymptotic stability.

P12.9   For the following characteristic equations in the discrete time domain, provide a Routh table and determine if the system is stable.

a. $z^3 - 2.898z^2 + 2.797z - 0.905 = 0$
b. $z^3 - 3z^2 - z + 4 = 0$
c. $z^3 - 3z^2 + 2z - 1 = 0$

P12.10   For the following characteristic equations, use r-transformation to determine the limits of gain for asymptotic stability.

a. $z^3 - z^2 + 2z - (K + 2) = 0$
b. $z^3 + 2z^2 + 3Kz - (K + 1) = 0$

P12.11 For the following first-order continuous system, use ZOH and plot the root locus in the z-plane. Determine the maximum gain for asymptotic stability. Determine any breakaway and break-in points.

$$G_P(s) = \frac{3}{s + 4}$$

P12.12 For the following second-order continuous system, use ZOH and plot the root locus in the z-plane. Determine the maximum gain, if such exists, for stability. Determine any breakaway and break-in points.

$$G_P(s) = \frac{s + 15}{s^2 + 7s + 10}$$

P12.13 For the following discrete second-order control system, plot the root locus and determine the maximum gain, if such exists, for asymptotic stability. Determine any breakaway and break-in points.

$$G(z) = \frac{0.212z - 0.063}{z^2 - 1.368z + 0.449}$$

P12.14 For the following open-loop discrete second-order system, plot the root locus and determine the maximum gain, if such exists, for asymptotic stability. Determine any breakaway and break-in points.

$$G(z) = \frac{0.081z - 0.066}{z^2 - 1.365z + 0.549}$$

P12.15 For the control system shown in Figure P12-15, convert the system to the z-plane and provide the root locus. Determine the maximum gain for asymptotic stability and the gain required for a system to have a damping ratio of 0.707. Provide the response of the system to a unit step input. (Use $T_s = 0.2$ seconds.)

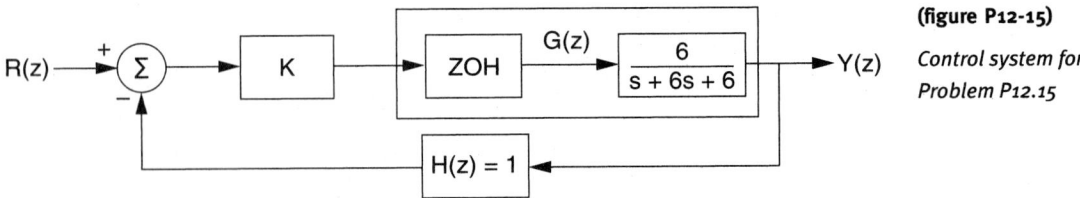

(figure P12-15)

Control system for Problem P12.15

P12.16 The following is the characteristic equation of a discrete control system. The system is to have a damped radian velocity of 2.8 rad/sec. Calculate the required gain. The system is being sampled at a rate of 0.1 seconds.

$$z^2 - (1.489 - .0246K)z + (0.549 + 0.0202K) = 0$$

MATLAB Problems

M12.1   Use **MATLAB** to verify your calculations for Problem P12.1.

M12.2   For Problem P12.3, use **MATLAB** to verify that the transfer function you calculated is correct. Produce the root locus of the system.

M12.3   For the following s-plane transfer function, use the **MATLAB c2d** command to produce the discrete transfer function for the 'zoh', 'matched', and 'foh' discrete functions.

$$G(s) = \frac{s + 6}{s^2 + 5s + 6}$$

a. Comment on any changes apparent in the transfer functions of each type.

b. From the root locus of each type, use **MATLAB** to determine if a damping ratio of 0.707 can be achieved.

M12.4   A z-plane transfer function can be input to **MATLAB** by using the following command:

$$\text{Sysd} = \text{tf([num], [den], } T_s)$$

Where the numerator and denominator are in the standard form and T_s is the sampling time. If no sampling time is required, enter -1 for T_s.

The z-plane open-loop transfer function to be investigated is

$$G(z) = \frac{K(0.05654z + 0.4946)}{z^2 - 1.564z + 0.6703}$$

a. Use the tf function to enter the transfer function with a sampling time of 0.1 seconds. Assume $K = 1.0$ for now.

b. Plot the root locus and determine the maximum asymptotic gain. Then replot the root locus and determine the required gain for a damping ratio of 0.4.

c. Compute the transfer function in the closed-loop form, using the gain found in part b. Assume a unity feedback system.

d. Plot the step response of the closed-loop transfer functions for both the discrete and continuous forms on the same graph. (Use the d2c command to determine the open-loop transfer function of the continuous system.)

Chapter 13 | Discrete PID, PD, and PI Control Systems

Chapter Objectives

After completing this chapter, you should be able to:

➤ State the algorithms used in the discrete implementation of the PID algorithm.

➤ Explain the differences between the trapezoidal and backward Euler implementation of the integral.

➤ Explain the PID algorithm and how it is implemented for PD, PI, and PID control systems.

➤ Explain the difference between the continuous and discrete PD and PI controllers.

➤ Show the use of **MATLAB** to implement the discrete control functions for all the types of controllers.

13.1 Introduction

Chapter 9 discussed the concepts of the PID controller. All of those concepts can be extended to discrete controller design. It is assumed that the system is to run in real time and the computer is capable of multitasking so that multiple loops can be implemented at the same time. From the concepts of the sampling times and gains of the system, it is clear that an improper choice of gains and sampling time can cause instability in the system.

13.2 Digital Implementation of the Discrete PID Algorithm

First, consider the concept of the PID controller in the time domain. The three gains required are related by

$$u(t) = K_P e(t) + K_I \int e(t) dt + K_D \frac{de(t)}{dt} \tag{13.1}$$

With all initial conditions set to zero. It is clear that the algorithm must cover both the integral and derivative portions of the compensator. For the integral portion, the *trapezoidal rule* for integration will be used; for the derivative portion, a *backward Euler* approximation will be used.

Considering a decreasing error and starting with the simpler proportional function

$$u_p(k) = K_P \cdot e_k \tag{13.2}$$

Figure 13-1 shows the graphical interpretation of Equation 13.2.

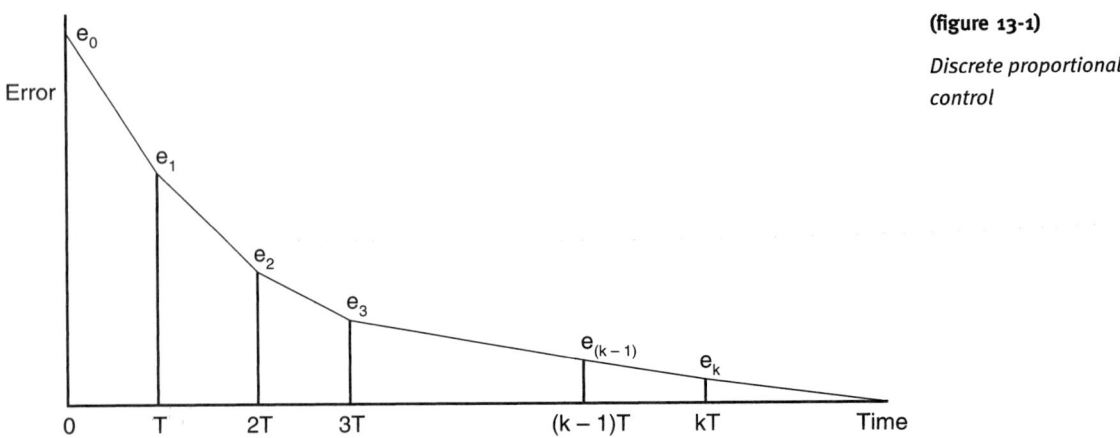

(figure 13-1)

Discrete proportional control

This form of control is often called the "present" form of control since it acts on the present sample of the error.

Trapezoidal Integration

To provide the integral control portion, the area under the curve of the error is required from zero to the present sample time (kT). Figure 13-2 shows that the total area is

$$A_0 + A_1 + A_2 + \cdots + A_{k-2} + A_{k-1} \tag{13.3}$$

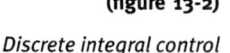

(figure 13-2)

Discrete integral control

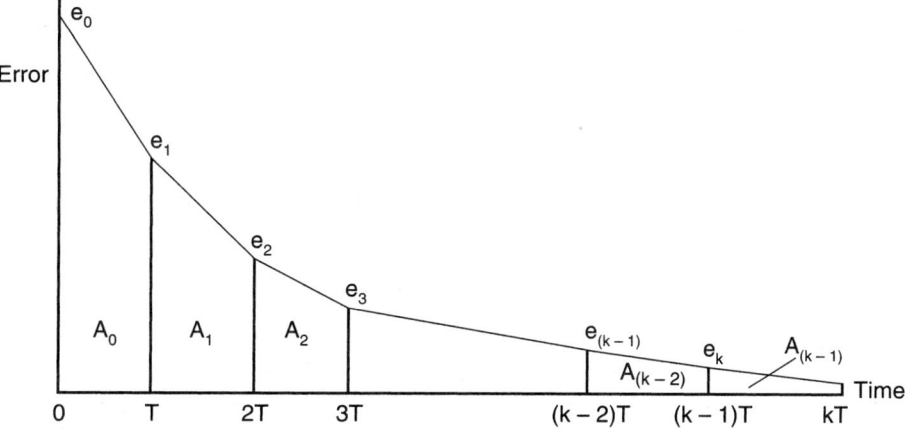

And the sum of the areas is given by

$$u_I(k) = \frac{K_I T}{2}\left[\left(e_0 + e_1\right) + \left(e_1 + e_2\right) + \cdots + \left(e_{k-2} + e_{k-1}\right) + \left(e_{k-1} + e_k\right)\right] \tag{13.4}$$

Since previous samples of the area are being stored, this type of control is often called the "past" form of control.

From the formula above, using the previous sample, you can find the z-transform of the discrete integrator from

$$u_k - u_{k-1} = \frac{K_I T}{2}\left(e_k + e_{k-1}\right)$$

$$U(z)\left(\frac{1 - z^{-1}}{T}\right) = \frac{K_I T}{2}\left(\frac{1 + z^{-1}}{T}\right)E(z) \tag{13.5}$$

$$\frac{U(z)}{E(z)} = \frac{K_I T}{2}\left(\frac{z + 1}{z - 1}\right)$$

Backward Euler Approximation

To provide derivative control, it is necessary to calculate only the slope of the error function at each sample, such as from e_k to e_{k-1}. The standard formula for slope can be used for this process. When transformed into the discrete form, the slope is

$$u_D(k) = \frac{K_D(e_k - e_{k-1})}{T} \tag{13.6}$$

The z-transform of the differentiator is found by taking the z-transform of both sides of Equation 13.6.

$$G_D(z) = K_D \frac{z - 1}{Tz} \tag{13.7}$$

PID Algorithm

The total response of the PID controller is found by summing the individual components as follows:

$$u(k) = K_P \cdot e_k + \frac{K_I T}{2}\left[(e_0 + e_1) + (e_1 + e_2) + \quad + (e_{k-2} + e_{k-1}) + (e_{k-1} + e_k)\right]$$
$$+ \frac{K_D(e_k - e_{k-1})}{T} \tag{13.8}$$

That formula is not in a form that is usable by a computer, however. If you examine the previous sample and replace k with $k-1$, you have

$$u(k - 1) = K_P \cdot e_{k-1} + \frac{K_I T}{2}\left[(e_0 + e_1) + (e_1 + e_2) + \quad + (e_{k-2} + e_{k-1})\right]$$
$$+ \frac{K_D(e_{k-1} - e_{k-2})}{T} \tag{13.9}$$

And subtracting Equation 13.9 from Equation 13.8 yields

$$u_k - u_{k-1} = K_P(e_k - e_{k-1}) + \frac{K_I T}{2}(e_k + e_{k-1}) + \frac{K_D}{T}(e_k - 2e_{k-1} + e_{k-2}) \tag{13.10}$$

This equation can be further simplified by collecting all of the terms in e_k, e_{k-1}, and e_{k-2} and multiplying them by an appropriate constant, as follows:

$$u_k - u_{k-1} = Ae_k + Be_{k-1} + Ce_{k-2}$$
$$u_k = u_{k-1} + Ae_k + Be_{k-1} + Ce_{k-2} \qquad (13.11)$$

Where

$$A = K_P + \frac{K_I T}{2} + \frac{K_D}{T}$$
$$B = \frac{K_I T}{2} - K_P - \frac{2K_D}{T} \qquad (13.12)$$
$$C = \frac{K_D}{T}$$

Once the constants A, B, and C are calculated, Equation 13.10 is easily implemented on any digital computer.

Discrete PD Controller

If K_I is zero, the proportional-derivative form of control is found from

$$G_C(z) = K_P - \frac{K_D}{T}\left(\frac{z - 1}{z}\right)$$

$$G_C(z) = \frac{K_P z + \frac{K_D}{T}z - \frac{K_D}{T}}{z} = \frac{\left(K_P + \frac{K_D}{T}\right)z - \frac{K_D}{T}}{z} \qquad (13.13)$$

Example 13-1

Consider the following antenna tracking system with an open-loop transfer function of

$$G(s) = \frac{6}{s^2 + 5s + 6}$$

Factoring the denominator, you find the system with real roots at $s = -2$ and $s = -3$. It is desired to convert this system to a digitally controlled system with a damping ratio of approximately 0.707, using a PD controller. Using a sample time of 0.1 seconds, generate the open-loop z-plane transfer function, using ZOH and a root locus, and determine the necessary gain for the damping ratio. Then provide the z-plane transfer function for the closed-loop system.

Example 13-1 (continued)

Solution

First, obtain the necessary inverse Laplace transforms of the system.

$$\mathcal{L}^{-1}\left[\frac{6}{s(s+2)(s+3)}\right] = 1 - 3e^{-2t} + 2e^{-3t}$$

Apply the necessary z-plane transforms to convert to the z-plane equation.

$$G_P(z) = \frac{1}{1-z^{-1}} - \frac{3}{1-e^{-0.2}z^{-1}} + \frac{2}{1-e^{-0.3}z^{-1}}$$

$$G_P(z) = \frac{1}{1-z^{-1}} - \frac{3}{1-0.8187z^{-1}} + \frac{2}{1-0.7408z^{-1}}$$

After simplification, the z-plane transfer function is

$$G_P(z) = \frac{0.02544z + 0.02154}{z^2 - 1.5595z + 0.6065}$$

Now apply the formula developed for the PD controller. Let the proportional gain be $K_P = 1.0$ to simplify the solution.

$$G_C(z) = \frac{\left(K_P + \dfrac{K_D}{T}\right)z - \dfrac{K_D}{T}}{z} = \frac{(1 + 10K_D)z - 10K_D}{z}$$

From which the open-loop transfer function in the z-plane is

$$G_{OL}(z) = \frac{(1 + 10K_D)z - 10K_D}{z} \cdot \frac{0.02544z + 0.02154}{z^2 - 1.5595z + 0.6065}$$

Upon simplification, the open-loop transfer function is

$$G_{OL}(z) = 0.2544K_D \cdot \frac{(z^2 - 0.1533z - 0.8467)}{z(z^2 - 1.5595z + 0.6065)}$$

The z-plane transfer function and root locus can be generated using **MATLAB's tf** and **rlocus** commands, as follows and as shown in Figure 13-3.

$$\text{sysd} = \text{tf}\left(\left[.2544 -.1533 -.2154\right], \left[1 - 1.5595\ .6065\ 0\right], .1\right)$$

$$\text{rlocus(sysd)}$$

Example 13-1 (continued)

This root locus leaves K_D as a variable. The gain for a damping ratio of 0.707 is found from the locus to be $K_D = 0.62$. The closed-loop characteristic equation of the system is

$$(z^3 - 1.5595z^2 + 0.6065z) + 0.2544K_D(z^2 - 0.1533z - 0.8467) = 0$$

(figure 13-3)

Root locus for Example 13-1

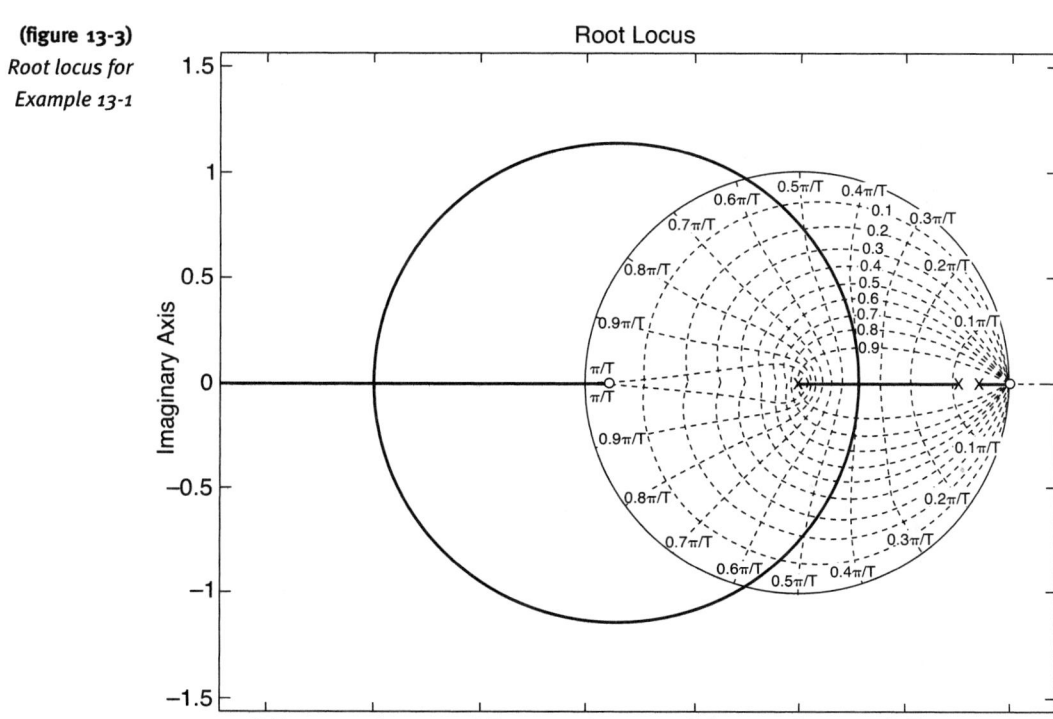

Discrete PI Controller

If K_D is zero, the discrete form of PI controller appears and, after simplification, is given by

$$G_C(z) = \frac{K_P z - (K_P - K_I T)}{z - 1} \qquad (13.14)$$

Example 13-2

This example will use the same plant as in Example 13-1. The same concept of placing the zero close to the pole at the origin of the s-plane still applies, but in this case, the zero is placed near the pole at $z = 1.0$.

Solution

You can refine the compensator equation by taking K_P out of the numerator and forming a new function.

$$G_C(z) = K_P \frac{z - \left(1 - \dfrac{K_I T}{K_P}\right)}{z - 1} \qquad (13.15)$$

Let the zero be placed at $z = 0.8$, and $K_P = 1.0$. All that remains, then, is to determine the value for K_I. You can create that problem by assigning the value 0.025 (the coefficient of z in Example 13-1) to a new variable K. The total transfer function then becomes

$$G_C(z)G_P(z) = KK_P \frac{z - \left(1 - \dfrac{K_I T}{K_P}\right)}{z - 1} \cdot \frac{(z + 0.8467)}{z^2 - 1.56z + 0.6065}$$

From which the value of K_I can be calculated.

$$1 - K_I T = 0.8$$

$$K_I = \frac{1 - .8}{T} = \frac{0.2}{0.1} = 2.0$$

And the zero is then located at $(0.0254z - 0.0204)$. The open-loop transfer function is then

$$G_C(z)G_P(z) = \frac{0.0254z^2 + 0.0011z - 0.0173}{z^3 - 2.56z^2 + 2.166z - 0.6065}$$

A root-locus plot showing the near cancellation of the pole at -0.8467 is shown in Figure 13-4. Note that this analysis is done without regard to any particular specifications.

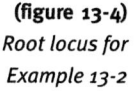

(figure 13-4)

Root locus for Example 13-2

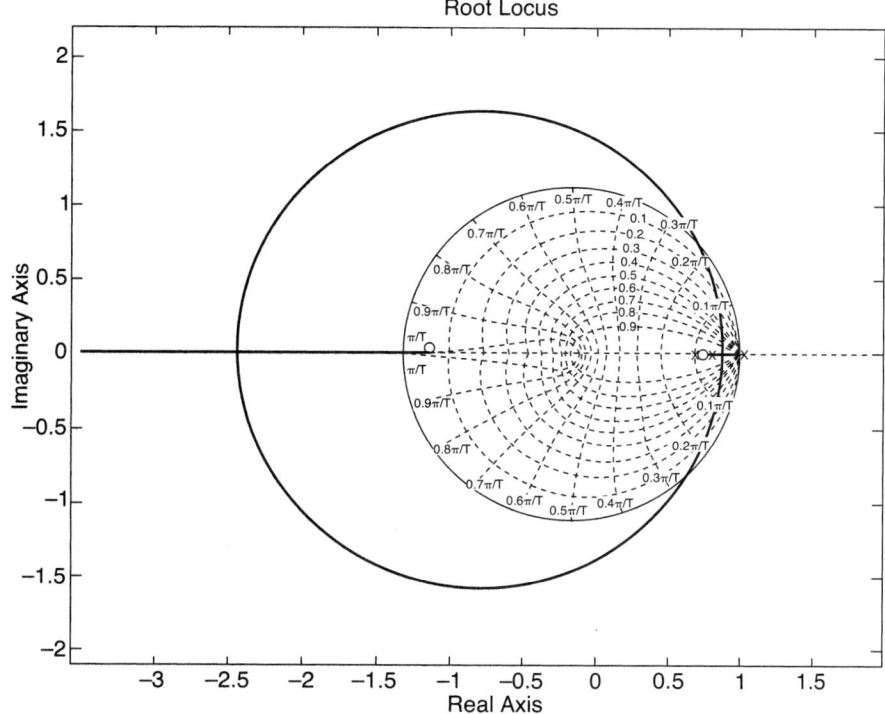

The entire system could have been done more simply in **MATLAB**, but it is important to understand the mechanisms involved first. It is left to you to use **MATLAB** to produce the same transfer function as in the example.

Summary

➤ The PI and PD controllers are derived from the PID algorithm.

➤ One method of implementation of the controllers is the trapezoidal method.

➤ The backward Euler integration and differentiation methods are also used for implementation of the controllers.

➤ Difference equations of the PID algorithm are readily developed for use in a digital computer.

➤ **MATLAB** may be used for producing the transfer functions and the root loci of the systems.

Problems

P13.1   Using Equation 13.12, show that the zero location of the PI controller is located at

$$z = -B/A.$$

P13.2   For the following discrete plant, design a PI controller such that the damping ratio is 0.707. The sampling time is 0.1 seconds.

$$G_P(z) = \frac{0.0906}{z - 0.819}$$

P13.3   The control system shown in Figure P13-3 is to be controlled by a PD controller. The settling time is to be less than 1 second, and the damping ratio is to be 0.707. Design the controller. Then put the system into the form of a discrete controller with a sampling time of 0.1 seconds. Compare the output of the continuous and digital systems. Based on the overshoot of each system, comment on the digital system response compared to that of the continuous system.

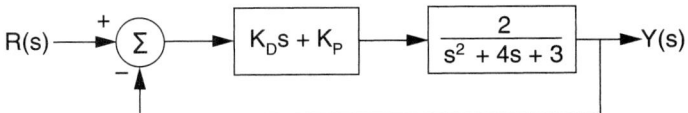

(figure P13-3)

Control system for Problem P13.3

P13.4   Using Equation 13.12, if $A \to K_D/T$ and $B \to -2K_D/T$, with $C = K_D/T$, show that the zeros of Equation 13.11 will approach the unit circle. Write the transfer function $G(z)$ for this condition.

P13.5   For the following continuous plant transfer function, determine the discrete domain transfer function. Then add a derivative controller to maintain a damping ratio of 0.707 and a settling time of 0.5 seconds with a sampling time of 0.05 seconds.

$$GP(s) = \frac{4}{s^2 + 7s + 6}$$

P13.6   A motor that is to be used in a discrete PID controller antenna rotator-system has the following parameters:

$$L_a = .01 \text{ H} \qquad\qquad K_b = 0.01 \text{ V/(rad/sec)}$$
$$R = 2 \ \Omega \qquad\qquad B = .03 \text{ oz-in(rad/sec)}$$
$$K_T = 10 \text{ oz-in/A} \qquad J = .002 \text{ oz-in-s}^2$$

a. Determine the motor transfer function.

b. Determine the open-loop transfer function in the z-plane, using the following gains: $K_I = 4.0$, $K_D = 0.00017$, and $K_P = 0.064$. The sampling time is 0.002 seconds.

c. Plot the poles and zeros in the z-plane and generate the root locus of the system.

d. Determine the closed-loop transfer function and plot the response to a step input.

P13.7   Using the transfer function of Problem P13.6, generate the difference equation for the PID controller.

P13.8   The closed-loop difference equation of a PI controller is given by

$$y_k = .00471u_{k-1} + 0.0002531u_{k-2} - 0.03505u_{k-3} + 2.424y_{k-1}$$
$$- 1.984y_{k-2} + 0.5488y_{k-3}$$

a. Determine the z-plane transfer function of the system. The sampling time is 0.1 seconds.

b. Plot the closed-loop root locus of the system. Determine the damping ratio of the second-order poles of the system and the maximum gain for stability.

P13.9   For the following continuous controller open-loop transfer function:

$$G(s) = \frac{24K}{s^4 + 10s^3 + 35s^2 + 50s + 24}$$

a. Determine the closed-loop transfer function.

b. Use the Routh table to determine the limits of stable gain.

c. Convert the system to the discrete domain; plot the open-loop root locus; and determine the limits of stable gain, leaving K as a variable.

d. Are the gain factors the same? Why or why not?

MATLAB Problems

M13.1   Use **MATLAB** to design the system in Problem P13.3.

M13.2   Use **MATLAB** to design the system in Problem P13.6.

M13.3   For the following open-loop continuous plant

$$G_p(s) = \frac{150}{(s^2 + 25s + 150)}$$

use **MATLAB** to design a PI controller in the discrete domain. Let $K_p = 1.0$ and place the zero at $z = 0.9$. The sampling time is 0.1 seconds.

M13.4   Use **MATLAB** to produce the transfer function of a PD controller for the following continuous open-loop transfer function:

$$G_p(s) = \frac{240}{s^2 + 100s + 2400}$$

The damping ratio is to be $\zeta = 0.707$. The proportional gain is $K_p = 25$. The sampling time is 0.0025 seconds.

M13.5   Use **MATLAB** to generate all of the functions in Problem P13.9. Do the values coincide with the values you calculated by hand? Explain any differences.

Chapter 14 | Discrete Compensation Methods

Chapter Objectives

After completing this chapter, you should be able to:

➤ Show by calculation the use of the matched pole-zero and bilinear transform methods as applied to the compensation of a control system.

➤ Explain the concepts of discrete lead, lag, and lead-lag compensation and the effect of the compensators on the response of the system.

➤ Explain the effect of a delay of n-samples on the phase response of a system.

14.1 Introduction

Chapter 8 described the concepts of lead, lag, and lead-lag compensation of continuous time systems. The methodology can be applied in the discrete domain as well. Further, two additional methods will be applied; both are borrowed from digital filter theory. They are the *matched pole-zero* and *bilinear transformation* techniques. You should understand that the compensation systems applied thus far are also filters, so knowledge of one can be applied to the other. The two techniques will aid in transforming the continuous time domain to the discrete domain.

In many cases, a design engineer first designs the compensation as a continuous time system and then converts the system to the discrete form, using the methods outlined thus far. This chapter will begin with the matched pole-zero technique.

14.2 Matched Pole-Zero Technique

Consider the following continuous time transfer function.

$$P(s) = K\frac{(s + r_1)(s + r_2)\cdots(s + r_n)}{(s + p_1)(s + p_2)\cdots(s + p_m)}$$ (14.1)

The transfer function is transformed to the z-plane, using the identity $z = e^{-sT}$. Any zeros located at infinity are mapped to $z = -1$, creating a matched number of poles and zeros in the form

$$P(z) = K_1 \cdot \frac{(z + 1)^{m-n}(z - e^{-r_1 T})(z - e^{-r_2 T})\cdots(z - e^{-r_n T})}{(z - e^{-p_1 T})(z - e^{-p_2 T})\cdots(z - e^{-p_m T})}$$ (14.2)

The gain K_1 is to be calculated such that the gain at the critical frequency of the original transfer function is preserved. Keeping that fact in mind, let's examine the technique with an example.

Example 14-1

Consider the following low-pass filter. Determine the digital filter equivalent and K_1, using the matched pole-zero method and a sampling time of 0.1 seconds.

$$P(z) = K_1 \cdot \frac{5}{s^2 + 2s + 5}$$

Example 14-1 (continued)

Solution

Since there are two poles and no zeros in the system, $m - n = 2$; and the numerator is $(z + 1)^2$. Then, applying Equation 14.2, the transfer function becomes

$$P(z) = K_1 \cdot \frac{(z + 1)^2}{z^2 - 1.774z + .8187}$$

Since this is a low-pass filter, the critical frequency is zero radians (DC). When s = 0, the final value is 1.0; and K_1 is calculated from the fact that $(z + 1)^2 = z^2 + 2z + 1$. With z set to 1, the numerator equals 4.0 and the denominator equals 0.0747. K_1 becomes

$$K_1 = \frac{0.0747}{4} = 0.0187$$

And the transfer function is

$$P(z) = 0.0187 \frac{z^2 + 2z + 1}{z^2 - 1.744z + 0.8187}$$

14.3 Bilinear Transformation Technique

Bilinear transformation involves the direct replacement of s in a continuous control system with a term in z, which is determined from the relation of z to s by $z = e^{-sT}$.

$$s = \frac{1}{T} \ln z \tag{14.3}$$

A series expansion of Equation 14.3 yields

$$\ln z = \frac{2}{T} \cdot \frac{z - 1}{z + 1} + \frac{1}{3T^3} \cdot \left(\frac{z - 1}{z + 1}\right)^3 + \cdots \tag{14.4}$$

From the expansion, the first term provides a good approximation to the function; and you can use just that term to approximate the function.

Example 14-2

Let's apply the method to the system of Example 14-1.

Solution

You start by replacing all s terms with the first term of the expansion as follows:

$$P(z) = \frac{5}{\left(\frac{2}{0.1} \cdot \frac{z-1}{z+1}\right)^2 + 2 \cdot \left(\frac{2}{0.1} \cdot \frac{z-1}{z+1}\right) + 5}$$

To simplify, you determine the common denominator and the coefficients of the system.

$$P(z) = \frac{5}{\dfrac{400z^2 - 800z + 400 + 40z^2 - 40 + 5z^2 + 10z + 5}{(z+1)^2}}$$

$$P(z) = \frac{5(z+1)^2}{445(z^2 - 1.775z + .0.8202)} = \frac{0.0112(z^2 + 2z + 1)}{z^2 - 1.775z + 0.8202}$$

Rather than do the tedious calculations, it is easier to construct the polynomial using **MATLAB**. The method is simple: construct the s-plane transfer function; then convert it to the discrete version, using the 'tustin' option for the **c2d** command as follows:

$$n = 5;\ d = \begin{bmatrix} 1 2 5 \end{bmatrix};$$
$$\text{sys} = \text{tf}(n,d)$$
$$\text{sysd} = \text{c2d}(\text{sys},.1,\ \text{'tustin'})$$

The original filter has a natural radian velocity of 2 rad/sec; the natural radian velocity of this filter is 2.45 rad/sec. That is due to the approximation to Equation 14.4. If you could increase the original continuous system frequency, the bilinear transformation would warp the frequency back to the original critical frequency.

From Equation 14.1, all critical frequencies are of the form

$$T(s) = 1 + \frac{s}{\omega_n} \tag{14.5}$$

When $s = j\omega$ and $\omega = \omega_n$ the frequency response is given by

$$T(j\omega) = 1 + j \tag{14.6}$$

You can prewarp the critical frequency to be

$$T_{PW}(s) = 1 + \frac{s}{\omega_{PW}} \tag{14.7}$$

Applying the bilinear transformation to Equation 14.7 provides

$$T_{PW}(z) = 1 + \frac{2}{\omega_{PW}T} \cdot \frac{z - 1}{z + 1}$$ (14.8)

Changing the z to its exponential equivalent causes the frequency response at the critical frequency to become

$$T_{PW}(e^{j\omega_n T}) = 1 + \frac{2}{\omega_{PW}T} \cdot \frac{e^{j\omega_n T} - 1}{e^{j\omega_n T} + 1}$$ (14.9)

It should be clear that Equations 14.9 and 14.6 are now equivalent and that

$$\frac{2}{\omega_{PW}T} \cdot \frac{e^{j\omega_n T} - 1}{e^{j\omega_n T} + 1} = j$$ (14.10)

Simplifying yields

$$\frac{2}{\omega_{PW}T} \cdot \left[\cos(\omega_n T) + j\sin(\omega_n T) - 1\right] = -\sin(\omega_n T) + j[\cos\omega_n T + 1]$$ (14.11)

Equating the real parts of Equation 14.11 yields

$$\frac{2}{\omega_{PW}T}\left[\cos(\omega_n T) - 1\right] = -\sin(\omega_n T)$$ (14.12)

$$\frac{2}{\omega_{PW}T}\sin(\omega_n T) = \cos(\omega_n T) + 1$$ (14.13)

Substituting Equation 14.12 into Equation 14.13 and simplifying yields

$$\left(\frac{\omega_{PW}T}{2}\right)^2 = \frac{1 - \cos(\omega_n T)}{1 + \cos(\omega_n T)}$$ (14.14)

Equation 14.14 is recognized as a trigonometric identity for the cosine squared function and can be reduced to

$$\omega_{PW} = \frac{2}{T}\tan\left(\frac{\omega_n T}{2}\right)$$ (14.15)

The standard form of a low-pass filter is given by

$$P(s) = \frac{\omega_n^2}{s^2 + 2\zeta\omega_n s + \omega_n^2}$$

And the prewarped form is

$$P(s) = \frac{\omega_{PW}^2}{s^2 + 2\zeta\omega_{PW}s + \omega_{PW}^2}$$

Let's examine how that works with an example.

Example 14-3

You will use the filter of Example 14-1 and determine the prewarped frequency and the new transfer function. The original natural frequency is 2.2367 rad/sec, and from Equation 14.15

$$\omega_{PW} = \frac{2}{T}\tan\left(\frac{2.2367 \cdot 0.1}{2}\right) = 2.2461\,\text{rad/sec}$$

Using the new frequency to rewrite the original transfer function gives

$$P_{PW}(s) = \frac{5.0448}{s^2 + 2.0088s + 5.0448}$$

Substitution of the bilinear transform yields

$$P_{PW}(z) = \frac{0.0113(z^2 + 2z + 1)}{z^2 - 1.774z + 0.8195}$$

There is not much difference between that transfer function and the bilinear transformation transfer function. Examination of the Bode plots of the original system, the bilinear transformed system, and the prewarped frequency system shows that there is approximately a 1 radian difference between the bilinear method and the original system; but the prewarped system returns the system to the correct critical frequency.

This system can be designed using **MATLAB**, with the 'prewarp' option in the **c2d** command as follows:

sysd = c2d(sys, 0.1, 'prewarp', 2)

The frequency to be used to prewarp the filter is the fourth parameter in the command, and it is required.

14.4 Discrete Lead Compensation

Chapter 8 covered the properties of lead compensation of continuous systems. Those concepts are extended to the discrete form of control, using the methods discussed below.

The original lead compensator had the form

$$G_C(s) = K\frac{s + a}{s + b} \qquad a < b \tag{14.16}$$

By applying the matched pole-zero technique, the z-plane equivalent becomes

$$G_C(z) = K_1 \frac{z - A}{z - B} \qquad A > B \qquad (14.17)$$

Where $A = e^{-at}$ and $B = e^{-bT}$, the value of K_1 is left to be calculated in the design process. An example will clarify the process.

Example 14-4

A control system using lead compensation has been designed in the s-domain. It is proper to cancel the pole closest to the origin to improve the system speed and radian velocity. The discrete system is to have a damping ratio of 0.707 and a radian velocity of 4 rad/sec. The system is shown in Figure 14-1.

Solution

For the system design, you will choose a sampling time of 0.1 seconds. Application of the ZOH process to the plant yields

$$G_P(z) = \frac{0.00864z + 0.00744}{(z - 0.9528)(z - 0.6692)}$$

The lead compensator must have a canceling pole at z = 0.9528, and the compensator will have the form

$$G_C(z) = K_1 \frac{z - 0.9528}{z - B}$$

(figure 14-1)

Control system for Example 14-4

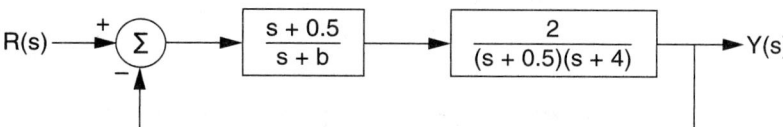

Example 14-4 (continued)

Use the relationship for ω, which is

$$\omega = \frac{1}{T}\cos^{-1}\left(\frac{b}{2\sqrt{c}}\right)$$

$$4 = \frac{1}{0.1}\cos^{-1}\left(\frac{b}{2\sqrt{c}}\right)$$

$$\frac{b}{2\sqrt{c}} = 0.9211$$

Using Equation 12.26, you arrive at a value of c.

$$\cos(-0.5\ln c) - 0.9211 = 0$$

The value of c is calculated to be

$$c = \exp\left[-\frac{\cos^{-1}(0.9211)}{0.5}\right] = 0.4494$$

The characteristic equation of the system is

$$z^2 - (B + .6692 - 0.00864K_1)z + (0.6692B + .00744K_1) = 0$$

From which you set up an equation in B and K_1.

$$0.6692B + 0.00744K_1 = 0.4494 \tag{14.18}$$

From the characteristic equation, you calculate the value of b.

$$\frac{b}{2\sqrt{0.4494}} = 0.9211$$

$$b = 1.235$$

Also from the characteristic equation, you have

$$B - 0.00864K_1 = 1.235 - 0.6692 = 0.5658 \tag{14.19}$$

Solving Equations 14.18 and 14.19 simultaneously yields values for K_1 and B. $K_1 = 5.352$ and B = 0.612.

The required discrete lead compensator is

$$G_C(z) = 5.352\frac{z - 0.9528}{z - 0.612}$$

While the preceding example provides some insight into the design of compensators in the discrete domain, most systems are designed in the s-domain and then are transferred to the z-plane.

14.5 Discrete Lag Compensation

The discrete lag compensator is a more complicated design in that two values for K_1 and b are arrived at and a decision about which one to use must be made.

The general form of the discrete lag compensator has the same form as the discrete lead compensator, but the requirements differ as follows:

$$G_C(z) = K_1 \frac{z - A}{z - B} \qquad A < B \qquad (14.20)$$

To appreciate the z-plane design techniques, you will see that a small change in the coefficients calculated produces a different type of compensation altogether. Once the design engineer is comfortable working in the z-plane, the complete design can be carried out using only the z-plane.

Let's carry out a design example for a discrete lag compensator.

Example 14-5

A proportional control first-order plant using a lag compensator is to be designed such that the error is no more than 10 percent and there is a damping ratio of 0.707. The system is shown in Figure 14-2.

It is decided that the system zero will be placed at 3 times the pole position. The desired pole will be placed near the origin of the s-plane to ensure a practical design, as outlined in Chapter 8.

Solution

The closed-loop transfer function of the system is

$$G(s) = \frac{K(s + 6)}{s^2 + (2 + b + K)s + (2b + 6K)}$$

(figure 14-2)

Control system for Example 14-5

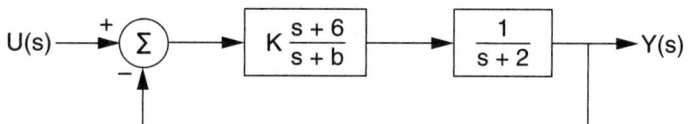

Example 14-5 (continued)

For the plant to exhibit a 10 percent error, the required error function is

$$\frac{6K}{2b + 6K} = 0.9$$

And the required value of K is

$$K = 3.0b \qquad (14.21)$$

The characteristic equation now becomes

$$s^2 + (2 + 4b)s + 20b = 0$$

For the system to have a 0.707 damping ratio

$$2(0.707)\omega = 2 + 4b \qquad (14.22)$$

And

$$\omega = \sqrt{20b} \qquad (14.23)$$

From Equations 14.22 and 14.23, you determine K and b.

$$2(0.707)\omega = 2 + 4b$$
$$39.9879b = 16b^2 + 16b + 4$$
$$16b^2 - 23.9879b + 4 = 0$$
$$b_1 = 0.1911 \qquad b_2 = 1.3081$$

And $K_1 = 0.5733$ and $K_2 = 3.9243$. The question is which one to use. In Chapter 8, you determined that a separation of 10 was sufficient to ensure a practical implementation. You will use the pole at $s = -1.3081$ and $K_2 = 3.9243$ for this design.

The design is now complete, and you can transfer it to the z-domain. The ZOH process for a 0.1-second sampling time for the plant is

$$G_C(z) = \frac{0.0906}{z - 0.8187}$$

You will apply the bilinear transformation to the compensator at the same sampling time to generate the required transfer function of the compensator.

$$3.9243 \frac{20\frac{z - 1}{z + 1} + 6}{20\frac{z - 1}{z + 1} + 1.3081} = 4.7884 \frac{z - 0.5385}{z - 0.8772}$$

The closed-loop characteristic equation of the system is

$$z^2 - 1.2621z + 0.4846 = 0$$

The damping ratio of the system is slightly lower than the desired 0.64, and the overshoot is about 14 percent. That is to be expected due to the ZOH process used for the plant.

If the gain is reduced, the system can be made to meet the specifications.

Discrete Lead-Lag Compensation

The continuous form of the lead-lag compensator is given by

$$G_C(s) = K\frac{(s + a_1)(s + a_2)}{(s + b_1)(s + b_2)} \qquad a_1 < b_1, a_2 > b_2 \qquad (14.24)$$

Application of the matched pole-zero technique gives the form of the compensator in the z-domain.

$$G_C(z) = K_1\frac{(s + A_1)(s + A_2)}{(s + B_1)(s + B_2)} \qquad A_1 > B_1, A_2 < B_2 \qquad (14.25)$$

Where

$$A_1 = e^{-a_1 T}, \quad A_2 = e^{-a_2 T}, \quad B_1 = e^{-b_1 T}, \quad B_2 = e^{-b_2 T}$$

Care must be taken in the design of lead-lag compensators because the zeros may or may not follow the s- to z-plane transfer property. You may try to calculate a pole to satisfy a requirement and end up with the pole outside the unit circle due to the selection of the zeros. K_1 will be calculated from the system requirements.

An example using the same plant as that in Example 14-5 will help clarify the design process.

Example 14-6

The control system of Example 14-5 is to exhibit zero error for a step input. The damped radian velocity and the damping ratio should remain the same. The error specification implies that an integrating pole be placed at z = 1.0. That creates a lag that slows the system down, and lead compensation will be needed to return the system to the original speed. For that reason, a lead-lag compensator is chosen. Figure 14-3 shows the system.

Solution

First, you must choose some of the parameters. The lag compensator pole will be placed at z = 1.0, and the lag zero can be safely placed to cancel the plant pole at z = 0.819 without jeopardizing the

(figure 14-3)

Control system for Example 14-6

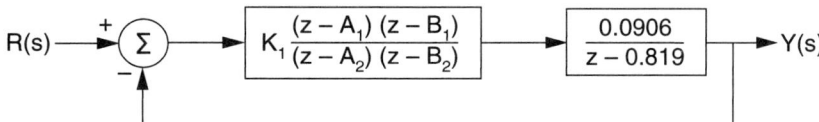

Example 14-6 (continued)

plant pole position. To obtain the required speed, the lead compensator position is chosen such that the speed reduction of the pole at $z = 1.0$ is offset; in this case, at $z = 0.25$. That leaves the lead compensator zero to be calculated. The transfer function of the lead-lag compensator is then

$$G_C(z) = K_1 \frac{(z - A_1)(z - 0.819)}{(z - 0.25)(z - 1.0)}$$

The closed-loop characteristic equation is then found from

$$z^2 - (1.25 - .0906K_1)z + (0.25 - 0.0906K_1A_1) = 0$$

You can now carry out an analysis based on the natural frequency of the system from Example 14-5. The natural frequency is 5.11 rad/sec. You start by calculating

$$\frac{b}{2\sqrt{c}} = \cos(5.11 \cdot 0.1) = 0.872$$

Substituting that value into Equation 12.26 yields

$$\cos(-0.5 \ln c) - 0.872 = 0$$

$$c = \exp\left(\frac{\cos^{-1} 0.872}{-0.5}\right) = 0.3595$$

And

$$\frac{b}{2\sqrt{0.3595}} = 0.872$$

$$b = 0.872 \cdot 2 \cdot \sqrt{.3595} = 1.0457$$

From the characteristic equation

$$1.25 - 0.0906K_1 = 1.0457$$

$$K_1 = \frac{1.25 - 1.0457}{.0906} = 2.255$$

And

$$0.25 - 0.906 \cdot 2.255A_1 = 0.3595$$

$$-A_1 = \frac{0.3595 - .25}{0.0906 \cdot 2.255} = -0.5360$$

Example 14-6 (continued)

Notice that the zero, A_1, is on the negative side of the z-plane. The completed closed-loop transfer function for the system is given by

$$G(z) = \frac{0.2043z + 0.1095}{z^2 - 1.0457z + 0.3595}$$

The damped radian frequency of this system has increased from 5 rad/sec to about 7 rad/sec. That is beneficial in terms of the speed of response of the system to a step input.

14.6 Discrete Transport Lag

The zero-order-hold process automatically inserts a delay of one sample time at the beginning of the analysis. In real systems, that is a major problem. However, the delay time works the same way in the continuous systems; that is, the amplitude due to e^{sT} is constant, but the phase varies widely depending on the number of sample time delays. In general, the delay is an integer number of samples and is given by

$$T_d = n \cdot T \tag{14.26}$$

The effect of the delay is the same as inserting a transfer block in the system with a transfer function of z^{-1}. Insertion of the block causes a phase change that can cause the system to tend toward instability.

Summary

➤ Matched pole-zero and bilinear transformation methods are used to transform the continuous transfer function into the z-plane.

➤ Bilinear transformation changes, or warps, the frequency response of the system.

➤ The problem of frequency warping can be alleviated by prewarping the continuous transfer function.

➤ The discrete forms of the lead, lag, and lead-lag compensation schemes as used with continuous systems can be applied to discrete systems.

➤ Application of lead, lag, and lead-lag compensators may reduce the system loop gain.

➤ Placement of the zeros can change the pole locations, placing one or more poles outside the unit circle.

➤ Transport lag in the terms of the ZOH process, and the fact that the delay could cause the system to be unstable may require extensive system redesign.

➤ MATLAB can be used to provide the matched pole-zero process using the Tustin method, as can the prewarp method for returning the frequency response of the system to its original value.

Problems

P14.1   The general form of the bandpass filter is

$$G(s) = \frac{2\zeta\omega_n s}{s^2 + 2\zeta\omega_n s + \omega_n^2}$$

If $\zeta = 0.25$ and $\omega = 1.0$ rad/sec, with $T_s = 1$ second, find the equivalent discrete form of the filter, using the following techniques:

a. ZOH method

b. Matched pole-zero technique

c. Bilinear transformation

d. Bilinear transformation with frequency prewarping

e. Continuous filter

P14.2   Plot the response of each of the filter types in Problem P14.1. Which form produces the best approximation to the continuous filter?

P14.3   Considering a critical frequency of 1 rad/sec, what is the benefit, if any, of using prewarping if the sampling frequency is made greater than 1 sample per second.

P14.4   The transfer function shown is for a prewarped bandpass filter. What is the original transfer function if the sampling time is 0.1 seconds?

$$G(s) = \frac{26.0974s}{s^2 + 1.0214s + 26.0974}$$

P14.5   The following is a continuous transfer function:

$$G(s) = \frac{5}{s^2 + 6s + 5}$$

a. Use the matched-pole technique to generate the discrete transfer function, using a sampling time of 0.1 seconds.

b. Plot both the continuous and discrete responses to a step input.

c. Change the sampling time to 0.5 seconds and plot the two responses. Is the new sampling time adequate to reproduce the filter response well?

P14.6   For the following closed-loop continuous transfer function, provide the discrete transfer function for a sampling time of 0.1 seconds.

$$G(s) = \frac{20(s + 1.8)}{s^3 + 8s^2 + 30s + 36}$$

P14.7   For the continuous system of Problem P14.6, provide the bilinear transformation method discrete transfer function.

P14.8   For the continuous system shown in Figure P14-8, use a lead compensator and determine a proper value of B and K. Select a suitable value of sampling time for the system. Then transform the system to the discrete domain, with a damping ratio of 0.707 and a damped radian velocity of 3.0 rad/sec. Plot the continuous and discrete responses to a step input. How well does your selected sampling time approximate the continuous function?

(figure P14-8)

Control system for Problem P14.8

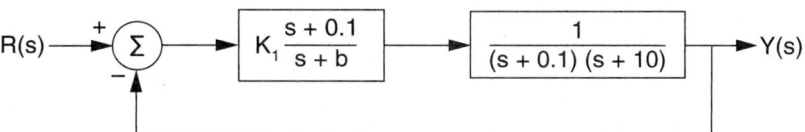

P14.9   A continuous plant has the following transfer function:

$$G_P(s) = \frac{9}{s^2 + 10s + 9}$$

The plant has a sluggish reaction for a step input due to the dominant pole at $s = -1.0$. It is desired to use a lead controller to cancel the dominant pole and to provide a damping ratio of 0.5, with a damped radian velocity of 5.0 rad/sec. Using a sampling time of 0.1 second, design the lead controller to provide the necessary response. Plot the continuous and discrete responses to a step input. Is the sampling time adequate to produce a good approximation for this system? If not, what should the sampling frequency be?

P14.10   The first-order plant shown below is to be proportionally controlled.

$$G_P(s) = \frac{1}{s + 5}$$

The specifications require that the error be no more than 10 percent, and the damping ratio is to be 0.6. A lag compensator is decided on to produce the required damping and error. Complete the design in the s-plane; then transfer the design to the z-plane. Use ZOH for the plant and bilinear transformation for the compensator. The sampling time should be a minimum of 0.1 seconds.

P14.11 The system of P14.10 is now specified to have a zero error for a step input. The damping ratio is to be 0.707, and the damped frequency is to remain the same. That implies an integrating pole as $z = 1.0$. Place the pole; then design the lead compensation portion to restore the speed of the system. Convert the system to the z-plane, using a sampling time of your choice.

P14.12 Consider the system shown in Figure P14-12.

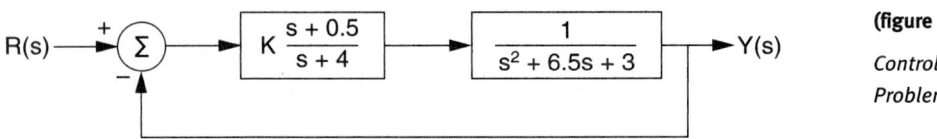

(figure P14-12)

Control system for Problem P14.12

    a. Calculate K for a damping ratio of 0.707.

    b. Convert $G_C(s)$ to $G_C(z)$, using the bilinear transformation method and a sampling time of your choice.

    c. Convert $G_P(s)$ to $G_P(z)$, using ZOH and the sampling time of part b.

    d. Plot the open-loop root locus, leaving K as a variable. Determine the gain for a damping ratio of 0.707. How does it compare to the gain found in part a?

    e. Plot the response of the system to a step input for the values of gain found in parts a and d.

P14.13 For the system shown in Figure P14-13, repeat Problem P14.12 for a damping ratio of 0.6.

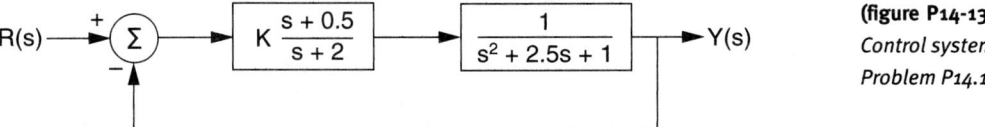

(figure P14-13)

Control system for Problem P14.13

P14.14 For the system shown in Figure P14-14, design the system to have a damping ratio of 0.707 and a 20 percent error for a unit step input. Select a sampling time that will provide a good approximation of the system.

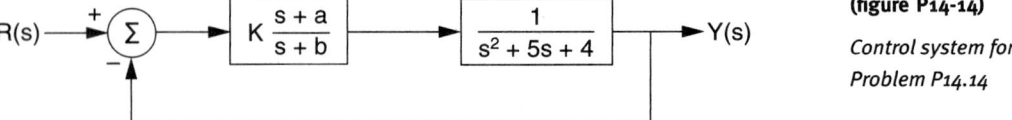

(figure P14-14)

Control system for Problem P14.14

P14.15 For the system shown in Figure P14-15, the design specifications are as follows:

Settling time ≤ 1 second
Error ≤ 10%
$\zeta = 0.707$

(figure P14-15)

Control system for
Problem P14.15

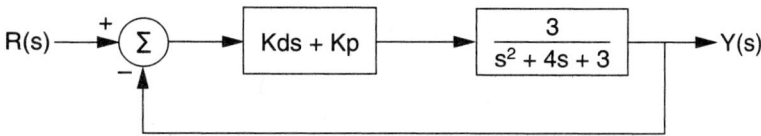

a. Design the linear system to meet the specifications.

b. Transfer the system to the z-plane, using the technique of your choice.

c. Plot the step response for the linear and discrete systems on the same graph. Does the response have the proper damping ratio? If not, what value of derivative gain will produce the correct damping ratio?

MATLAB Problems

M14.1 Use **MATLAB** to repeat Problem P14.1. Does **MATLAB** produce the same results as yours? Explain any differences.

M14.2 For Problem P14.4, show that the prewarped frequency maintained the critical frequency of the continuous system. Use the Bode plot function.

M14.3 The ZOH produces an inherent one sample delay in the response of a system. Using Problem P14.9, show that the Tustin (bilinear) approximation eliminates this delay.

M14.4 Use **MATLAB** to design the system of Problem P14.12.

M14.5 For Problem P13.6, use **MATLAB** to produce the necessary transfer function and to design the system to the specifications in the problem.

M14.6 Using Example 14-6, show that if the zero in the lead-lag compensator is placed at $z = 0.8$ and the pole is calculated, the pole will lie outside the unit circle, making the system unstable. Show the response of the system to a unit step input.

M14.7   For the following s-plane transfer function, use **MATLAB**'s **c2d** command to convert the system to the discrete domain. Then investigate the effect of the sampling time on the risetime, peak value, and settling time of the system for a unit step input. Use sampling times of 0.5, 0.25, 0.1, and 0.05 seconds and plot the continuous and discrete functions (Tustin form) on the same graph, using the step command with both systems, such as *step(sys, sysd)*. Discuss the effect of the sampling time on the parameters.

$$G(z) = \frac{12s + 13}{s^3 + 5s^2 + 17s + 13}$$

M14.8   For the following open-loop discrete transfer function, plot the root locus and select a value of gain K that will produce a damping ratio of 0.707, with a sampling time of 0.1 seconds. Then determine the closed-loop transfer function and plot the response to a unit step input showing that the gain is correct.

$$G(z) = K \cdot \frac{z - 0.55}{z^2 - 1.81z + 0.818}$$

M14.9   A low-pass filter has the following transfer function:

$$G(s) = \frac{1}{s^2 + .2s + 1} \qquad T_s = 1.0 \text{ second}$$

a. Convert the filter to the discrete domain for ZOH, Tustin, and pre-warp forms.

b. Plot all three forms on the same graph and comment on why the ZOH method produces an unstable system.

c. What would you change to make ZOH stable?

M14.10 For Problem M14.9, determine the sampling time for which the ZOH produces a stable system.

| # Discrete State-Variable Methods

Chapter Objectives

After completing this chapter, you should be able to:

➤ Write the general form of the discrete state-vector equation.

➤ Show the discrete form of the state-transition matrix.

➤ State the differences between the concept of the transfer function in discrete state-space and the continuous time state-transition matrix.

➤ Use the controllability and observability methods as applied to discrete state-space.

15.1 Introduction

Chapter 10 developed the concept of state variables and their application to continuous systems. You found that constant coefficient differential equations are easily shown in the form of state-vector equations and then are solved using numerical methods. The concepts of controllability and observability were introduced. Those concepts can be applied in the discrete domain, as well.

Let's begin by investigating the solution of the discrete state-vector equation.

15.2 Solution of the Discrete State-Vector Equation

Considering initial conditions presented in Chapter 10, you know that the state equation is

$$X(s) = [sI - A]^{-1} x(0) + [sI - A]^{-1} BR(s) \tag{15.1}$$

For which the time domain solution is

$$x(t) = \Phi(t)x(0) + \int_0^t \Phi(t - \tau) \cdot Br(\tau)\, d\tau \tag{15.2}$$

Where $\Phi(t)$ is the state transmission matrix. If you now consider Figure 15-1a, needing the output at time t_2, Equation 15.2 becomes

$$x(t) = \Phi(t_2 - t_1)x(t_1) + \int_0^t \Phi(t_2 - \tau) \cdot Br(\tau)\, d\tau \tag{15.3}$$

Where t_1 represents the initial condition time.

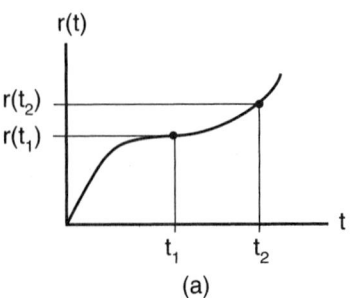

(a)

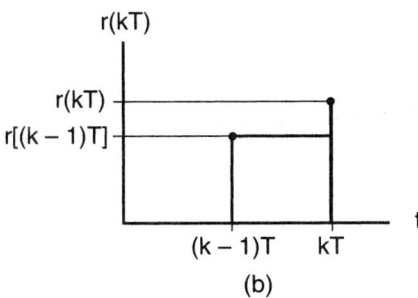

(b)

(figure 15-1)

System outputs:
(a) continuous
(b) with ZOH

In applying a ZOH to the equation, Figure 15-1b represents the discrete input to the system, and Equation 15.3 becomes

$$x(kT) = \Phi(T)x[(k-1)T] + \int_{(k-1)T}^{kT} \Phi(kT-\tau) \cdot Br[(k-1)T]d\tau \qquad (15.4)$$

Upon simplification, Equation 15.4 becomes

$$x_k = \Phi(T)x_{k-1} + \int_{(k-1)T}^{kT} \Phi(kT-\tau) \cdot B \, d\tau \cdot r_{k-1} \qquad (15.5)$$

Consider that $r[(k-1)]T$ is a constant in that it represents sampling. You can define a new constant (λ) such that

$$\lambda = kT - \tau \qquad (15.6)$$

Equation 15.5 is then written as

$$x_k = \Phi(T)x_{k-1} + \int_{(k-1)T}^{kT} [\Phi(\lambda) \cdot B \, d\lambda]r_{k-1} \qquad (15.7)$$

From Equation 15.7, you have

$$x_k = Ax_{k-1} + Br_{k-1} \qquad (15.8)$$

Where

$$A = \Phi(T) \qquad (15.9)$$

$$B = \int_0^T \Phi(\lambda) \cdot B \, d\lambda \qquad (15.10)$$

Equations 15.9 and 15.10 provide a means of converting the continuous state equations to the discrete domain.

The output equation in the discrete domain is

$$y_{k-1} = Cx_{k-1} \qquad (15.11)$$

Let's use Example 10-1 to determine the discrete state equations of the system.

Example 15-1

Example 10-1 is to be converted into the discrete state form. The sampling time will be 0.1 seconds. The continuous form of $\Phi(t)$ is repeated here for convenience.

$$\Phi(t) = \begin{bmatrix} \left(-e^{-t} + 2e^{-2t}\right) & \left(e^{-t} - e^{-2t}\right) \\ \left(-2e^{-t} + 2e^{-2t}\right) & \left(2e^{-t} - e^{-2t}\right) \end{bmatrix}$$

Solution

The **A** matrix is a function of $\Phi(T)$ and is

$$A = \Phi(0.1) = \begin{bmatrix} \left(-e^{-0.1} + 2e^{-0.2}\right) & \left(e^{-0.1} - e^{-0.2}\right) \\ \left(-2e^{-0.1} + 2e^{-0.2}\right) & \left(2e^{-0.1} - e^{-0.2}\right) \end{bmatrix}$$

You can obtain numeric values for the matrix, which are

$$A = \begin{bmatrix} (0.7326) & (0.0861) \\ (-0.1722) & (0.9909) \end{bmatrix}$$

The **B** matrix is then given by

$$B = \int_0^T \begin{bmatrix} \left(-e^{-\lambda} + 2e^{-2\lambda}\right) & \left(e^{-\lambda} - e^{-2\lambda}\right) \\ \left(-2e^{-\lambda} + 2e^{-2\lambda}\right) & \left(2e^{-\lambda} - e^{-2\lambda}\right) \end{bmatrix} \cdot \begin{bmatrix} 4 \\ -5 \end{bmatrix}$$

After integration and simplification, the **B** matrix is

$$B = \begin{bmatrix} 0.3218 \\ -0.5347 \end{bmatrix}$$

The discrete state-vector equation is then

$$\begin{bmatrix} x_1(k) \\ x_2(k) \end{bmatrix} = \begin{bmatrix} 0.7326 & 0.0861 \\ -0.1722 & 0.9909 \end{bmatrix} \begin{bmatrix} x_1(k-1) \\ x_2(k-1) \end{bmatrix} + \begin{bmatrix} 0.3218 \\ -0.5347 \end{bmatrix} r_{k-1}$$

From the state-vector matrices, the system of difference equations are

$$x_1(k) = 0.7326x_1(k-1) + 0.08611x_2(k-1) + 0.3218r_{k-1}$$
$$x_2(k) = -0.1722x_1(k-1) + 0.9909x_2(k-1) - 0.5347r_{k-1}$$

The discrete output equation is given by

$$y_{k-1} = \begin{bmatrix} 1 & -1 \end{bmatrix} \begin{bmatrix} x_1(k-1) \\ x_2(k-1) \end{bmatrix}$$

Example 15-1 (continued)

From which the output difference equation is

$$y_{k-1} = x_1(k-1) - x_2(k-1)$$

The equations as developed in this example can easily be simulated by a computer.

15.3 Discrete Transfer Functions in State-Space

The discrete transfer function is found using a matrix that is similar to the transfer function matrix of Chapter 10. The use of the identity matrix to provide similar matrix dimensions for the state-transition matrix is also necessary.

If you take the z-transform of Equation 15.8, you have

$$X(z) = Az^{-1}X(z) + Bz^{-1}R(z) \tag{15.12}$$

And

$$zX(z) = AX(z) + BR(z) \tag{15.13}$$

$X(z)$ is solved similarly to the way it was solved in Chapter 10.

$$X(z) - AX(z)z = BR(z)$$

$$X(z) = \left[zI - A \right]^{-1} BR(z) \tag{15.14}$$

The output Equation 15.11 has the z-transform form

$$z^{-1}Y(z) = z^{-1}CX(z)$$

Simplifying yields

$$Y(z) = CX(z) \tag{15.15}$$

Substitution of Equation 15.13 into Equation 15.15 yields

$$Y(z) = C \cdot \left[zI - A \right]^{-1} \cdot BR(z) \tag{15.16}$$

Since Equation 15.16 represents an output for a given input, the output-input relationship $Y(z)/R(z)$ is given by an $n \times m$ matrix of the form

$$P(z) = C\left[zI - A\right]^{-1} \cdot B \tag{15.17}$$

The transfer function matrix is

$$P(z) = \begin{bmatrix} P_{11} & P_{12} & \cdots & P_{1m} \\ P_{21} & P_{22} & \cdots & P_{2m} \\ \vdots & & & \\ P_{n1} & P_{n2} & \cdots & P_{nm} \end{bmatrix} \tag{15.18}$$

The characteristic equation of the transfer function is also given by the determinant of the $P(z)$ matrix, as in Chapter 10.

$$\det\left[zI - A\right] = 0 \tag{15.19}$$

The similarity of the continuous and discrete forms of transfer function matrix is evident.

Example 15-2

Determine the discrete transfer function of Example 15-1.

Solution

For the discrete state-space system in Example 15-1, determine the discrete transfer function matrix and the response to a unit step input. Find the location of the poles in the z-plane.

$$\left[zI - A\right] = \begin{bmatrix} z & 0 \\ 0 & z \end{bmatrix} - \begin{bmatrix} 0.7326 & 0.0861 \\ -0.1722 & 0.9909 \end{bmatrix}$$

$$= \begin{bmatrix} z - 0.7326 & -0.0861 \\ 0.1722 & z - 0.9909 \end{bmatrix}$$

The determinant of [zI – A] is

$$z^2 - 1.7235z + 0.7407$$

The inverse matrix is

$$\left[zI - A\right]^{-1} = \left(\frac{1}{z^2 - 1.7235z + 0.7407}\right) \cdot \begin{bmatrix} z - 0.9909 & 0.0861 \\ -0.1722 & z - 0.7326 \end{bmatrix}$$

Example 15-2 (continued)

Combining that into the equation for **P(z)**, you have

$$P(z) = \left(\frac{1}{z^2 - 1.7235z + 0.7407} \right) \cdot [1 - 1] \cdot \begin{bmatrix} z - 0.9909 & 0.0861 \\ -0.1722 & z - 0.7326 \end{bmatrix} \cdot \begin{bmatrix} 0.3218 \\ -0.5347 \end{bmatrix}$$

Performing the indicated operations yields

$$P(z) = \frac{0.8565(z - 0.818)}{z^2 - 1.7235z + 0.7407}$$

The pole locations are found by factoring the characteristic equation; they are $z_1 = 0.9055$ and $z_2 = 0.818$.

$$P(z) = \frac{0.8565(z - 0.818)}{(z - 0.9055)(z - 0.818)} = \frac{0.8565}{z - 0.9055}$$

Applying a unit step to the system, the output becomes

$$Y(z) = \frac{0.8565}{(1 - z^{-1})(z - 0.9055)}$$

Simplifying, the result is

$$y(k) = 9 - 9(0.9055)^{k-1} = 9\left[1 - (0.9055)^{k-1}\right]$$

15.4 Controllability of Discrete Systems

The concepts of controllability apply in the discrete domain as well as in the continuous time domain. The rules are the same for the discrete domain; therefore, the matrix is

$$M_C = \left[B \middle| AB \middle| \cdots \middle| A^{n-1}B \right] \tag{15.20}$$

The matrix is completely controllable if the matrix is of full rank. If not, one or more states are not controllable. Let's apply that theorem to the matrices of Example 15-1.

Example 15-3

Determine the controllability of the system of Example 15-1.

Solution

Since the system is second-order, the controlling matrices are

$$A = \begin{bmatrix} 0.7326 & 0.0861 \\ -0.1722 & 0.9909 \end{bmatrix} \qquad\qquad B = \begin{bmatrix} 0.3218 \\ -0.5347 \end{bmatrix}$$

Example 15-3 (continued)

And

$$M_C = \begin{bmatrix} B & AB \end{bmatrix}$$

$$M_C = \begin{bmatrix} 0.3218 & 0.1897 \\ -0.5347 & -0.5852 \end{bmatrix}$$

The determinant is −0.0869 and is nonzero; therefore, the discrete system is completely controllable.

15.5 Observability of Discrete Systems

The definition of *observability* is the same as it is for the continuous system. The observability matrix is the same, but the individual matrices are the discrete system matrices. The observability matrix must be nonzero and of full rank for the system to be completely observable. The matrix form is repeated here for convenience.

$$M_O = \begin{bmatrix} C \\ \hline CA \\ \hline \vdots \\ \vdots \\ CA^{n-1} \end{bmatrix} \tag{15.21}$$

Example 15-4

Show the observability of the system of Example 15-1.

Solution

The observability matrix is

$$M_O = \begin{bmatrix} C \\ CA \end{bmatrix}$$

$$CA = \begin{bmatrix} 1 & -1 \end{bmatrix} \begin{bmatrix} 0.7326 & 0.0861 \\ -0.1722 & 0.9909 \end{bmatrix} = \begin{bmatrix} 0.9048 & -0.9048 \end{bmatrix}$$

$$M_O = \begin{bmatrix} 1 & -1 \\ 0.9048 & -0.9048 \end{bmatrix}$$

The determinant of the matrix is zero, and its rank is one. Therefore, the system is unobservable.

Summary

➤ The methodology for producing the discrete form of state variables of continuous systems uses the discrete form of the state-transition matrix.

➤ The discrete form of the state-transition matrix is similar to the state-transition matrix in the continuous time domain.

➤ The discrete characteristic equation is the determinant of the $P(z)$ matrix.

➤ The controllability and observability of the discrete systems is essentially the same as those developed in Chapter 10.

Problems

For all problems, use **MATLAB** or another computer program to check your answers.

P15.1   For the following time domain matrices, find the $[zI - A]^{-1}$ matrix and the characteristic equation for the sampling times shown.

a. $\Phi(t) = \begin{bmatrix} \left(1.4e^{-2t} - 0.4e^{-7t}\right) & \left(-0.2e^{-2t} + 0.2e^{-7t}\right) \\ \left(2.8e^{-2t} - 2.8e^{-7t}\right) & \left(-0.4e^{-2t} + 1.4e^{-7t}\right) \end{bmatrix}$, $Ts = 0.25\ s$

b. $\Phi(t) = \begin{bmatrix} \left(3e^{-t} - 3e^{-2t} + 3e^{-3t}\right) & \left(-3e^{-t} + 6e^{-2t} - 3e^{-3t}\right) & \left(3e^{-t} - 12e^{-2t} + 9e^{-3t}\right) \\ \left(2.5e^{-t} - 4e^{-2t} + 1.5e^{-3t}\right) & \left(-2.5e^{-t} + 8e^{-2t} - 4.5e^{-3t}\right) & \left(-2.5e^{-t} - 16e^{-2t} + 13.5e^{-3t}\right) \\ \left(0.5e^{-t} - e^{-2t} + 0.5e^{-3t}\right) & \left(-0.5e^{-t} + 2e^{-2t} - 1.5e^{-3t}\right) & \left(0.5e^{-t} - 4e^{-2t} + 4.5^{e-3t}\right) \end{bmatrix}$,

$Ts = .3\,s$

P15.2   For the plant simulation shown in Figure P15-2, draw a minimum phase
diagram, choose a suitable sampling time, and generate the discrete state
equations of the system.

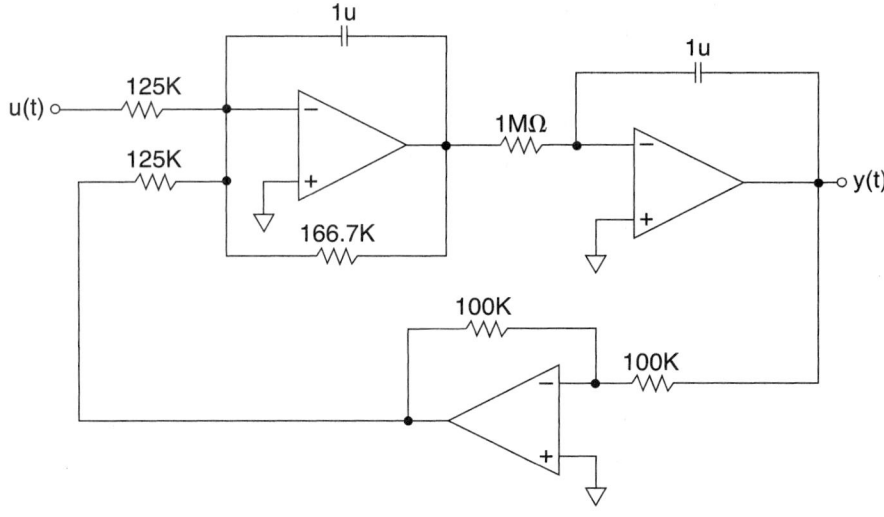

(figure P15-2)
*Control system simulation
circuit for Problem 15.2*

P15.3   For the following continuous state equations, select a suitable sampling time
and generate the discrete transfer function. Determine the damping ratio of
the discrete system and compare it to the continuous damping ratio.

$$\begin{bmatrix} x_1 \\ x_2 \end{bmatrix} = \begin{bmatrix} -6 & 0 \\ -4 & -1 \end{bmatrix}\begin{bmatrix} x_1 \\ x_2 \end{bmatrix} + \begin{bmatrix} 1 \\ -2 \end{bmatrix} u(t)$$

$$y(t) = \begin{bmatrix} 1 & -1 \end{bmatrix}\begin{bmatrix} x_1 \\ x_2 \end{bmatrix}$$

P15.4   Transform the following continuous state equations to a discrete system.
Determine the $\Phi(T) = [zI - A]^{-1}$ matrix and the characteristic equation of
the system for a sampling time of 0.1 seconds.

$$\begin{bmatrix} x_1 \\ x_2 \end{bmatrix} = \begin{bmatrix} 0 & 1 \\ -3 & -4 \end{bmatrix}\begin{bmatrix} x_1 \\ x_2 \end{bmatrix} + \begin{bmatrix} 0 \\ 1 \end{bmatrix} u(t)$$

$$y(t) = \begin{bmatrix} 3 & 0 \end{bmatrix}\begin{bmatrix} x_1 \\ x_2 \end{bmatrix}$$

P15.5   For the minimum phase diagram shown in Figure P15-5, determine the
continuous time state equations and transform them to the discrete domain.
Test the transfer function for pole-zero cancellations and provide the step
output for the closed-loop system.

(figure P15-5)

Minimum phase diagram
for Problem 15.5

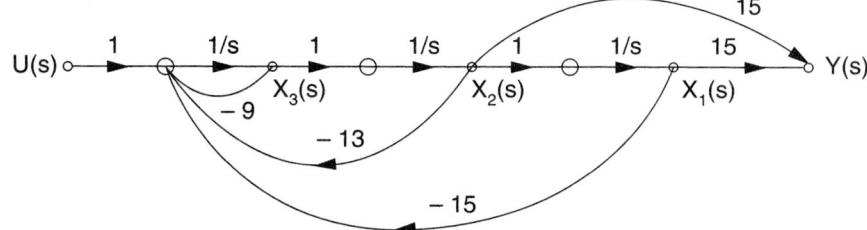

P15.6   For the following state matrices, determine the A and B matrices for discrete time. Then determine the transfer function, using Equation 15.17.

$$\Phi(0.1) = \begin{bmatrix} 1.5e^{-0.2} - 0.5e^{-0.6} & 0.25e^{-0.2} - 0.25e^{-0.6} \\ -3e^{-0.2} - 3e^{-0.6} & -0.5e^{-0.2} + 1.5e^{-0.6} \end{bmatrix}$$

$$B = \begin{bmatrix} 0 \\ 1 \end{bmatrix} \qquad C = \begin{bmatrix} 1 & 0 \end{bmatrix}$$

P15.7   For the following state matrices, use Equation 15.17 to determine the characteristic equation and the transfer function. The sampling time is 0.1 seconds.

$$[zI - A]^{-1} = \begin{bmatrix} (z - 0.1) & 0.5 & 0.25 \\ -0.25 & (z - 0.2) & 1.5 \\ -0.05 & -0.2 & (z - 0.3) \end{bmatrix} \qquad B = \begin{bmatrix} 0 \\ 0 \\ 1 \end{bmatrix} \qquad C = \begin{bmatrix} 1 & 0 & 0 \end{bmatrix}$$

P15.8   Determine if the following state matrices provide a completely controllable and observable system.

$$A = \begin{bmatrix} 0.2 & 0 & -0.2 \\ 0 & 0.4 & -1 \\ 0.4 & 0 & 0.6 \end{bmatrix} \qquad B = \begin{bmatrix} 0 \\ 0 \\ 0.5 \end{bmatrix} \qquad C = \begin{bmatrix} 1 & 0 & 0 \end{bmatrix}$$

P15.9   For Problems P15.6 and P15.7, determine if the systems are completely controllable and observable.

P15.10   For the following discrete state matrices, determine if the system is controllable and observable.

$$A = \begin{bmatrix} 0.8187 & 0 & 0 \\ 0 & 0.7408 & 0 \\ 0 & 0 & 0.6703 \end{bmatrix} \qquad B = \begin{bmatrix} 0 \\ 1 \\ 1 \end{bmatrix} \qquad C = \begin{bmatrix} 1 & 1 & 1 \end{bmatrix}$$

Using MATLAB

MATLAB FEATURES

MATLAB has numerous features that help in analyzing the performance of a system. While most computations and analyses of systems can be done with the basic computing program, MATLAB has several application-specific toolboxes that can be used to perform some of the functions with a minimum of programming. Toolboxes that are of particular use in control systems engineering are as follows:

Control System Toolbox

System Identification Toolbox

Fuzzy Logic Toolbox

Robust Control Toolbox

Model Predictive Control Toolbox

More toolboxes are available, and each toolbox has dedicated functions that allow complex computational functions to be implemented with a single command. In this book, the most used toolbox is the Control System Toolbox.

Further, MATLAB has a block diagram-simulating program called Simulink that allows a system to be input in the form of a common block diagram. The results of all of the analysis types can be displayed in high-resolution graphic format.

Starting with MATLAB

MATLAB is matrix-oriented. Most of the time, the data to be manipulated is input in the form of matrices or matrix vectors. All of the basic matrix manipulations of addition, subtraction, multiplication, and inversion are available.

Matrices are entered using square braces. If there is more than one row or column the rows and columns are separated using semicolons. If the matrix is a single row vector or column vector, the matrix is entered as

$$A = \begin{bmatrix} 1 & 6 & 4 & 9 & 6 \end{bmatrix}$$

When the Enter key is pressed, the row is shown as

A =

    1   6   4   9   6

If the matrix is a column vector, the single quote is used at the end of the vector for the transpose of the matrix as follows:

A = [1 6 4 9 6]'

And the column is shown as

A =

    1

    6

    4

    9

    6

If the matrix is not a single row or column vector, the matrix can be input in one of two ways. The first way is as follows:

A = [1 3 5; 6 0 9; 11 −3 5]

Note the use of the semicolons to separate the row vectors. Alternatively, the matrix may be entered as

A = [1   3   5

6 0 9

11 −3 5]

Press the Enter key after each row is entered. When the Enter key is pressed after the last row is entered, the resultant matrix is as follows:

    1   3   5
    6   0   9
   11   −3   5

If the transpose of the matrix is required, the (') is used; and the matrix is

A = [1 3 5; 6 0 9; 11 −3 5]'

The matrix is transposed, and the output is

```
1    6    11
3    0    -3
5    9    5
```

When a semicolon is included at the end of the matrix definition, the matrix is not shown on the screen.

A = [1 3 5; 6 0 9; 11 –3 5];

The determinant of the matrix is determined using

det(A)

The determinant of the matrix, A, is 144.

When two or more matrixes are defined, all of the matrix operations are available as follows:

B = [0 9 –5; 6 5 3; 8 –5 5];

A + B

```
1    12    0
12    5    12
19    -8    10
```

A – B

```
1    -6    10
0    -5    6
3    2    0
```

A * B

```
58    -1    29
72    9    15
22    59    -39
```

The inverse of the matrix is calculated in the following manner: A^{-1}, or inv(A). Assigning the matrix to a different variable allows the variable to be used in calculations.

C = A^ – 1

```
0.1875    -0.2083    0.1875
0.4792    -0.3472    0.1458
-0.1250    0.2500     0.1250
```

Multiplication of the C and A matrices returns the identity matrix.

```
D = C * A
    1.0000      0          0
    0           1.0000     0
   −0.0000      0.0000     1.0000
```

A (−0) may occur on one or more of the values due to the accuracy limitations of the computation.

MATLAB is case-sensitive. For the matrices above, the matrix A is not the same as another matrix, a. If the matrix is entered using a capital letter for its name, identifying the matrix with a lowercase letter will result in an error.

If the roots of a characteristic or other equation are required, **MATLAB** will provide them. To factor an equation into its component roots, enter the equation as a single row vector as follows:

```
A = [1   6   11   12];
```

```
roots(A)
```

Returns -4.0, $1.0000 + 1.4142i$, $-1.0000 - 1.4142i$. Alternatively, you can enter

```
roots([1   6   11   12])
```

Which returns the same answers.

Complex numbers are allowed in all operations and functions in **MATLAB**. For the roots of A above, two are complex and the form shown is $-1 + 1.4142i$. However, the complex numbers may also be entered as $-1 + 1.4142j$.

If for any reason an expression is complicated and cannot fit on one line, an ellipsis consisting of three or more periods may be used to indicate that the statement continues on the next line.

```
x = 1 + 1/2 + 1/3 + 1/5 + 1/7 + 1/9 ...
      1/11 + 1/13 + 1/15;
```

When the Enter key is pressed, the value of x will be calculated but not shown.

Variables and Expressions in MATLAB

The general method for input of a variable is *variable = expression*. When a variable is used, the value is stored and can be used in any subsequent calculations.

```
Y = exp(−2)
```

Which returns the value of e^{-2}, or

0.1353

Similarly, an expression can be formed for other functions. The general form of an incremental function is *variable = (start value: increment: stop value)*, as follows:

$t = (0{:}0.01{:}1)$

t can be used in any analysis as a variable for the time of analysis or another requirement. (*t* is not required to be used for time, but it is convenient.)

Logarithms can be entered into **MATLAB** using one of two forms: *log(expression)* returns the natural logarithm of the expression, while *log10(expression)* returns the logarithm to the base 10.

The colon is important in **MATLAB** for the generation of vectors and tables of numbers. Consider

$r = 1{:}5$

The row vector produced is

$r =$

    1   2   3   4   5

To produce a table of values of *r* and *w*, enter the values in the following way:

$r = (1{:}5)';$

$w = 2 * pi * r;$

$[r\ w]$

The table produced is

```
ans =
        1.0000      6.2832
        2.0000     12.5664
        3.0000     18.8496
        4.0000     25.1327
        5.0000     31.4159
```

In **MATLAB**, polynomials are represented as row vectors containing coefficients of the polynomial ordered in descending powers. The command *poly(matrix name)* returns the polynomial of the matrix. For example, consider the matrix of Example 10-6.

$$A = \begin{bmatrix} 0 & 1 & 0 \\ 0 & 0 & 1 \\ -24 & -26 & -9 \end{bmatrix}$$

The characteristic equation of the matrix can be obtained by entering

poly(A)

The returned values are

Ans =

    1   9   26   24

Which are equivalent to $\det(\lambda I - A)$

$$\det(\lambda I - A) = \lambda^3 + 9\lambda^2 + 26\lambda + 24$$

Using MATLAB Functions to Model Systems

To use **MATLAB** to model a system, the system must first be defined; then the constraints of the analysis are defined. Finally, the plotting expressions are defined when a plot of the input/output relationship is needed. The following short example shows the method:

$$\frac{Y(s)}{U(s)} = \frac{12}{s^3 + 4s^2 + 12s + 12}$$

The input to the system is a ramp of $tu(t) = 0.5t$. The first step is to define the ramp. The next step is to define the system and the method to use for analysis and then the plot, as follows:

$t = (0:0.01:5)$;	% t sets up the time limits.
$r = 0.5 * t$;	% r defines the rate of change of t.
$n = 12; d = [1 \quad 4 \quad 12 \quad 12]$;	% Definition of the transfer function
$[y, x] = lsim(n, d, r, t)$;	% Simulation of the system
plot(t, y, t, r)	% Plot the transfer function as a function of time.

t defines the time of analysis, r defines a ramp input of $0.5t$, n and d are the numerator and denominator of the system, *lsim* is a linear simulation of the system, and *plot* defines two plots to be placed on the same chart with the same time and amplitude constraints. The plot of the system is shown in Figure A-1.

(figure A-1)

*Output of linear
simulation*

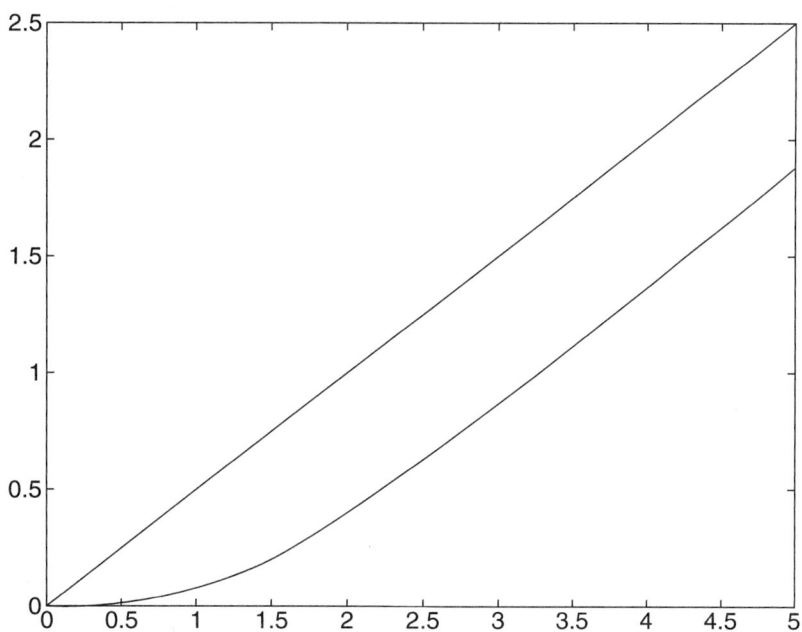

To obtain a hard copy of the graph, select Print from the File menu of the plot.

Plotting Outputs

Plotting the outputs of systems is one of the most important things engineers do to obtain information about how a system will respond to a stimulus. Often, the system will be driven by a step input, and the analysis is usually adequate for the system response. However, sometimes engineers want to change the time of analysis to show more detail.

It is sometimes desirable to plot more than one curve on a single graph, such as a plot of the step and frequency response of a closed-loop system.

Let's consider the example shown in the following system:

$$\frac{Y(s)}{U(s)} = \frac{4s + 6}{s^3 + 3s^2 + 8s + 6}$$

First, you will examine putting more than one graph on the same plot. Assume you want to examine the effect of changing the position of the zero from 1.5 to 1.0 to 0.5. To do that, you must define three transfer functions in the following way:

sys1 = tf([4 6],[1 3 8 6]); sys2 = tf([6 6],[1 3 8 6]);

step(sys1, sys2)

Two different colored graphs will be displayed on the same graph in two predefined colors: sys1 is blue, and sys2 is green.

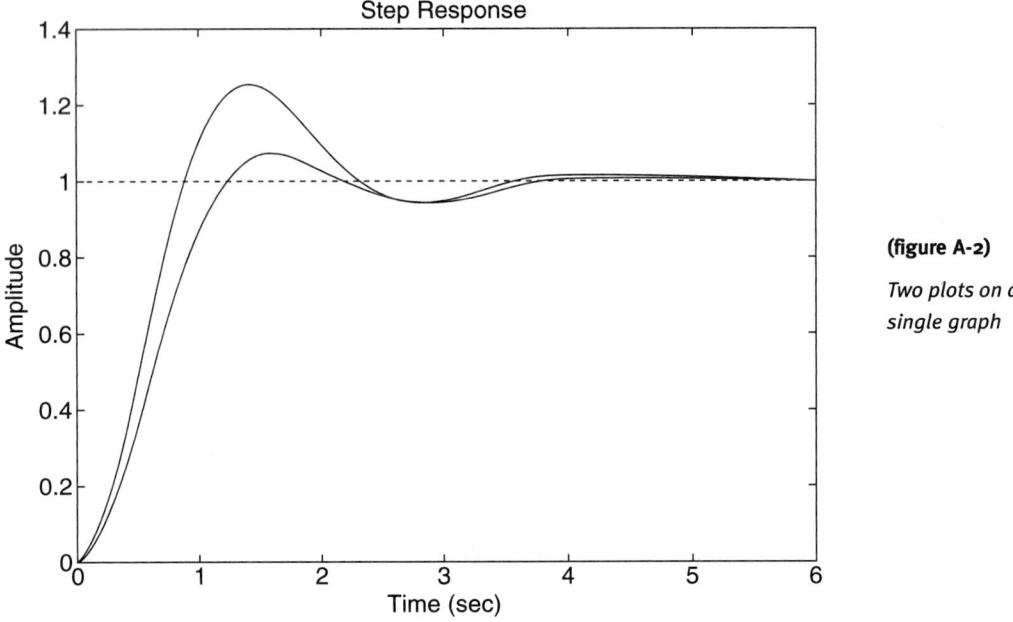

(figure A-2)

Two plots on a single graph

If another form of data is required, the same graph window can be used to show more than one type of data. The procedure to do that is controlled using the *subplot* command.

subplot(*m,n,p*)

That command partitions the figure window into an *m* by *n* matrix of smaller subplots and selects the *pth* subplot for the current plot.

To place both the step and Bode responses on the same window, use the following method:

sys1 = tf([4 6],[1 3 8 6]); sys2 = tf([6 6],[1 3 8 6]);

subplot(2,1,1), step(sys1,sys2), subplot(2,1,2), bode(sys1,sys2)

The result is shown in Figure A-3.

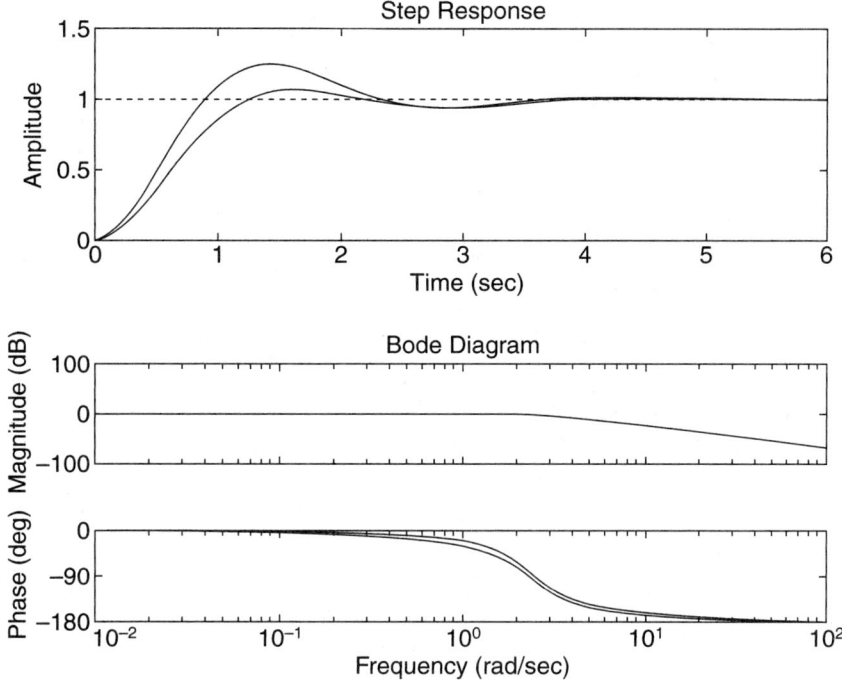

(figure A-3)

Different data types on a single graph

Adding Delay Time to a System

When a Bode plot of a system is run, the phase plot gives the maximum delay time for the system input or output when the stability margins are displayed. If the system you are working with has a delay, the effect of the delay on the phase margin can be readily shown.

To add a delay to a system, the following method is used:

   sys = tf(num, den, 'inputdelay', delay time)

Or

   sys = tf(num, den, 'outputdelay', delay time)

Consider the system

$$G(s) = \frac{25}{s(s^2 + 6s + 25)}$$

A Bode plot of the system shows the phase margin to be 75.8° and the phase crossover to be 5 rad/sec with a maximum delay margin of 1.31 seconds. If there is a 0.5 second delay in measurement time and signal input, the effect of the delay can be found using the following instructions:

$n = 25; d = [1 \quad 6 \quad 25 \quad 0];$

sys = tf(n, d, 'inputdelay', .5)   % Alternatively sys = tf(n, d, 'outputdelay', .5)

bode(sys)

The effect of the input delay on the original system is shown in Figure A-4.

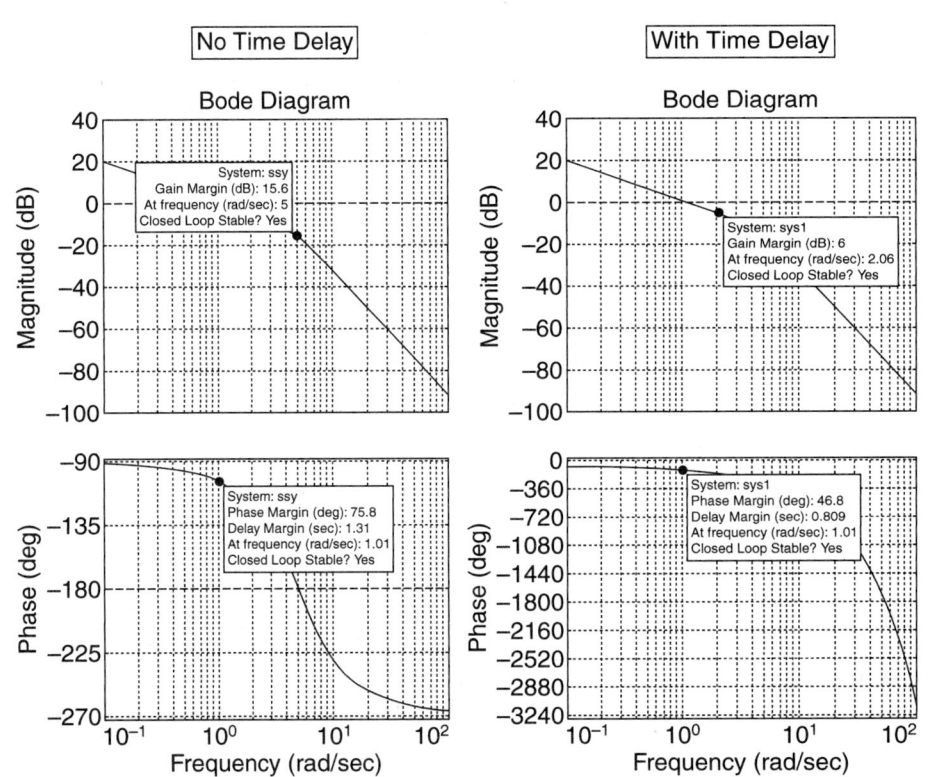

(figure A-4)

Effect of a time delay on the phase of a system

It is clear that the delay has affected the phase margin, reducing it by nearly 30°.

Some State-Space Functions of MATLAB

At times, when you have the matrices for a system, you need to obtain the transfer function of the system from the matrices. You can, of course, do the necessary math to obtain the transfer function; but it may be more convenient to allow **MATLAB** do the work. For instance, given the matrices

$$A = \begin{bmatrix} 0 & 1 & 0 \\ 0 & 0 & 1 \\ -6 & -9 & 12 \end{bmatrix} \quad B = \begin{bmatrix} 0 \\ 1 \\ 1 \end{bmatrix}$$

$$C = \begin{bmatrix} 1 & 1 & 0 \end{bmatrix} \quad D = 0$$

You can either do the $\begin{bmatrix} sI - A \end{bmatrix}$ function and then calculate the transfer function or you can enter the matrices in **MATLAB** and use the *ss(A,B,C,D)* function as follows:

a = [0 1 0; 0 0 1;–6 –9 –12]; b = [0 1 1]';

c = [1 1 0]; d = 0;

sys = ss(*a,b,c,d*), which gives

a =

	x1	x2	x3
x1	0	1	0
x2	0	0	1
x3	-6	-9	-12

b =

	u1
x1	0
x2	1
x3	1

c =

	x1	x2	x3
y1	1	1	0

d =

	u1
y1	0

The transfer function is easily found using the *tf* command

sysdc = tf(sys)

Which produces

sys1 = tf(sys)

Transfer function:

$$\frac{s^2 + 14s + 13}{s^3 + 12s^2 + 9s + 6}$$

The zero-pole-gain form of the system can also be found directly by invoking the *zpk(sys)* command as follows.

gzpk = zpk(sys)

Zero/pole/gain:

$$\frac{(s + 1)\ (s + 13)}{(s + 11.25)\ (s^2 + 0.7528s + 0.5335)}$$

Creating MATLAB M-files

An M-file is a reusable file that can contain variables, expressions, and analysis types. If you make certain calculations often, the M-file is a convenient way to store the equations for calculation. There are two types of M-files:

1. Scripts, which do not accept input arguments or return output data. They operate in the workspace environment.
2. Functions, which do accept input arguments and return output data. Any variables in the function are local to the function.

Scripts

A script file contains all of the parameters required for analysis. This type of file executes when the filename is typed in the command window. To invoke the internal editor, type either

edit

Or

edit filename

If the filename does not exist, you are prompted to create a new M-file. M-files can be entered using any standard text editor. If you do not use the internal editor, which is NotePad, be sure to use *filename.m* when you save the file. An example of a simple script that produces the amplitude and frequency response of a third-order system saved by the name *xferfunc.m* produced using the **MATLAB** editor is:

% Script M-file to produce step and Bode plots of the

% output of closed-loop transfer function of a system

num = 15; den = [1 6 11 15];

sys = tf(num,den)

subplot(2,1,1), step(sys), subplot(2,1,2), bode(sys)

The first two lines of the file are the help lines for the script. If the keyword *help filename* is used, the first two lines are displayed in the command window as follows:

help xferfunc

To run the file, make sure you have the proper path name for the file. If the file is not in the workspace, change the path, using the following method:

path(path,'c:/pathname/')

Or

cd('c:/pathname/')

Type the filename into the command window. The output of this file is shown in Figure A-5.

(figure A-5)

Graphics output of script file

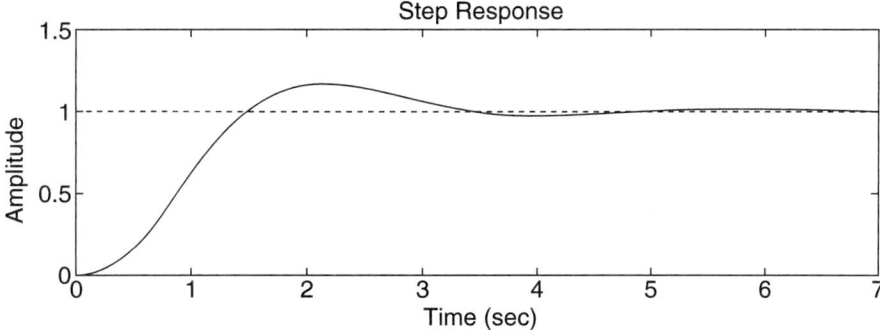

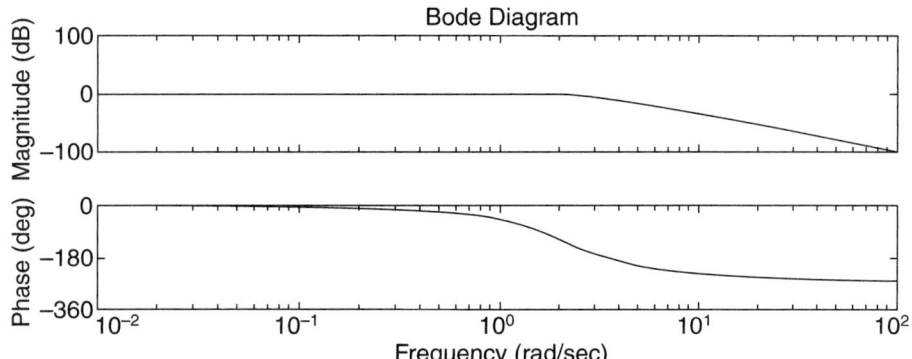

To save the file, it is best that you change the path to a folder in which you can keep M-files. If you don't, the next time **MATLAB** is updated, the files are likely to be erased.

Of course, when you run the file, it will always produce the same curves unless you edit the file to have the transfer function and graphics you need.

An example of a slightly more complex script file with four plots on the same graphics screen is

```
% Script M-file to produce step and Bode plots of

% output of closed-loop transfer function of a system


k = 3;

z = -5;

p = [0 - 3 + j * 1.4142 - 3 - j * 1.4142]';

[n,d] = zp2tf(z,p,k);

subplot(2,2,1), rlocus(n,d), subplot(2,2,2), bode(n,d)

subplot(2,2,3), nyquist(n,d), subplot(2,2,4), nichols(n,d)
```

The output of this file is shown in Figure A-6.

(figure A-6)

Graphics output of script file

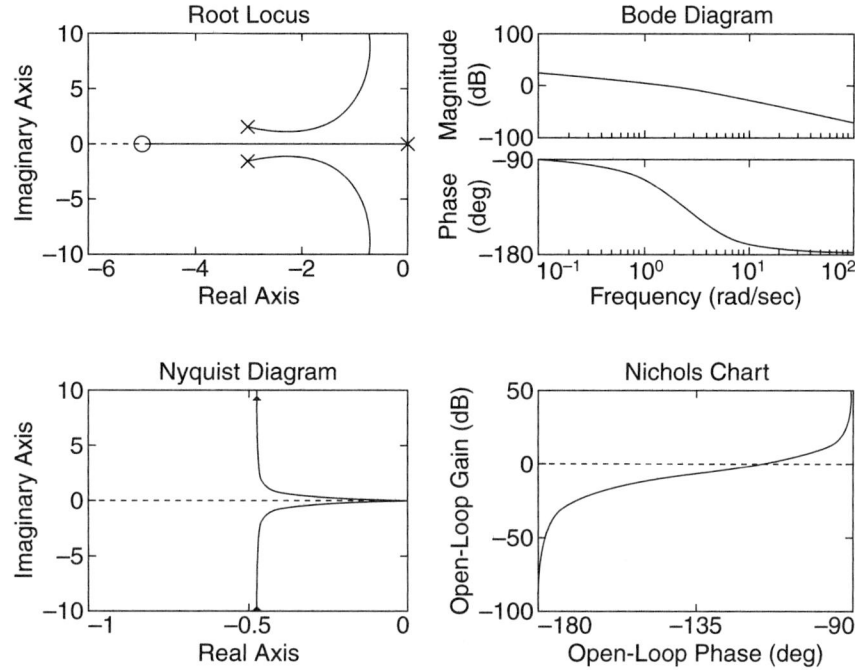

Functions

The use of a function in an M-file allows you to enter parameters directly in the command window and then execute the file. The keyword *function* is usually the opening line for this type of M-file. The help lines follow the function keyword but can come before the function argument, as shown below.

function   output argument = function name(input arg1, input arg2, ...)

The name of the M-file should be the same as the function name.

The following example shows a function M-file that allows you to enter the numerator, denominator, and sampling time of a transfer function in the continuous time form and calculates the continuous and discrete forms of transfer function and then plots both step and frequency responses on a single graph.

function sysc = xfer(n,d,t)

% M-file to generate a transfer function in both

% continuous and discrete time

% For the discrete system, only the "ZOH" method is used.

% Enter the numerator and denominator in continuous form.

% Then enter the sampling time, $t > 0$. The step and frequency

% response of both systems is then plotted on a single graph.

sys = tf(n, d)

sysd = c2d(sys, t)

subplot(1, 2, 1), step(sys, sysd), subplot(1, 2, 2), bode(sys, sysd)

To call this M-file, first enter n, d, and t and then use

xfer(n, d, t)

Both forms of transfer function are displayed on the command window and then the output graphs. The plots of a system with $n = \begin{bmatrix} 1 & 15 \end{bmatrix}$, $d = \begin{bmatrix} 1 & 6 & 11 & 15 \end{bmatrix}$, and $t = 0.1$ are shown in Figure A-7.

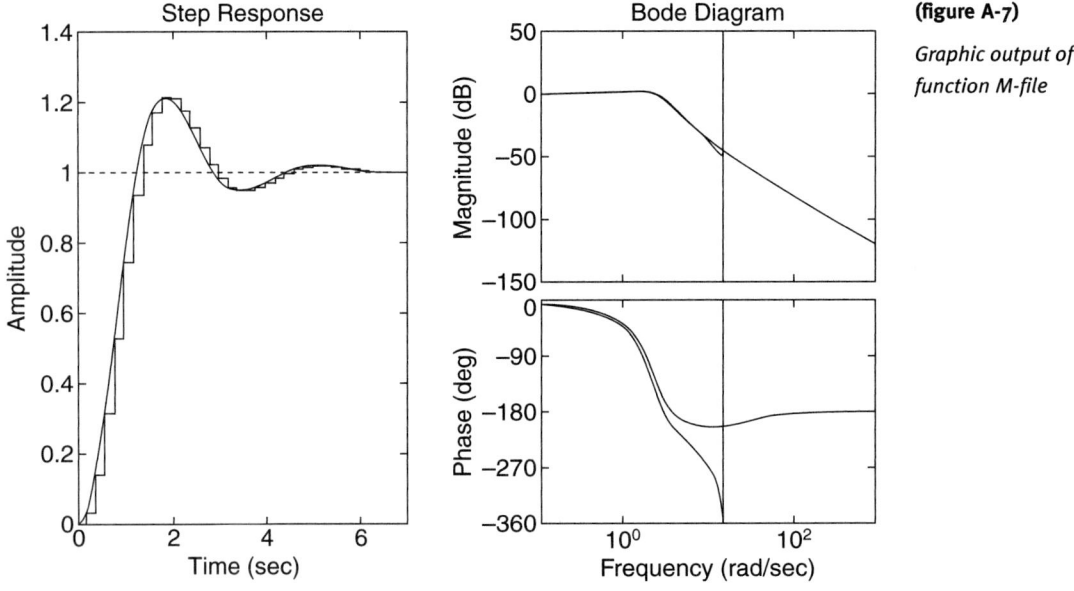

(figure A-7)

Graphic output of function M-file

If you type *help xfer,* the help lines will be displayed on the command window screen.

The M-file can also prompt for keyboard input. To prompt for input from the user, use the following method:

name = input('prompt string')

For the above file, enter the prompted information at each prompt. Enter only the vector for the requested information, not the name of the vector. Note that the function is now empty at the calling line. The file should be modified as follows:

function sysc = xfer()

%M-file to generate a transfer function in both

%continuous and discrete time.

%For this M-file, only the "ZOH" method is used.

%Enter the numerator, denominator in continuous form.

%Then enter the sampling time, $t > 0$.

n = input('Enter numerator ')

d = input('Enter denominator ')

t = input('Enter sampling time ')

sys = tf(n,d)

sysd = c2d(sys,t)

subplot(1,2,1), step(sys, sysd), subplot(1,2,2), bode(sys,sysd)

To activate the file, first call the file using

xfer()

Then follow the prompts as follows:

Enter numerator [1 15]

n =

     1   15

Enter denominator [1 6 11 15]

d =

     1   6   11   15

Enter a sampling time of 0.1 second.

t = 0.1

The output is, of course, the same as that in Figure A-6 for the original M-file.

Generating Data

MATLAB has its own form of the FOR and WHILE loops found in most computer programming languages. Using those functions, you can generate a fixed number of data points for any given system.

The FOR loop has the following syntax:

FOR x = expression

     statements

END

The FOR loop executes a fixed number of times and may be nested as in other computer languages.

For instance, consider Example 2-6 whose difference equation is repeated here.

$$u_k = \frac{Te_k + u_{k-1}}{1 + 2T}$$

From the example, $T = 0.1$ second and $e_k = 1.0$. The equation can be programmed into a FOR loop to generate the values given in Table 2-2. Since you know that the total analysis time is 3 seconds, you can eliminate the setting of a time variable and set a variable for the time increments

$k = .1:.1:3;$

Then two variables, u_k and u_{k-1}, must be defined and set to zero to produce the proper value at 0.1 seconds.

$u\_k = 0;$

$u\_k\_1 = 0;$

Then set up the FOR loop as follows:

for $k = .1:.1:3$

u_k_1 must be equated to u_k for each succeeding calculation. The loop will not execute until the end statement is entered.

$u\_k = (.1 + u\_k\_1)/1.2$

$u\_k\_1 = u\_k$

end

WHILE loops execute an indeterminate number of times as long as a given condition is true. That being the case, care must be taken when designing the conditions so that an infinite loop is not created. The WHILE loop has the following form:

WHILE expression

    statements

END

As an example, you can generate the factorial of the first ten integer numbers. First, define an integer value as n (or any variable name you want)

$n = 1$;

Then set the WHILE loop to produce the factorial values.

while $n <= 10$

n

factorial(n)

$n = n + 1$;

end

Note that the program simply prints n and the number being calculated to the screen. It is important that $n = n + 1$ come after the factorial. If not, the factorial for 11 will be produced.

MATLAB also has provisions for conditional branching in the form of the IF, ELSEIF, and ELSE commands. The general format for the commands is

IF expression

    statements

ELSEIF

    statements

ELSE

    statements

END

A simple example of the IF, ELSEIF, and ELSE commands are in the factoring of the roots of the quadratic equation $Ax^2 + Bx + C = 0$. The three root conditions of the quadratic equation are real and unequal roots, real and equal roots, and complex or imaginary roots. You will make a function M-file for this example. Only the discriminant will be considered in this function. Start a new function file by opening the editor, as follows:

function Roots = Quad(A,B,C)

% File calculates the discriminant of the quadratic equation and

% displays the root conditions of the quadratic.

A = input(' Enter Coefficient A = ')

B = input(' Enter Coefficient B = ')

C = input(' Enter Coefficient B = ')

disc = sqrt(B^2 – 4 * A * C)

if disc > 0

    'Discriminant > 0. Roots are real and unequal.'

elseif disc = 0

    'Discriminant = 0. Roots are real and equal.'

else

    'Discriminant < 0. Roots are complex or imaginary.'

end

Notice the spacing on the input commands. That is not required and is done only for cosmetic purposes.

MATLAB has many more functions than can be included here. You should refer to the literature that is included with your version of the software and to the help available on the Internet.

Useful Laplace and z–Transforms

Useful Laplace Transforms

$f(t)$	$F(s)$
1 (unit impulse)	1
$u_s(t)$ (unit step)	$\dfrac{1}{s}$
$tu_s(t)$ (unit ramp)	$\dfrac{1}{s^2}$
$\dfrac{t^2 u_s(t)}{2}$ (parabolic input)	$\dfrac{1}{s^3}$
$u(t) - u(t - a)$ (rectangular pulse)	$\dfrac{1}{s}(1 - e^{-as})$
e^{-at}	$\dfrac{1}{s + a}$
te^{-at}	$\dfrac{1}{(s + a)^2}$
$\dfrac{1}{(n - 1)!}t^{n-1}e^{-at}$ (n is a positive integer)	$\dfrac{1}{(s + a)^n}$
$\dfrac{1}{a}(1 - e^{-at})$	$\dfrac{1}{s(s + a)}$
$1 - et^{-at}$	$\dfrac{a}{s(s + a)}$
$\sin \omega t$	$\dfrac{\omega}{s^2 + \omega^2}$
$\cos \omega t$	$\dfrac{s}{s^2 + \omega^2}$

f(t)	F(s)
$e^{-at} \sin \omega t$	$\dfrac{\omega}{(s + a)^2 + \omega^2}$
$e^{-at} \cos \omega t$	$\dfrac{s + a}{(s + a)^2 + \omega^2}$

Useful z-Transforms

f(t)	F(z)
1 (unit impulse)	1
$u_s(t)$ (unit step)	$\dfrac{z}{z - 1}$
$tu_s(t)$ (unit ramp)	$\dfrac{Tz}{(z - 1)^2}$
$\dfrac{t^2 u_s(t)}{2}$ (parabolic input)	$\dfrac{T^2 z(z + 1)}{2(z - 1)^3}$
e^{-at}	$\dfrac{z}{z - e^{-aT}}$
te^{-at}	$\dfrac{Tze^{-aT}}{(z - e^{-aT})^2}$
$1 - e^{-at}$	$\dfrac{z(1 - e^{-aT})}{(z - 1)(z - e^{-aT})}$
$\sin \omega t$	$\dfrac{z \sin \omega T}{z^2 - 2z \cos \omega T + 1}$
$\cos \omega t$	$\dfrac{z(z - \cos \omega T)}{z^2 - 2z \cos \omega T + 1}$
$e^{-at} \sin \omega t$	$\dfrac{ze^{-aT} \sin \omega T}{z^2 - 2ze^{-aT} \cos \omega T + e^{-2aT}}$
$e^{-at} \cos \omega t$	$\dfrac{z(z - e^{-aT} \cos \omega T)}{z^2 - 2ze^{-aT} \cos \omega T + e^{-2aT}}$

Index

A

acceleration error, 187–190
A/D (analog-to-digital) converter, 5
  accuracy, 375
  aperture time, 375
  bandwidth, increasing, 377
  conversion time, 375
  resolution, 375
  sample and hold, 374–377
  zero-order hold, 374–377
  z-plane and pole locations, 379–381
adding matrices, 30, 55–56
adjoint of matrices, 31–32
advance theorem, 46
algebra
  block diagrams, 122–128
  SFGs (signal flow graphs), 137–138
all-pole model, 135–136
amplitude response, 248
angle criterion, 204–205
angles of departure and arrival, 206
asymptotic properties, 206
asymptotic stability
  definition, 152
  frequency domain methods, 152
  methods for determining, 152
  s-domain methods, 152

B

backlash, 82
backward Euler approximation, 412, 414
bandwidth
  Bode's analysis, 243
  control systems. *See* PD (proportional-derivative) control; PID (proportional-integral-derivative) control.
  increasing, 377
bilinear transformation
  discrete compensation methods, 425–428
  discrete systems, 396
block diagrams
  algebra, 122–128
  canonic forms, 121–122
  characteristic equations, 130
  disturbance input, 128–129
  feed forward gain, 126
  gain blocks, 120–121
  graphical output. *See* **LTI Viewer**.
  identities, 129–130
  reference input, 121
  setpoint, 121
  summing junctions, 121
  transfer functions, 130–131
  types of blocks, 120–121
Bode, Hendrik W., 242

Z